AF864387

Lecture Notes in Bioengineering

Advisory Editors

Nigel H. Lovell, Graduate School of Biomedical Engineering, University of New South Wales, Kensington, NSW, Australia

Luca Oneto, DIBRIS, Università di Genova, Genova, Italy

Stefano Piotto, Department of Pharmacy, University of Salerno, Fisciano, Italy

Federico Rossi, Department of Earth, University of Salerno, Fisciano, Siena, Italy

Alexei V. Samsonovich, Krasnow Institute for Advanced Study, George Mason University, Fairfax, VA, USA

Fabio Babiloni, Department of Molecular Medicine, University of Rome Sapienza, Rome, Italy

Adam Liwo, Faculty of Chemistry, University of Gdansk, Gdansk, Poland

Ratko Magjarevic, Faculty of Electrical Engineering and Computing, University of Zagreb, Zagreb, Croatia

Lecture Notes in Bioengineering (LNBE) publishes the latest developments in bioengineering. It covers a wide range of topics, including (but not limited to):

- Bio-inspired Technology & Biomimetics
- Biosensors
- Bionanomaterials
- Biomedical Instrumentation
- Biological Signal Processing
- Medical Robotics and Assistive Technology
- Computational Medicine, Computational Pharmacology and Computational Biology
- Personalized Medicine
- Data Analysis in Bioengineering
- Neuroengineering
- Bioengineering Ethics

Original research reported in proceedings and edited books are at the core of LNBE. Monographs presenting cutting-edge findings, new perspectives on classical fields or reviewing the state-of-the art in a certain subfield of bioengineering may exceptionally be considered for publication. Alternatively, they may be redirected to more specific book series. The series' target audience includes advanced level students, researchers, and industry professionals working at the forefront of their fields.

Indexed by SCOPUS, INSPEC, zbMATH, SCImago.

Syed Faris Syed Omar ·
Mohd Hasnun Arif Hassan · Alexander Casson ·
Alan Godfrey · Anwar P. P. Abdul Majeed
Editors

Innovation and Technology in Sports

Proceedings of the International Conference on Innovation and Technology in Sports, (ICITS) 2022, Malaysia

Springer

Editors
Syed Faris Syed Omar
National Sports Institute of Malaysia
Kuala Lumpur, Malaysia

Alexander Casson
University of Manchester
Manchester, UK

Anwar P. P. Abdul Majeed
School of Robotics
Xi'an Jiaotong – Liverpool University
Suzhou, China

Mohd Hasnun Arif Hassan
Faculty of Mechanical and Automotive Engineering Technician
Universiti Malaysia Pahang
Pekan, Pahang, Malaysia

Alan Godfrey
Northumbria University
Newcastle upon Tyne, UK

ISSN 2195-271X ISSN 2195-2728 (electronic)
Lecture Notes in Bioengineering
ISBN 978-981-99-0296-5 ISBN 978-981-99-0297-2 (eBook)
https://doi.org/10.1007/978-981-99-0297-2

© The Editor(s) (if applicable) and The Author(s), under exclusive license to Springer Nature Singapore Pte Ltd. 2023
This work is subject to copyright. All rights are solely and exclusively licensed by the Publisher, whether the whole or part of the material is concerned, specifically the rights of translation, reprinting, reuse of illustrations, recitation, broadcasting, reproduction on microfilms or in any other physical way, and transmission or information storage and retrieval, electronic adaptation, computer software, or by similar or dissimilar methodology now known or hereafter developed.
The use of general descriptive names, registered names, trademarks, service marks, etc. in this publication does not imply, even in the absence of a specific statement, that such names are exempt from the relevant protective laws and regulations and therefore free for general use.
The publisher, the authors, and the editors are safe to assume that the advice and information in this book are believed to be true and accurate at the date of publication. Neither the publisher nor the authors or the editors give a warranty, expressed or implied, with respect to the material contained herein or for any errors or omissions that may have been made. The publisher remains neutral with regard to jurisdictional claims in published maps and institutional affiliations.

This Springer imprint is published by the registered company Springer Nature Singapore Pte Ltd.
The registered company address is: 152 Beach Road, #21-01/04 Gateway East, Singapore 189721, Singapore

Preface

The International Conference on Innovation and Technology in Sports (ICITS 2022) is the first edition of the conference organised by the Malaysian Sports Technology Association (MySTA) and the National Sports Institute, Malaysia, which is supported by the Malaysian Board of Technology (MBOT). This conference is one of the platforms to bring together those involved in the sports industry including researchers, engineers and technologists to share the latest research and innovation in the field of sports technology to further empower the sports industry. This volume hosts 24 papers that have been thoroughly reviewed by the appointed technical review committee that consists of various experts in the field of sports.

A sincere thanks to all members of the organising committee for making the conference a success. Not forgetting our sponsors, Reveal DNA from Genomas as the main sponsor of this conference and other sponsors consisting of Futurise, The Institution of Engineering (IET) Malaysia Local Network, Gatorade, Bleu, Goodday, Yakult, ATF Sport and also Richard Wee Chambers for their kind gesture and continuous support. We also would like to extend our appreciation to the authors for contributing valuable papers to the proceedings. We hope this book will intensify the knowledge sharing amongst colleagues in the field of sports technology and innovation.

Kuala Lumpur, Malaysia	Syed Faris Syed Omar
Pekan, Malaysia	Mohd Hasnun Arif Hassan
Manchester, UK	Alexander Casson
Newcastle upon Tyne, UK	Alan Godfrey
Suzhou, China	Anwar P. P. Abdul Majeed

Contents

Kinematic of Body Segments, the Force Growth, and the Speed of the Rowing

Ab Aziz Mohd Yusof, Hazim Sharudin, Wan Muhammad Syahmi Wan Fauzi, and Muhamad Noor Harun

Abstract The rowing performance depends on how fast the boat moves during the race, contributed by the generated forces and rower kinematic. Based on that interest, the objective of the study is to investigate the kinematics during a stroke by including the generated forces and related speed. The study was carried out using a dynamic rowing simulator which simplified the rowing boat and mimicked the rowing biomechanics and the hydrodynamic condition. The experimental result found that extending the legs generated a higher force to boost the speed. Using legs, trunk and arms in their overlapping motion helps the rower instantly reach the peak effort and maintain the pressure. About 496–539 N of the handle forces were captured by the simulator, which generated a blade force peak between 222 and 234 N, with the average peak force achieved for these three strokes being 46% of the drive. The oar angular speed captured was from 1 to 1.5 rad/s. In return, accelerate the boat at two m/s for 25–27 stroke/min rowing stroke. In conclusion, the combination of the kinematic, force and speed related to the rowing during the stroke is important to consider during the study. The effect of rower kinematic can be observed directly in the forces generated and associated speeds.

Keywords Rowing · Kinematic · Force · Speed

1 Introduction

The rowing race involves propelling the boat using an oar to generate the hydrodynamic force on the fully submerged blade. The race intends to move forward faster by maximising the generated force, which pushes the boat to a higher speed. During the beginning, the rower rows with precision strokes and produces high horsepower

A. A. Mohd Yusof · H. Sharudin · W. M. S. Wan Fauzi
School of Mechanical Engineering, College of Engineering, Universiti Teknologi MARA, UiTM, Johor Branch, Masai, Malaysia

M. N. Harun (✉)
Faculty of Engineering, School of Mechanical Engineering, Universiti Teknologi Malaysia, Johor Bahru, Malaysia
e-mail: mnoorharun@utm.my

© The Author(s), under exclusive license to Springer Nature Singapore Pte Ltd. 2023
S. F. Syed Omar et al. (eds.), *Innovation and Technology in Sports*, Lecture Notes in Bioengineering, https://doi.org/10.1007/978-981-99-0297-2_1

to develop force and speeds as quickly as possible. The performance highly depends on the rower's kinematic, the generated force and speed during the stroke. These three aspects are important to consider since they will help the rower move forward and be comfortable manipulating the rest of the race.

During the stroke, rowers transfer the generated energy to biomechanics capability through body kinematics using three body segments, arms, trunk and legs. The rower moves the body and generates force as the handle force. The force rapidly grew to achieve a peak at an average of 40% of the drive event. Biomechanics can be improved through anthropometry, body weight, stroke rate, and rowing style [1, 2]. V. Kleshnev reported that the average amount of power applied during the stroke was 46.4 ± 4.5% from the legs, 30.9 ± 5.2% from the trunk, and 22.7 ± 5.2% from the arms [3]. Changing the percentage of power from any of these body segments consequently changes the handle force pattern known as stroke style [4].

The rowing stroke contains four phrases repeated in the same sequence over the race event. The phases are begun with a catch, followed by the drive, finish, and end with recovery before it is repeated [3, 5–7]. The drive phase is the main contributor to hydrodynamic propulsion force, and the other phases are only complementary to the stroke [8]. During the catch phase, the blade is initially submerged, and then it is followed by the drive phase, where the blade travels through the arc path. The finish phase is achieved when the blade reaches the maximum travelling arc and is prepared to be lifted. The final phase is recovery, where the blade feathers and travels in the air, then return to the catch position. The blade force is developed based on the blade-boat relative speed. In this aspect, the oar blade moves faster than the boat, and the relative speed is obtained by minus the oar blade speed with boat speed. If the oar blade is faster than the boat, negative pressure is generated on the front of the blade to propel the boat. But if the oar blade is slower than the boat, negative pressure is generated at the back of the blade, consequently slowing the boat down.

This study was carried out to investigate how rower kinematics relates to force and speed growth. The study focused on the start phase, where force and speed gradually develop as the rower tries to move the boat from the stationary condition before extending up to 3 strokes before force and speed fully develop. As far as authors consent, very limited literature investigates kinematic, force and speed in one study to elucidate the dependency between them.

2 Method

2.1 Experimental Approach

The study mimics the rowing biomechanics and the hydrodynamic condition using a dynamic rowing simulator shown in Fig. 1. This approach eliminated the skin drag on the rowing shell, heave and pitch motion, balancing effect experienced by the rower, and bow surges the boat move. Furthermore, as the boat was fully controlled, the blade's hydrodynamic force was efficiently utilised to propel the boat.

The dynamic rowing simulator used in the experiment consisted of three main parts: (1) the simplified boat. (2) boat rail. (3) water tank. The simplified boat moved freely along the rail during the experiment using roller bearings. An oar with a Macon blade was attached to the boat through the oar lock placed at the right wall.

The size of the simplified boat was 1.50 × 0.69 × 0.36 m for length, width and height, respectively. The simplified rowing boat and oar weight was 35 kg and 2.5 kg. The rail size was 15.00 × 0.95 × 0.90 m, designed to fit the simplified rowing boat. The water tank size was 15.00 × 5.00 × 0.90 m for length, width and height.

Handle and blade hydrodynamic forces were sensed according to the shaft strain bending [9]. Four foil strain gauges FLA-6–350-17 manufactured by Tokyo Sokki Kenkyujo were used in the experiment. Strain gauges were glued on the shaft using a general-purpose Cyanoacrylate adhesive material. Two strain gauges were located on the outboard shaft, one on the tension and the other on the compression side, 24 cm away from the oar collar. The other two strain gauges were placed inboard at the region experiencing tension and compression. Both inboard strain gauges were located 15 cm from the oar collar.

The oar rotational angle and boat velocity were monitored using the rotary encoder DFS60A-S4PC65536, manufactured by SICK and attached to a simplified rowing boat. The one encoder was on top of the gate, and the second one was on the roller.

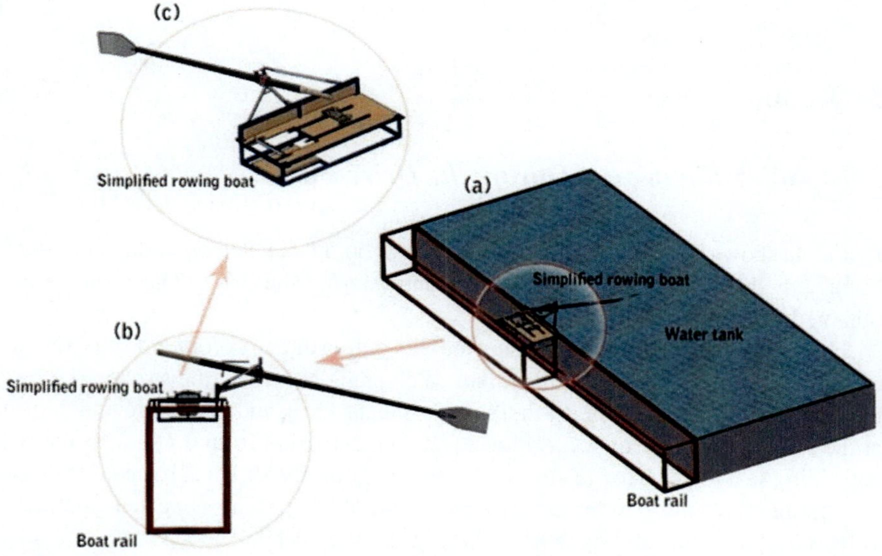

Fig. 1 Experimental equipment used in the study; **a** complete setup of rowing equipment which consisted of the simplified boat, rail, and water tank, **b** model of the rail and how the simplified boat was placed and **c** model of the simplified boat that replicated the real model of the rowing boat

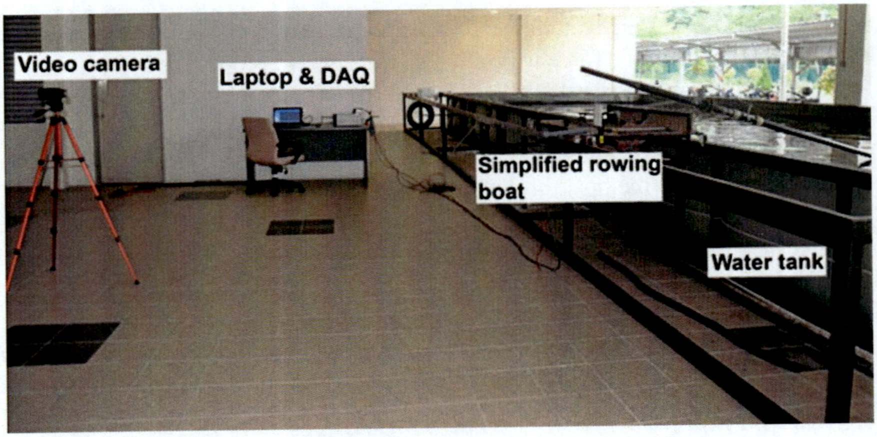

Fig. 2 Experimental setup for the study

Strain gauges and encoders were connected to a National Instrument data acquisition system with NI 9219 and NI 9401 modules and slotted into the NI9174 chassis.

The experiments began by pulling the oar handle towards the rower. Next, the motion of the rower body segments was captured using a camera (Sony FDR-AXP35) at 25 frames per second, as shown in Fig. 2. The experimental procedure was used to simultaneously capture rower kinematic, handle force, and blade hydrodynamics.

3 Result

3.1 Rower Kinematics During the Drive Phase

Figure 3a shows how the rower situated his body in the hatched position at the start. Figure 3b shows the body segment kinematic during the drive. The results were obtained from video analysis.

The stroke is started by compressing the legs, tilting the trunk forward and straightening the arms to achieve the maximum catch angle to initiate the stroke. The legs then extended backwards for the first 0.4 s, while the trunk and arms were maintained at the same pose as the initial. An increasing leg extension from 0.4 to 0.6 s caused the sliding seat to move for 12 cm. After 0.6 s, a significant leg trunk happens to cause the sliding seat to move to the maximum distance of 23 cm. The stoke then continues by bending the arms and increasing the oar angular speed. The leg extension stop at 0.8 s. Instead, the oar handle moved due to the increasing trunk tile angle, where the angle changed about 68° relative to the original position. Besides that, after half of the drive, the arms also moved quicker by moving the handle faster towards the body, which made the total elbow bending equally to 83°. By the end of the drive,

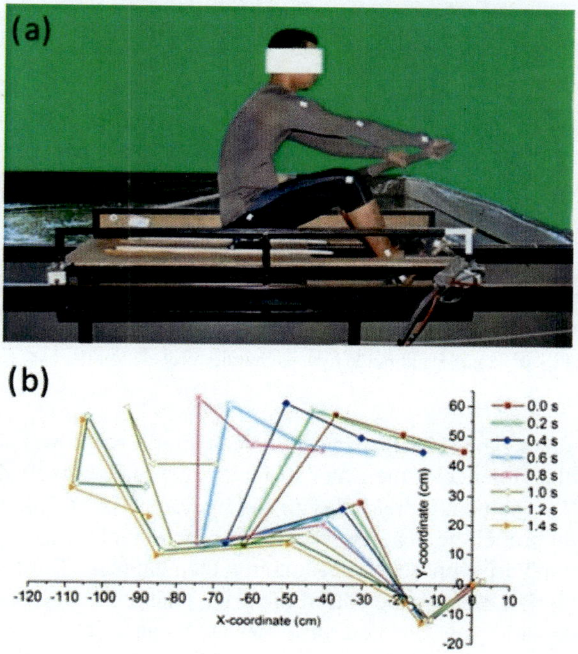

Fig. 3 **a** Rower equipped with the marker for video analysis **b** Kinematic of rower for the first stroke

these three body segments coordinated to achieve the finish, which can be seen at 1.0 and 1.2 s. At the end of the drive at 1.4 s, only the rower's arm removed the blade from the water, whereas other body segments stopped as they reached the maximum reachable distance.

3.2 *Force Profile of the Drive Phase*

The handle and the corresponding blade hydrodynamic force were recorded using strain gauges, shown in Fig. 4. The force increased gradually during the early drive phase due to rower kinematic. As the force magnitude reached 46% of the drive, it decreased gradually to zero as the blade was lifted from the water. The maximum handle forces applied by the rower at the starting phase are 496 N, where 539 N for the second stroke and 501 N for the third stroke.

The blade force is the hydrodynamic force that acts on the blade due to the relative velocity between the oar rotation and boat translation. As soon as the stroke started, the blade hydrodynamic force increased gradually to the highest peak of 222 N at 46% of the first drive. Meanwhile, for the second stroke, a peak force of 227 N was achieved at 42% of the second drive. The last peak of 234 N was reached at 50% of the drive. The average peak force achieved for these three strokes is 46% of the

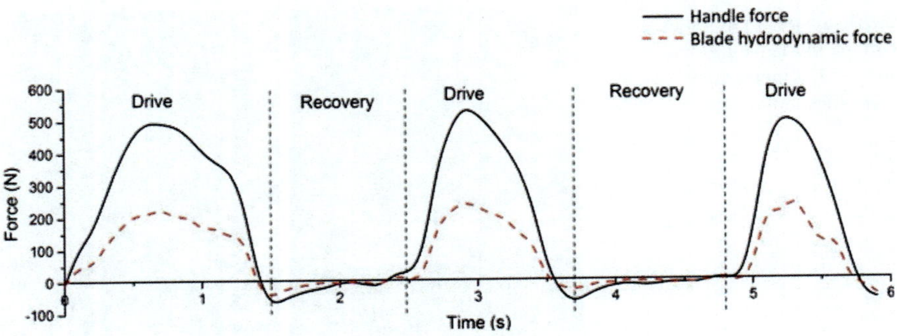

Fig. 4 Handle and blade hydrodynamic force act on the oar

drive. The smoothness of the blade graph was lower than the handling graph due to the violence of the water surface as the blade travelled in water.

The boat increased with the increment of the blade force at power was applied to propel the boat. However, once the magnitude of handle force passes 50% of the total drive phase, force reduction then happens. But the boat continues to gain speed until it reaches the end of the drive phase. The increasing boat speed influences the situation due to the cruise inertia. Meanwhile, the reduction in the relative speed between the blade and the boat speed affected the blade force [10]. As the blade force reduced, the oar was rotated easily, which is represented by a higher angular speed at the end of the drive but contains less power.

3.3 *Speed Profile of the Oar and Boat*

The best three speeds and oar angular speeds were captured as in Fig. 5. The oar angular speed grows gradually up to 1 rad/s for 80% of drive time. Once the oar reaches the finishing phase, the blade overturns, illustrated as negative notation. For most overturn or recovery phases, the average travelling velocity is at 1.5 rad/s, 33% faster than drive. Similar to the first stroke, the oar angular speed of the second stroke increased gradually from zero to the maximum angular speed. The number was 1.4 rad/s higher and 0.3 s shorter in duration than the first stroke. Meanwhile, the third stroke was 0.1 s faster than the second and maintained the peak angular speed.

Since the rowing boat was initially stationary, most of the force applied was used to overcome the stationary inertia. As a result, the boat moved slower before accelerating to achieve two m/s, as shown in Fig. 5. However, during the translation stage between the first and second stroke, speed reduction happened as the rower returned to the hatch position to start the second stroke as the rower moved in the opposite direction of the boat. Since the boat had already moved and the oar was rotated faster than the initial stroke, the boat seemed to move even faster related

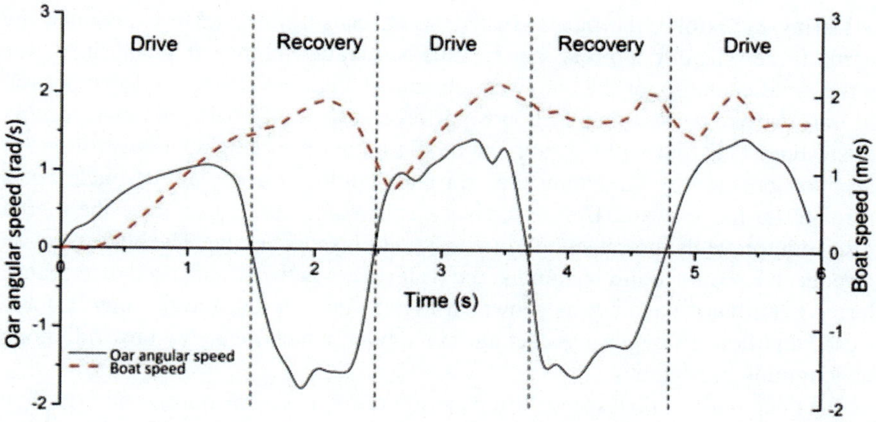

Fig. 5 Angular speed of the oar during the start and rowing boat velocity profile

to the increment of the stroke rate from 25 strokes/min at the initial stroke to 26 strokes/min for the second stroke and 27 strokes/min for the third stroke.

4 Discussion

Leg, trunk and arm body segments were used during the stroke. Most of the power is contributed by the leg and trunk. The stroke began when the rower extended the legs and generated a large and boost force. The use of the leg is important as it contributes $46.4 \pm 4.5\%$ of the stroke power, as reported by Kleshnev [3]. The stroke was sustained with simultaneous emphasis on the leg extension, trunk alternation and the arms bending to extend the sweeping angle. The use of the trunk provided $30.9 \pm 5.2\%$ of the stroke power, and the arm provided $22.7 \pm 5.2\%$ of the stroke power [3]. There are several advantages of joining the use of legs, trunk and arms in their overlapping motion. The rower can reach the peak effort instantly and maintain the pressure on the blade to provide the propulsive force in accelerating the boat.

The average maximum handle force is 512 N with a standard deviation of 23 N, and the average corresponding blade force is 231 N with a standard deviation of 13 N. The handle and blade forces show a pattern as in Fig. 4. This study confirms that the oar is working according to the first-class lever. It shows the load, fulcrum and effort lever element arrangement, representing the handle force, oarlock and blade force. This concept agrees with Kleshnev study [6]. The ratio of the outboard to the inboard length of the oar is defined as the mechanical advantage. In this experimental study, the oar's mechanical advantage was 0.45. However, the value was small, less than one. However, the advantage of the first-class lever of the oar produced a larger arc sweeping distance and moved the boat forward within a short time with a small force on the blade as an anchor.

During the stroke, the oar is rotated faster, causing the oar to experience the inertia force. The effect of oar inertia force is seen during the catch and finish due to the rapid changing of the oar sweep direction. Figure 4 shows the force peak at the early stroke and the higher opposing force at the end of the drive. According to Kleshnev [11], during the catch, the inertia force effect helps the rower improve the body position by stretching the arm and shoulder muscle as the blade tends to rotate further forward. Before recovery, it is suggested to maximise the inertia force to improve the propulsion through a good finish technique. The blade rotation introduced a momentum-changing at the boat side, which caused speed disturbance during catch and finishing, as shown in Fig. 5. In addition, the oar inertia force caused additional moving mass acting on the boat, which made the boat roll along the longitudinal boat axis.

The fluctuation of boat speed during the stroke decreased the boat speed in motion disturbance. According to this study, three factors that influence the fluctuation of the boat were identified: first, fluid drag acting on the blade once the blade submerged during the catch. Once the blade is submerged for the catch, the blade's rotational speed is zero. This has caused drag resistance at the back of the blade, which resists the boat's motion. Second, changing the stroke phase from water support during the drive phase to unsupported water condition during the recovery phase. Between these phases, the energy to accelerate the boat is cut as the blade is lifted from the water. Third, the rower moved in the opposite direction compared to the boat velocity while returning to the catch position. It created additional momentum, which disturbed the boat's momentum. The fluctuation of boat speed can be reduced by increasing the stroke rate, especially during returning from the finish to the next catch position, because it reduces the time interval and maintains the power supply to the boat. During returning to the hatch body position, a smooth and soft movement should be envisaged by the rower to reduce the momentum of altering to avoid any disturbance of the boat velocity.

5 Conclusion

In conclusion, the handle force and blade hydrodynamic force were important in the kinematic or stroke style investigation to obtain the accurate boat speed for the performance evaluation. The right catch timing the controlled by the kinematic, which can be achieved by moving the blade at a suitable speed matching the boat speed. The performance can be increased by applying the right kinematic in the form of stroke style. Response of the kinematic happed directly in the form of force and speed. The fluctuation of the moving boat affects rowing efficiency. The right timing of blade placement can improve during the drive, better-finished phase and proper movement during returning to the catch position.

References

1. Doyle MM, Lyttle A, Elliott B (2010) Comparison of force-related performance indicators between heavyweight and lightweight rowers. Sports Biomech 9(3):178–192
2. Kleshnev V (2006) Rowing biomechanics. Biorower
3. Kleshnev V (2002) Segments contribution to the stroke length and rower power. Higher percentage of the trunk power helps to achieve higher boat speed. Rowing Biomech Newslett 2(2)
4. Kleshnev V (2005) Experimental data on comparison of rowing on-water with Concept-II and RowPerfect ergometers. Rowing Biomech Newslett 5(3)
5. Caplan N, Coppel A, Gardner T (2010) A review of propulsive mechanisms in rowing. Proc Inst Mech Eng Part P J Sports Eng Technol 224(1):1–8
6. Kleshnev V (2002) Moving the Rowers: biomechanical background, vol 25. Australian Rowing, Carine, WA, pp 16–19
7. Sliasas A, Tullis S (2011) Modelling the effect of oar shaft bending during the rowing stroke. Proc Inst Mech Eng Part P J Sports Eng Technol 225(4):265–270
8. Baudouin A, Hawkins D (2004) Investigation of biomechanical factors affecting rowing performance. J Biomech 37(7):969–976
9. Caplan N, Gardner TN (2007) Optimization of oar blade design for improved performance in rowing. J Sports Sci 25(13):1471–1478
10. Sliasas A, Tullis S (2009) Numerical modelling of rowing blade hydrodynamics. Sports Eng 12(1):31–40
11. Kleshnev V (2015) Oar inertia forces. Rowing Biomech Newsl 170

Mechanical Testing of Futsal Footwear: Friction Coefficient Under Different Sliding Direction

Shariman Ismadi Ismail, Hiroyuki Nunome, Filip Gertz Lysdal, Uwe Gustav Kersting, Ahmad Faizal Salleh, and Hosni Hasan

Abstract This study aimed to clarify the differences on friction coefficient of footwear used in futsal when mechanically measured in two different sliding direction. Available Friction Coefficient (AFC) and Traction Force (TF) of three futsal footwear with different outsole design (S1, S2 and S3) were measured using a novel six-degree of freedom mechanical test in anteroposterior (AP) and mediolateral (ML) sliding direction. Results have shown differences of AFC value when measured in different sliding direction (AP and ML) for all three shoes. In addition, it was observed that S2 shoe was the least affected in terms of reduction of AFC value when compared between AP and ML direction. It was also observed that among the three shoes tested, S2 has produced the highest TF in both AP and ML direction as compared with other shoes. From these findings, it can be suggested that traction performance of sports footwear should be evaluated by multi-directional sliding approach, and conventional one directional footwear evaluation standard such as BE EN ISO 13287 is most likely not adequate to analyse sports footwear–sports playing surface traction performance in real world.

S. I. Ismail (✉) · H. Hasan
Faculty of Sports Science & Recreation, Universiti Teknologi MARA, Shah Alam, Malaysia
e-mail: shariman_ismadi@uitm.edu.my

H. Nunome
Faculty of Sports and Health Science, Fukuoka University, Fukuoka, Japan

F. G. Lysdal
Faculty of Sport, Allied Health and Performance Sciences, St Mary's University, Twickenham, London, UK

U. G. Kersting
Institute of Biomechanics and Orthopaedics, German Sport University Cologne, Cologne, Germany

Sport Sciences, Department of Health Science and Technology, Aalborg University, Aalborg, Denmark

A. F. Salleh
Sports Engineering Research Centre, Centre of Excellence (SERC), Universiti Malaysia Perlis, 02600 Arau, Perlis, Malaysia

© The Author(s), under exclusive license to Springer Nature Singapore Pte Ltd. 2023
S. F. Syed Omar et al. (eds.), *Innovation and Technology in Sports*, Lecture Notes in Bioengineering, https://doi.org/10.1007/978-981-99-0297-2_2

Keywords Friction coefficient · Football codes · Shoes · Mechanical · Futsal · Force platform · Traction · Interaction

1 Introduction

There are several studies investigated the influence of outsole tread grooves on the friction coefficient [1–4]. Those researchers commonly used mechanical testers which examined the friction of footwear in a quasi-static condition in one (anteroposterior) sliding direction such as BE EN ISO 13287 (2007) which is an international standard on test methods for slip resistance in footwear to ensure prevention against slipping. However, one previous study [5] pointed that mechanically measured frictional values do not necessarily represent actual traction performance in sports-specific human movements. There are inconsistent findings between mechanically measured traction coefficient and human traction performance commonly observed in previous studies [6–9]. It can be assumed that these inconsistent findings are potentially due to the fact that test conditions used in conventional mechanical tests are far different from footwear—surface interaction made by players in real-world, sporting situation. This implies the limitation of conventional mechanical tests on the veracity to anticipate actual footwear traction performance in sports-specific scenarios. We aimed to clarify the differences on friction coefficient of footwear used in futsal when mechanically measured in anteroposterior and mediolateral sliding direction.

2 Methodology

2.1 *Shoes and Playing Surface*

Three types of futsal shoes (Fig. 1: S1, S2 and S3) with different outsole tread grooves were selected for the test. Three pairs of shoes (size 27.0 cm) as shown in Fig. 1, were selected for mechanical test. Each shoe properties and features are described in Table 1. For the sample of playing surface (futsal court material), an area-elastic hardwood surface was selected for this study (Fig. 1). The technical specification (shock absorption and sliding coefficient properties) of the hardwood surface is described in Table 1 as well.

2.2 *Mechanical Testing*

Traction force (TF) was measured using a mechanical test consisting a 4-degrees of freedom force platform that is controlled by a hydraulic system [10]. Above the platform, an artificial foot made from nylon was statically secured to a profile steel

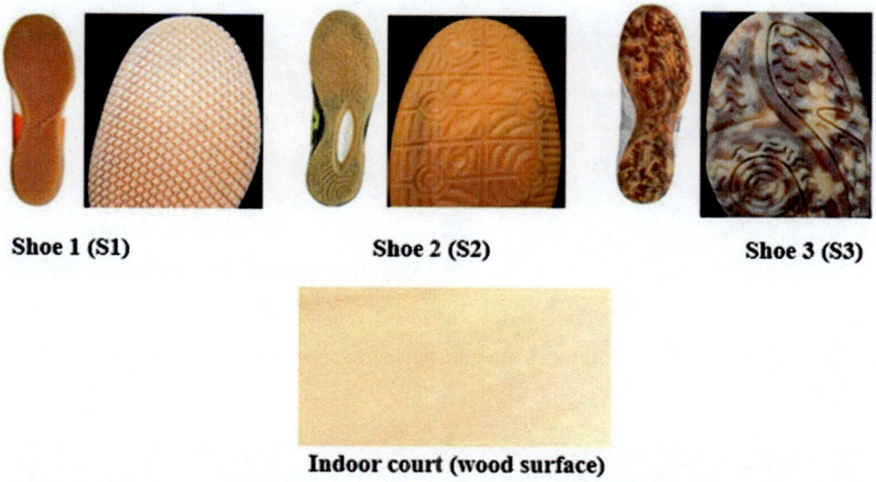

Fig. 1 The shoes and indoor court material (futsal court) selected in the study

Table 1 Shoes and playing surface properties

Shoe	Mass per shoe (gram)	Sole hardness, Shore A (degree) [manufacturer data]	Playing surface properties [manufacturer data]	
S1	311	57	Shock absorption: 40–55%	Sliding coefficient: 80–110
S2	232	60		
S3	276	60		

frame structure, to which the three types of footwear were tightly secured. To avoid any unwanted movements between the shoe and artificial foot, the shoe was bolted on the anterior and posterior tips of the artificial foot (Fig. 2). The ground reaction forces (Fx, Fy and Fz) were recorded using a force platform (AMTI) that is synchronised to a Qualisys motion analysis system (Qualisys AB, Gothenburg, Sweden) as shown in Fig. 3.

Futsal playing surfaces were tightly bonded on top surface of the force platform using a strong double-sided tape. The moving force platform was controlled using a customized system (Mr. Kick, v. 2.030, Knud Larsen, Aalborg University, Denmark). The platform starts at an initial position below the footwear and upon activation, it will move towards the static footwear to apply the normal load. Upon applying the certain magnitude of normal load, the force platform will move horizontally, creating a backward sliding motion between the top layer of the playing surface and the footwear outsole.

TF was calculated from the mean traction force during 125 ms of steady-state condition. Using TF values and the offset normal force (500 N), available friction coefficient (AFC) was computed. The calculation of AFC is shown in Eq. 1, where

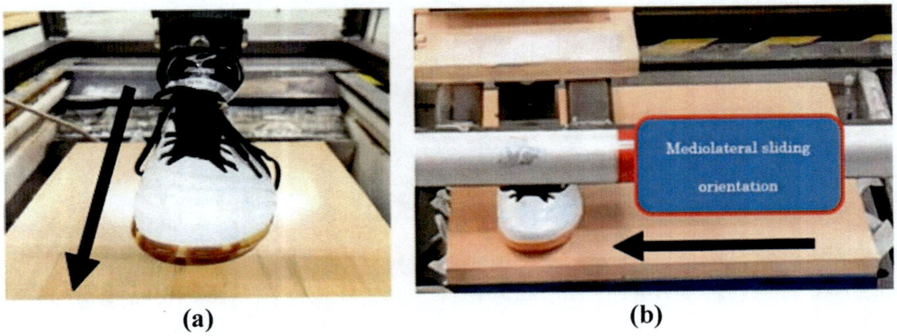

Fig. 2 Illustration of shoe attachment to the artificial foot, **a** anteroposterior, **b** mediolateral sliding orientation

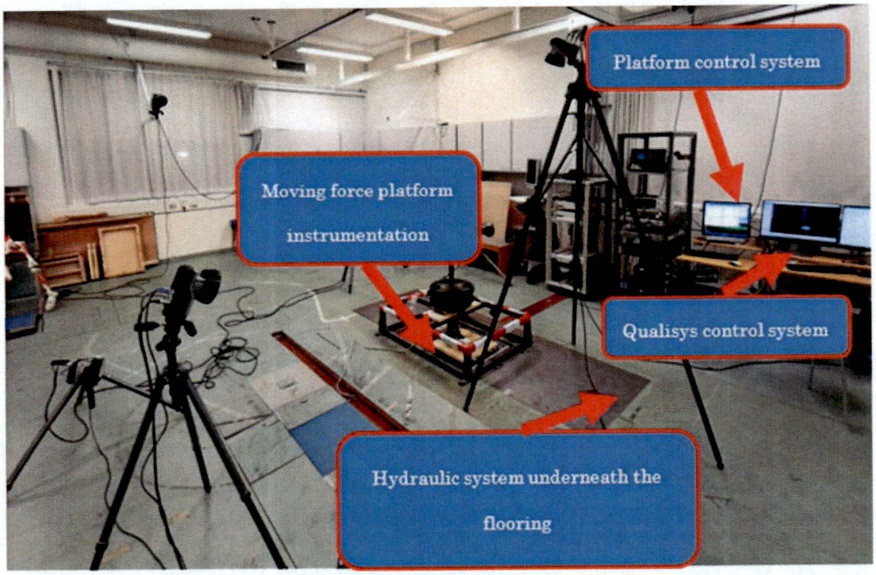

Fig. 3 Test setup for the moving force platform that is synchronized to a Qualisys motion analysis system

AFC was defined as the mean value of friction coefficient during 125 ms of the steady-state condition of the friction coefficient curve [11]. Raw data of the force platform were filtered by 4th order low-pass Butterworth filter at 50 Hz cut-off frequency [12].

$$\mathrm{AFC} = \frac{\sqrt{Fx^2 + Fy^2}}{|Fz|} \tag{1}$$

AFC and TF measurements were carried out with three footwear—playing surface (three shoes, and one playing surface) under dry friction condition. The test condition was set as the followings: (1) the offset normal load = 500 N, (2) the sliding velocity of the force platform was 0.3 m/s and (3) the force platform slides against footwear outsole in two directions (anteroposterior and mediolateral) with contact angle at 0° in accordance to the BE EN ISO 13287 test standard. After anteroposterior sliding tests were completed, the artificial foot is rotated 90° from the original position to allow the test to be conducted to the mediolateral side footwear.

2.3 *Statistical Analysis*

Statistical analyses were performed using IBM SPSS Statistics 22 (SPP Inc., Chicago, IL, USA). The mean TF and AFC of tested conditions between AP and ML sliding orientations for each shoe were compared using paired sample t-test (two-tailed) and one-way ANOVA repeated measure was used to identify differences between the threeshoes TF and AFC properties. Statistical significance was set at $p < 0.05$.

3 Results

3.1 *Traction Force (TF)*

There was a significant difference ($p < 0.05$) of TF of S2 shoe in both AP (722 ± 2 N) and ML (664 ± 22 N) direction as compared with other shoes [S1: AP- 672 ± 10 N, ML- 622 ± 30 N; S3: AP- 700 ± 2 N, ML- 605 ± 5 N] as shown in Table 2. It was observed that among the three shoes tested, S2 has produced the highest TF in both sliding orientations ($p < 0.05$). It was also noticed that TF for ML sliding orientation were significantly lower ($p < 0.05$) than AP orientation for all three shoes.

3.2 *Available Friction Coefficient (AFC)*

Similar to results found in TF, it was observed that among the three shoes tested, S2 has produced the highest AFC in both AP and ML sliding orientations ($p < 0.05$). There was a significant difference ($p < 0.05$) of AFC value when measured in different sliding direction (AP and ML) for all three shoes [S1: AP ($M = 1.25$, $SD = 0.006$) and S1: ML ($M = 1.00$, $SD = 0.002$), $p = 0.0001$; S2: AP ($M = 1.34$, $SD = 0.009$) and S2: ML ($M = 1.31$, $SD = 0.004$), $p = 0.0001$; S3: AP ($M = 1.30$, $SD = 0.006$) and S3: ML ($M = 1.11$, $SD = 0.008$), $p = 0.001$]. In addition, it was observed that S2 shoe was the least affected in terms of reduction of AFC value when compared between AP and ML direction [AFC reduction AP vs. ML: S1- 20% reduction; S2- 2% reduction; S3- 14.6% reduction].

Table 2 Traction force and available friction coefficient mechanical test results under AP and ML sliding orientations

		Traction force [N]		Available friction coefficient (AFC)	
Shoe	Orientation	Mean	SD	Mean	SD
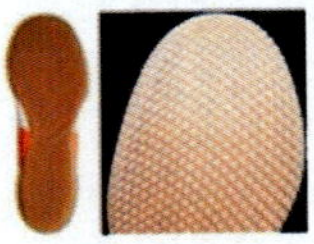 Shoe 1 (S1)	AP	672 **A***	10	1.25 **A***	0.006
	Ml	622 **A***	30	1.00 **A***,†	0.002
		T*		**T***	
 Shoe 2 (S2)	AP	722 **A***,‡	2	1.34 **A***,‡	0.009
	ML	664 **A***,‡	22	1.31 **A***,‡	0.004
		T*		**T***	
 Shoe 3 (S3)	AP	700 **A**‡	2	1.30 **A**‡	0.006
	ML	605 A‡	5	1.11 A*,†	0.008
		T*		**T***	

Standard deviation: SD; Paired sample t-test AP versus ML ($p < 0.05$): T*; One-way ANOVA: A; S1 versus S2: *, S1 versus S3: †, S2 versus S3: ‡ ($p < 0.05$)

4 Discussion

This study examined whether there is a significant difference in traction force (TF) and available friction coefficient (AFC) when mechanically measured under different sliding orientation. Three futsal shoes with different shoe outsole tread grooves were mechanically tested in accordance to the BE EN ISO 13287 test standard (AP sliding orientation). In addition, we also tested the shoe and surface interaction under ML sliding orientation, which is not included in the ISO standard. It was found that there was a significant difference in TF as well as AFC properties for shoe and surface interaction measured under different sliding orientation and different shoes. It was also observed that among the three shoes tested, S2 shoe has produced significantly greater TF and AFC as compared to S1 and S3 shoes. In addition, it was identified that there were AFC performance reduction under ML sliding orientation, in which S2 shoe had the least AFC reduction as compared to S1 and S3 shoes.

This study has confirmed that shoe-surface interaction can produce different outcome under different test condition. The shoe outsole property possessed by each shoe may have contributed to these findings. Our mechanical test results are also in line with other studies in the past that involved with human subject functional test, in which have reported that shoe and surface interaction has shown to produced different outcome under different movement task and different shoe properties involving different movement orientation such as straight-line sprint and change of direction movement [5, 13]. In addition, sport such as futsal does require the players to perform various movement [14] and the aspects of traction performance has been shown to significantly influence the functionality and performance of the player [15]. These movements typically will produce different magnitude of the ground reaction force, in which would command different shoe-surface interaction requirement or demand. As frictional properties of surfaces are specific to particular loading patterns [16], the observed discrepancy can be explained by different loading patterns applied during the tests. From the given findings, it should be highlighted that the mechanical test used in this study succeeded in discriminating concealed differences of frictional properties among the three shoes. Finding from this study also has highlighted the limitation of current footwear test standard in relation to the sports footwear. Sports movements require multi-directional task, and therefore require a more specific test condition as compared to footwear that is use for daily activity.

In terms of the shoe design, it was found that the outsole tread grooves design may have contributed to TF and AFC properties (S2 > S1 and S3). These findings are also in line with other footwear studies that have found the influence of outsoles properties on traction performance which could also affects sports performance [17, 18]. In addition, the range of AFC found in this study (1.00–1.34) was in line with finding from previous study which suggested the traction coefficient for sports shoes are required to achieve minimum threshold of 0.7 to allow enough confidence to perform dynamics task such as change of direction in futsal [19].

However, in this study, only three types of futsal shoes and one playing surface have been tested. Therefore, the study results may not represent the general behaviour of sports footwear and sports surface interaction. More investigation is required for different type of shoes and different materials of playing surface to understand more about sports shoe-surface interaction.

5 Conclusion

Different outcomes of TF and AFC for these sliding directions could potentially due to the different outsole properties which dictates the ground reaction force produced during shoe–surface interaction under different sliding direction. From these findings, it can be suggested that traction performance of sports footwear should be evaluated by multi-directional sliding approach, and conventional one directional evaluation is most likely not adequate to mimic footwear–playing surface traction performance in real world.

Acknowledgements The authors would like to thank Mizuno Inc., Japan for partially supporting this study.

References

1. Beschorner KE, Iraqi A, Redfern MS, Cham R, Li Y (2019) Predicting slips based on the STM 603 whole-footwear tribometer under different coefficient of friction testing conditions. Ergonomics 62(5):668–681
2. Blanchette MG, Powers CM (2015) The influence of footwear tread groove parameters on available friction. Appl Ergon 50:237–241
3. Iraqi A, Cham R, Redfern MS, Beschorner KE (2018) Coefficient of friction testing parameters influence the prediction of human slips. Appl Ergon 70:118–126
4. Yamaguchi T, Katsurashima Y, Shibata K, Hokkirigawa K (2015) Study on rubber tread block design for high slip resistance on liquid-contaminated condition. In: Proceedings 19th triennial congress of the IEA. vol 9, pp 14. (Aug 2015)
5. Sterzing T, Müller C, Hennig EM, Milani TL (2009) Actual and perceived running performance in soccer shoes: a series of eight studies. Footwear Sci 1(1):5–17
6. Milani TL, Hennig EM (2002) Biomechanische Testverfahren und Laufschuhforschung. Medizinisch-Orthopädische Technik 122(3):68–75
7. Nigg BM (1990) The validity and relevance of tests used for the assessment of sports surfaces. Med Sci Sports Exerc 22(1):131–139
8. Pedroza A, Fernandez S, Heidt R, Kaeding C (2010) Evaluation of the shoe-surface interaction using an agility maneuver. Med Sci Sports Exerc 42(9):1754–1759
9. Sterzing T, Muller C, Schwanitz S, Odenwald S, Milani TL (2008) Discrepancies between mechanical and biomechanical measurements of soccer shoe traction on artificial turf. In: 26th symposium of the international society of biomechanics in sports. Seoul National University, pp 339–342. https://ojs.ub.unikonstanz.de/cpa/article/view/2103
10. van Doornik J, Sinkjær T (2007) Robotic platform for human gait analysis. IEEE Trans Biomed Eng 54(9):1696–1702
11. Morio C, Sissler L, Guéguen N (2015) Static vs. dynamic friction coefficients, which one to use in sports footwear research? Footwear Sci 7(sup1):S63–S64
12. De Clercq D, Debuyck G, Gerlo J, Rambour S, Segers V, Van Caekenberghe I (2014) Cutting performance wearing different studded soccer shoes on dry and wet artificial turf. Footwear Sci 6(2):81–87
13. Ismail SI, Nunome H, Tamura Y (2021) Does visual representation of futsal shoes outsole tread groove design resemblance its mechanical traction, dynamic human traction performance, and perceived traction during change of direction and straight sprint tasks? Footwear Sci 13(1):79–89
14. Ismail SI, Nunome H (2021) The effect of different tournament stages on the movement dynamics of futsal players while in ball possession. Hum Mov 22(4):68–77
15. Ismail SI, Nunome H, Tamura Y (2021) The influence of forefoot bending stiffness of futsal shoes on multiple V-cut run performance. Front Psychol 12:625079
16. Clarke J, Carré M, Damm L, Dixon S (2012) Understanding the influence of surface roughness on the tribological interactions at the shoe–surface interface in tennis. Proc Inst Mech Eng Part J J Eng Tribol 226(7):636–647
17. McGhie D, Ettema G (2013) Biomechanical analysis of traction at the shoe-surface interface on third-generation artificial turf. Sports Eng 16(2):71–80

18. Serrano C, Sánchez-Sánchez J, López-Fernández J, Hernando E, Gallardo L (2020) Influence of the playing surface on changes of direction and plantar pressures during an agility test in youth futsal players. Eur J Sport Sci 20(7):906–914
19. Morio C, Bourrelly A, Sissler L, Gueguen N (2017) Perceiving slipperiness and grip: a meaningful relationship of the shoe-ground interface. Gait Posture 51:58–63

Physical Fitness Profile and Match Analysis of Elite Junior Badminton Players: Case Studies

Wei Sheng Wei Kui, Hui Yin Ler, and Mei Teng Woo

Abstract The purpose of this study was to determine the physical fitness profile, heart rate responses, technical and timing analysis of elite junior badminton players. Four elite junior badminton players (2 males & 2 females; Average age = 17.5 years) with a minimum of 12 h of training per week were included in this study. The players completed a series of fitness testing (cardiorespiratory fitness, speed, agility, upper strength & lower strength, and power) to obtain their physical fitness profile. Heart rate responses were recorded during the matches. Videos of both genders were analyzed using a customized badminton tagging panel with video analysis software. Results showed the fitness levels of male players were better than female players. During the matches, female players demonstrated a greater average % HR compared to male players. Match analysis demonstrated that male single favourite shots were net (31.68%) and smash (20.04%), while female favourite shots were clear (25.43%) and drop (19.58%). The winning shots for the males were mostly from smash, while the females' winning shots were from drop. The results of the study provide an insight to the coaches to develop effective training plans for junior players based on gender. (197 words).

Keywords Technical analysis · Heart rate · Timing motion analysis

W. S. W. Kui (✉) · H. Y. Ler
Faculty of Applied Sciences, Department of Sport Science, Tunku Abdul Rahman University of Management and Technology, Kuala Lumpur, Malaysia
e-mail: kuiws-wa16@student.tarc.edu.my

H. Y. Ler
e-mail: lerhy@tarc.edu.my

M. T. Woo
School of Sports, Health and Leisure, Republic Polytechnic, Singapore, Singapore
e-mail: woo_mei_teng@rp.edu.sg

© The Author(s), under exclusive license to Springer Nature Singapore Pte Ltd. 2023

S. F. Syed Omar et al. (eds.), *Innovation and Technology in Sports*, Lecture Notes in Bioengineering, https://doi.org/10.1007/978-981-99-0297-2_3

1 Introduction

Badminton, one of the most popular sports in Malaysia after football. The major achievement was winning the Thomas Cup back in 1992. However, they have not been able to get the same performance results in the Thomas Cup ever since. The unbalanced team, which relies on specific singles and doubles could be the major disadvantage for the team event at Thomas Cup. To strengthen the team event in the major badminton games such as Thomas Cup and SEA Games, it is particularly important to build the pool of badminton athletes through an effective junior development programme. This allows greater transition and wider net for athlete's selection for team event (e.g., Thomas Cup).

In a typical development programme and coaching plan, it is important for coaches and sports scientists to consider the physiological requirements, and the tactical and technical aspects of a badminton match. Heart rate monitoring is important from a physiological perspective because it is an indicator that reflects body metabolism and is widely applied to sports training practice. It provides a comprehensive reflection of the physiological conditions in vivo during sport movement and of the real-time sports intensity during training and competition (Chi, 2014). A few researchers reported that an average HR_{max} value of 190.5 beats.min^{-1} in elite males and 193 beats.min^{-1} in elite females during tournaments [1–5]. While male juniors have an average of 198 beats.min^{-1} of HR_{max} and female juniors have an average of 199 beats.min^{-1} of HR_{max} during the bleep test. To the best of our knowledge, no study has reported the heart rate responses of junior players during tournaments.

A typical badminton match consists of repeated bouts of high intensity movements with short periods of rest [5]. The primary energy system involved during the match is the anaerobic system, the alactic anaerobic system [5–8]. Moreover, Leong and Krasilshchikov [9] found that elite players required a higher aerobic capacity than junior elite players in the international tournament (1449.2 s vs. 1066.3 s) with higher intensity (12.3 vs. 8.7 shots per rally). The understanding of physiological demand during badminton matches would help coaches develop better strategies for both tactical plays between the sets as well as between the games.

Match analysis has been commonly used in analyzing the technical aspects of team sports such as water polo, handball [10], basketball [11] and soccer [12]. Recently, an increasing interest has been observed in using video-based analysis in badminton [13]. Technical analysis and timing analysis, also known as temporal structure, are the typical match analysis used in sports. Timing motion analysis is known as playing time, rally time, pause time and effective playing time in a match. While the technical analysis corresponds to the quantification of the offensive and defensive actions performed by the athletes [14]. The overall match related analysis data could provide insights to enhance coaches' understanding on the baseline fitness and important stroke requirements of a match [13, 15]. Previous studies showed that Men singles' most favorite shots in the last shot of the rallies was smash (29.1%) and drive (6.3%) [16–18]; while for women, singles drop was the favorite shot of the last rallies

(9.0) [16, 19]. These indicated that both genders tend to use more offensive strategies than defensive strategies, which leads to greater physical demand. However, there have been limited studies published on the junior playing style and physiological demand during matches that provide development coacher with information for training planning.

2 Objectives

To determine the physical fitness profile, heart rate responses, and technical and timing analyses of both junior men's and women's badminton matches.

3 Method

3.1 Samples

Four elite junior badminton players (2 males & 2 females; age = 17.5 years old; Height = 168.75 cm; Weight = 65 kg) with minimum 12 h of training per week were included in this study. Two matches (Male 1 (M1) vs. Male 2 (M2); Female 1 (F1) vs. Female 2 (F2)) were recorded during the National Junior Internal Ranking 2021. All of them had international tournament experiences. The study was reviewed and approved by Tunku Abdul Rahman University College Ethics Committee (TARUC/EC/2022/02-7).

3.2 Procedures

Subjects were required to perform a series of physical tests to determine their cardiorespiratory fitness, speed, agility, upper strength & lower strength, and power.

3.2.1 Cardiorespiratory Fitness

Beep test [20] was used to evaluate the players cardiorespiratory fitness. Two markers were placed in the starting and finishing line with 20 m distance. The player started when the beep test audio instructed. A successful completion of a shuttle counts if the player can reach the end of the 20 m shuttle before the sound of the next beep which indicates the start of the next shuttle. The process was repeated until the player failed to reach the marker three times before the beep sound. The maximum heart rate of each player was recorded during the beep test.

3.2.2 Speed

A 20 m sprint test [21] was used to evaluate the player's speed ability. The Fusion Smart Speed device was placed in the starting and finishing line with 20 m distance. The player started with a ready position behind the Fusion Smart Speed. The command of '1, 2, 3, Go' was given by the BAM exercise physiologist. After the command, the player was instructed to run as fast as possible to the finishing line and pass through the SmartSpeed Timing Gate System. Subjects were given three trials and the fastest time was recorded.

3.2.3 Agility

The badminton specific agility test [22] was used to determine the players' agility performance. The Fusion Smart Speed device was placed in the front and back of the badminton court and 4 shuttlecocks were placed in four corners of the badminton court. The player starts when he/she is ready from the middle court. The players moved to the backhand front court first then followed by the forehand front court, then to the backhand back court and finished with the forehand backcourt.

3.2.4 Upper Body Strength and Lower Body Strength

The 5RM Bench Press test was used to evaluate upper body strength and the 5RM Squat test was used to evaluate lower body strength. The players started warming up with 10 repetition 50% of previous 1RM, then 8 repetition 75% of previous 1RM. After the warmup, players started their first attempts at 5RM. A total of four attempts were done and the last attempt's result was recorded to calculate an estimated 1RM of each player [22]. The result was presented in relative strength (total lifted weight divided by body weight).

3.2.5 Power

A countermovement jump test [22] was used to determine the players' power ability. Players were instructed to jump on the Fusion Smart Jump mat. During the jump, the players must place both hands on the hips throughout the jump. Once received then command of '1, 2, 3, Jump' given by the researcher, the players jumped as high as possible with both hands placed on the hip. They were given three trials and the highest jump height was recorded.

3.2.6 Heart Rate Analysis

Prior to the competition, permission from the coach was obtained to measure the heart rate responses during the match. The selected junior badminton players were instructed to secure the heart rate strap surrounding their chest. The player turned on the heart rate monitor just before the game started. Heart rate responses were recorded every second by telemetry throughout the matches, using the Polar Team Pro system. The Polar Team Pro App was used to determine parameters such as the maximum and average heart rate of each player, the intensity during the matches based on % HRmax during the matches.

3.2.7 Video Analysis

The footage was recorded with a video camera (Gopro Hero 4) with a focal length of 3.45–41.4 mm and which was placed behind the court. The court (13.40 × 6.10 m) was registered in its entirety to ensure a complete view of all the actions during the National Junior Internal Ranking 2021. The temporal structure was obtained from subsequent analysis of the videotaped matches by calculation of the rally time, rest time (seconds), and work density (ratio of performance time to rest time; a non-dimensional variable). Meanwhile, performance rates were analyzed by watching the video of each match and summarizing them as follows [2]: (i) Unforced errors: errors committed by the player in a situation where an error is not expected; (ii) Winning shots: shots that, on account of their effective execution, score a point; (iii) Number of shots that occur in each point (number of shots per rally), in each of the sets and matches (total shots); (iv) Total number of rallies: number of interventions occurring throughout the sets and whole match. All the videos were analyzed using the video analysis software (Dartfish Pro S).

3.3 Statistical Analysis

Statistical analyses were performed on the data collected using the IBM Statistical Package for Social Science for Windows (IBM SPSS version 24.0). The descriptive statistic of timing characteristic was presented in mean and standard deviation. The heart rate data and technical analysis were presented in percentage.

Table 1 Physical fitness tests on four elite juniors badminton players

Type of test	Players			
	M1	M2	F1	F2
Predicted VO_{2max} (ml/kg/min)	58	59	45	48
Speed (s)	3.01	3.05	3.76	3.53
Four corner agility (s)	6.9	6.89	8.1	7.36
Upper body strength (kg/bw)	1.1	1.2	0.7	0.7
Lower body strength (kg/bw)	2.5	2.9	2.2	2.3
Highest jump height (cm)	49.2	49.3	31.1	34

4 Results

4.1 *Physical Fitness Tests on Four Elite Juniors Badminton Players*

Table 1 showed that M1 and M2 have similar fitness levels, whereas F2 performed better in aerobic fitness, speed, agility and jump height as compared to F1.

4.2 *Heart Rate*

Figure 1 shows M1 and M2 heart rate responses during the match. Both players' maximum and minimum heart rates gradually increased throughout the match. The average heart rate responses were higher in the 2nd and 3rd sets as compared to the 1st set.

Figure 2 shows F1 and F2 heart rate responses during the match. F1 maintained her average heart rate responses during the 2nd and 3rd sets, whereas F1's heart rate responses increased throughout the match.

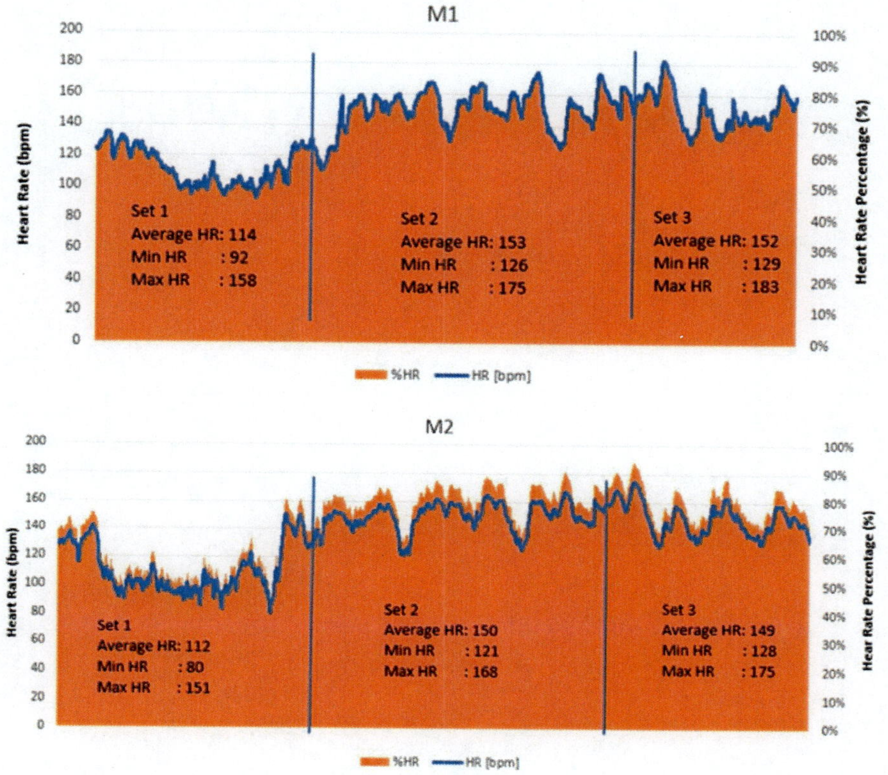

Fig. 1 Heart rate responses of M1 and M2 during the match

4.3 *Offensive Shots*

Figures 3 and 4 show the offensive shots for both males and females. Overall, the commonly used offensive shot by both male players is Net, followed by push, smash and drop. M2 playing style was towards offensive where more smash shots (22.22%) in the 1st set, however M1 played more net shots (31.03%) in the 2nd set compared to the first set (29.63%) in order to create a higher chance of smash (25.29%). In the third set, M1 used the same strategy as the second set.

F1 controlled the game by using clear shots (20.99%) and net (18.52%) to create chances to smash (18.52%). F2 increased her clear shot from 23.26 to 28.57% during the second set, and she tried to play more netting (22.86%). In the last set, F2 played a combination of clear, drop, and net to create a 15% of smash.

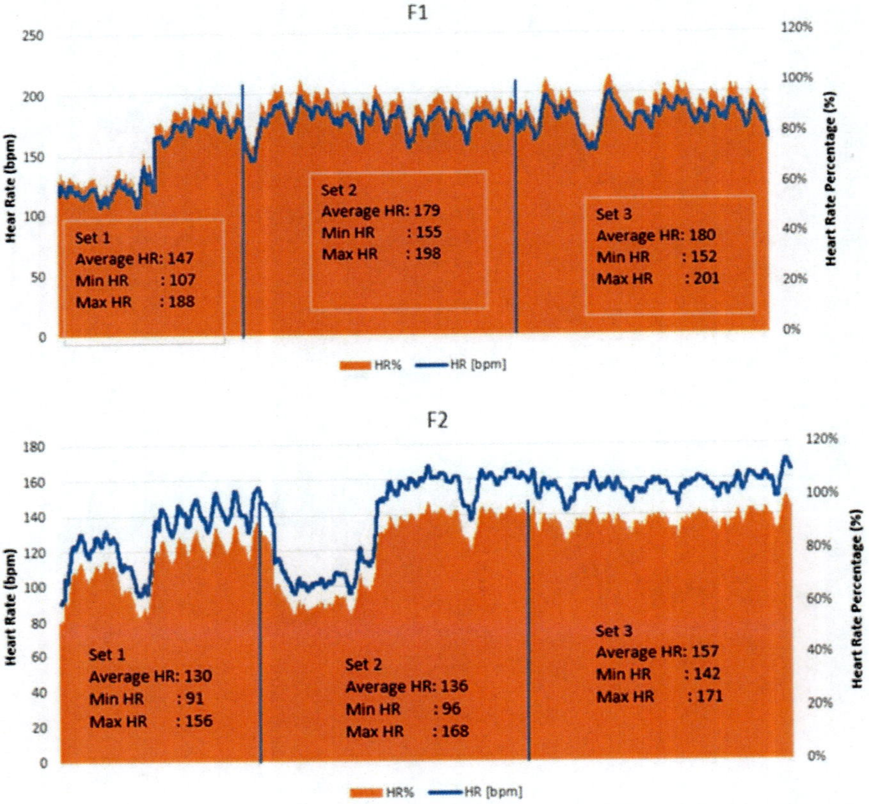

Fig. 2 Heart rate responses of F1 and F2 during the match

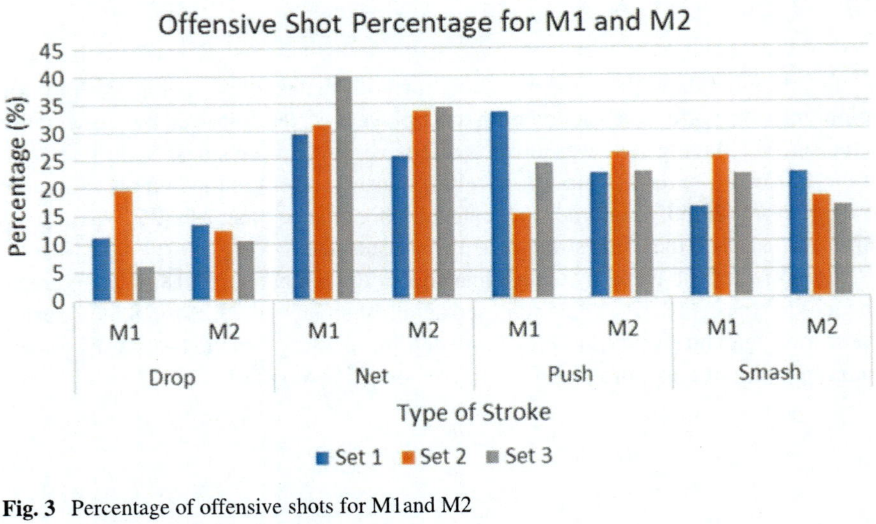

Fig. 3 Percentage of offensive shots for M1and M2

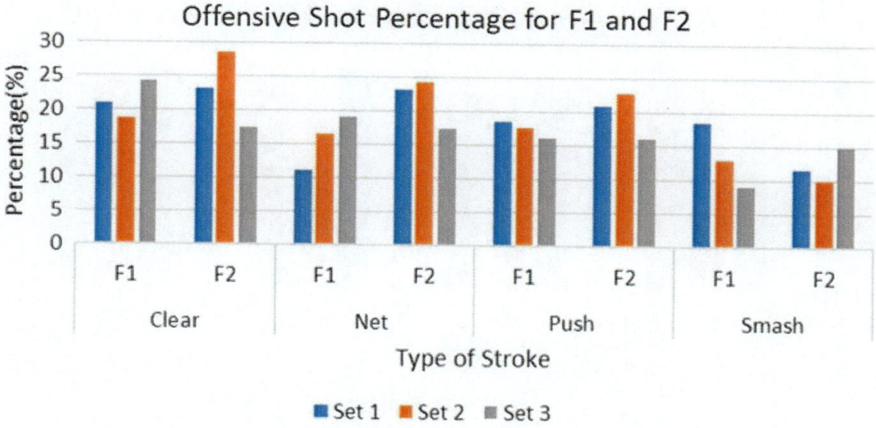

Fig. 4 Percentage of offensive shots for F1 and F2

4.4 Defensive Shots

Figures 5 and 6 show the defensive shots for both males and females. Overall, females have lower straight blocks than male because they have lower percentage of smash as well.

Males showed a higher straight block than females, although they have a lesser percentage of net lifting because their opponents performed a higher smash shot.

For females, F1 increased her net lifting from set 1 (48%), set 2 (57%) to set 3 (62%), while F2 decreased her net lifting from set 1 (55%), set 2 (40%) to set 3 (38%). It resulted in a decrease in straight blocks when F2 decreased her net lifting shot.

4.5 Timing Characteristic

Table 2 shows that males a have higher average shot frequency, average shot per rally, average number of rest than females.

5 Discussion

The objective of this study was to describe the general physical fitness characteristics, heart rate responses, temporal and timing analysis during the games of four elite junior badminton players (2 males and 2 females). The main findings of this study were (i) fitness levels of male players were better than female players. The average heart rate of females was 10.12% higher than male players. Match analysis demonstrated

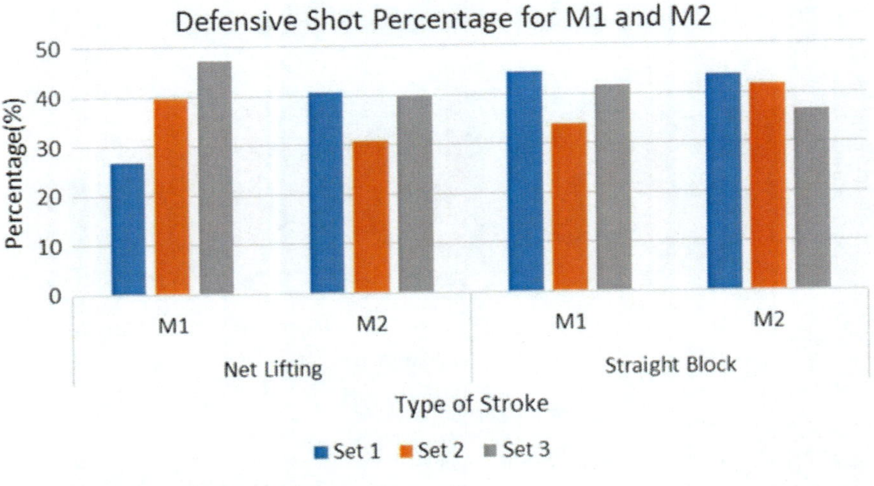

Fig. 5 Percentage of defensive shot for M1 and M2

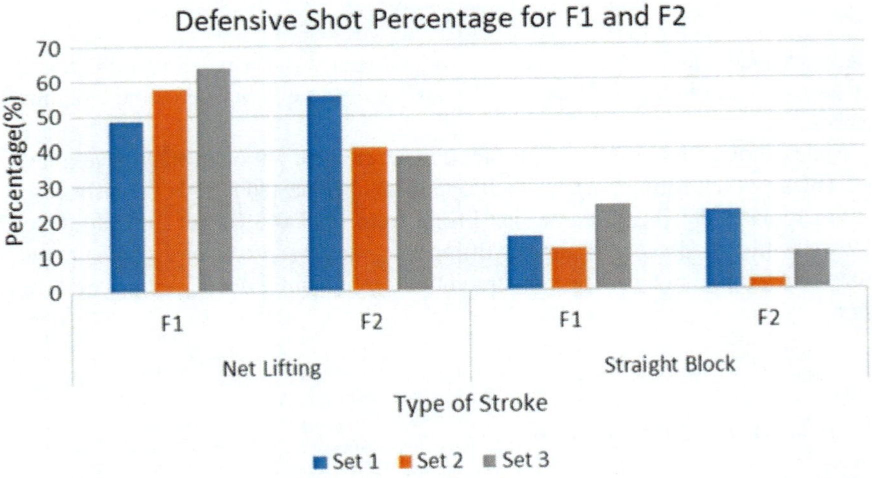

Fig. 6 Percentage of defensive shot for F1 and F2

that male single favorite shots were net (31.68%) and smash (20.04%), while female favorite shots were clear (25.43%) and drop (19.58%).

5.1 *Physical Fitness*

The result showed that both male have similar physical fitness which is better compared to females. Our result showed similar fitness results with Ooi et al. [22]

Table 2 Timing characteristics for male and female matches

Players/set	Timing characteristic				
	Average shot frequency (s)	Average shot per rally	Total playing time (min)	Average rally time (s)	Average number of rest (s)
Male					
Set 1	1.04	6.93	16.50	6.88	17.5
Set 2	0.99	8.56	17.13	8.11	24.97
Set 3	0.94	6.36	14.35	5.94	20.29
Female					
Set 1	0.75	5.41	14.45	7.2	10.94
Set 2	0.84	6.54	17.01	7.77	18.11
Set 3	0.87	8.21	19.40	9.42	13.86

study which also used elite badminton athletes. Beside that, F2 has slightly better results in aerobic fitness, speed, agility and jump height than F1. This may give advantage to F2 to win the match against F1 because greater aerobic fitness can have faster recovery between rallies [23] and better agility and speed can put her in a better position when receiving shuttle from F1.

5.2 *Heart Rate (HR) Analysis of Badminton Players*

The result showed that M1's average HR was at 69% of HR_{max} and his peak HR at 91% of HR_{max} during the 48.38 min of playing time, whereas M2's average HR reaches 73% and peak HR reaches 94% of HR_{max} in 48.38 min of playing time. This indicated that M1.

In comparison to the Alcock and Cable [24] study, these results showed a slightly lower percentage in both average and peak HR (88.8% of HR_{max}, 96.8%HR_{max}, respectively). This may be due to previous study subjects having lower aerobic capacity than current study subjects. Another reason could be the junior plays shorter rally time and fewer shots per rally, therefore resulting in a lower average heart rate. However, our result indicated that junior match intensity is similar to senior level based on heart rate.

A greater aerobic capacity has a higher efficiency of oxygen transport, which results in a lower HR at submaximal load [25]. In this present study, the average HR and HR_{peak} of female players were greater than those of male players. The average heart rate of F1 was at 87% of HR_{max} and the peak heart rate was at 103% HR_{max} in 41.46 min of playing time. F2's average HR was at 83% HR_{max} and the peak HR was at 100% HR_{max}. This can be explained by the fact that females have lower aerobic capacity, which means the heart needs extra work to deliver oxygen to the body [25].

Therefore, it resulted in a higher heart rate in the match. The results showed that aerobic capacity is important to maintain heart rate at submaximal intensity.

5.3 Temporal Analysis

In the male junior internal ranking, M1 won M2 with a score of 19–21, 21–15, and 21–12. M1 lost the first set because he played more defensively where a higher percentage of defensive shots (53% of straight block and defensive lift) were used during 1st set. In contrast, M2 playing style was towards offensive where more smash shots (22.22%) with a 40% successful rate was observed. During the second set, M1 played more net shots (31.03%) compared to the first set (29.63%) in order to create a higher chance of smash (25.29%). When M1 increased his smash shots in the second set, it created a lot of pressure on M2.In the final game, M1 continued using the same strategy (created chances to smash through the netting) which led him to winning the set. On the other hand, M2 used the same strategy as M1 but he was only able to secure 16% of the successful smash in the whole third set. rs. From this match, it shows the importance of smash as it could lead to the opponent making more errors which lead to winning points [15, 26]. Based on our results, men single, the highest rate of winning shots is from smash M1 (42.50%) and M2 (28.57%), followed by net and push (M1:17.5%, 15%; M2: 22.86%, 25.7%), which is in line with other researchers [16, 17, 18].

For women's single internal ranking, F2 won F1, 15–21, 21–18, 21–15. In the first set, the main reason for F1 winning the set was that F1 controlled the game by using clear shots (20.99%) and net (18.52%) to create attacking chances. Secondly, F2 adopted the wrong strategy by giving W a lot of net lifting (55,56%), which gives an advantage to F1to attack.. In the second set, F1's play was more defensive as she had more net lifting (57.69%) compared to the first set (48.48%) which could be the reason that led to her loss. In contrast, F2 increased her clear shot and reduced net lift to (40.54%) percentage., and she tried to play more netting (22.86%) to create attacking opportunities, particularly on sharp drop technique. This has resulted in 38.5% of her winning points were from drop. In the final game, F1 used the same strategy as in set 1 by doing clear and drop shots to control the game. However, F2 continued her approach in set 2 by doing less net lifting, which reduced the chances of smash by F1. In addition, F2 used a combination of clear, drop, and net to force F1 into a defensive position which limited F1's attacking..). As a result, F2 successfully increased the percentage of smash (23.1%) in the 3rd set compared to the first set (7.14%) and second set (7.69%). Our result showed that clear and drop are the favorite shots for women, which is also in line with previous research [16, 27], where female's winning shots mostly come from drop and smash.

From the match analysis, we observed that females used different strategies compared to male players. In this case, the main difference between male and female is the female's strategy is clear, then drop or smash to get a point [16, 27], while the male's favorite shot is smash, followed by net and push skills. This may be due to physical differences, males tend to play more aggressive shots than females. We can see that both M1 and M2 have higher upper body strength than F1 and F2 which can explain why smashing will be the best choice to finish the point. In conclusion, although different strategies of play, both genders still use more offensive strategies than defensive strategies.

5.4 *Timing Analysis*

Torres-Luque et al. [28] reported that both men and women single in the Olympic games 2012 and 2016 have a higher average rally time than our current study average rally time. This may be due to seniors having a better consistency in stroke which produces longer rally time [28]. The result also showed that our juniors players have fewer shots per rally compared to the olympic game player [13] which can indicate that juniors have less consistency and more errors made in a rally. Moreover, we found that the average rest time in every set was more than 12 s, which is similar to other studies [2, 13, 15, 16]. This indicates that players need more rest time to recover as the intensity in junior and senior is higher [15].

6 Conclusion

In conclusion, for male junior players, winning shots mostly come from smashing, while female winning shots mostly come from drop and net. This can be the main training focus for the coaches when it comes to skill training or the coaches can increase other skills to improve the playing style. The results of the study provide an insight to the coaches to develop effective training plans for junior players based on gender. Moreover, the intensity in junior was as high as in senior badminton players, with higher peak heart rate and average heart rate. This suggests that cardiovascular fitness is important in order to compete in such a high-intensity game. Besides that, the junior player should improve the consistency of the stroke, as senior players play longer rallies and more shots per rally.

7 Limitation/Practical Implications

The study's limitation was that data was collected only during internal ranking and not during other periods due to the COVID-19 pandemic. The results of the study help coaches understand the playing style of each player and can help them plan training programmes based on their strengths and weaknesses. Furthermore, it provides information to the sports scientist about the physical demands of a badminton match.

Acknowledgements The authors would like to express their gratitude to the Badminton Association of Malaysia (BAM) for allowing our research team to take video recordings during the National Junior Internal Ranking 2021.

References

1. Cabello D, Padial P, Lees A et al (2004) Temporal and physiological characteristics of elite women's and men's singles badminton. Int J Appl Sport Sci 16(2):1–26
2. Manrique DC, Gonzalez-Badillo JJ (2003) Analysis of the characteristics of competitive badminton. Br J Sports Med 37(1):62–66
3. Gowitzke BA, Waddell DB (1979) Technique of badminton stroke production: science in badminton. In: Terauds J (ed) Racket Sport. Academic Publishers, Del Mar, pp 17–41
4. Heller J (2010) Physiological profiles of elite badminton players aspects of age and gender. Br J Sports Med 44(Suppl I):17
5. Faude O, Meyer T, Rosenberger F, Fries M, Huber G, Kindermann W (2007) Physiological characteristics of badminton match play. Eur J Appl Physiol 100(4):479–485
6. Chin MK, Wong AS, So RC, Siu OT, Steininger K, Lo DT (1995) Sport specific fitness testing of elite badminton players. Br J Sports Med 29(3):153–157
7. Van Lieshout KA, Lombard AJ (2003) Fitness profile of elite junior South African badminton players. African J Phys Act Health Sci 9(3):114–120
8. Andersen LL, Larsson B, Overgaard H, Aagaard P (2007) Torque–velocity characteristics and contractile rate of force development in elite badminton players. Eur J Sport Sci 7(3):127–134
9. Leong KL, Krasilshchikov O (2016) Match and game performance structure variables in elite and youth International badminton players. J Phys Edu Sport 16(2):330
10. Bilge M (2012) Game analysis of Olympic, World and European Championships in men's handball. J Hum Kinet 35(1):109–118
11. Csataljay G, O'Donoghue P, Hughes M, Dancs H (2009) Performance indicators that distinguish winning and losing teams in basketball. Inter J Perform Anal Sport 9(1):60–66
12. Dellal A, Chamari K, Wong DP, Ahmaidi S, Keller D, Barros R, Bisciotti GN, Carling C (2011) Comparison of physical and technical performance in European soccer match-play: FA premier league and La Liga. Eur J Sport Sci 11(1):51–59
13. Phomsoupha M, Laffaye G (2015) The science of badminton: game characteristics, anthropometry, physiology, visual fitness and biomechanics. Sports medicine 45(4):473–495
14. Lees A (2003) Science and the major racket sports: a review. J Sports Sci 21(9):707–732
15. Chiminazzo JGC, Barreira J, Luz LS, Saraiva WC, Cayres JT (2018) Technical and timing characteristics of badminton men's single: comparison between groups and play-offs stages in 2016 Rio Olympic Games. Int J Perform Anal Sport 18(2):245–254
16. Abian-Vicen J, Castanedo A, Abian P, Sampedro J (2013) Temporal and notational comparison of badminton matches between men's singles and women's singles. Int J Perform Anal Sport 13(2):310–320
17. Abián P, Castanedo A, Feng XQ, Sampedro J, Abian-Vicen J (2014) Notational comparison of men's singles badminton matches between Olympic Games in Beijing and London. Int J Perform Anal Sport 14(1):42–53
18. Oswald E (2006) A computer-aided comparison of the playing pattern of the world's top male players and Austrian top male players in single badminton. In: Proceedings of the IV world congress of science and racket sports. Alcoy: Alto Rendimiento, pp 1–10
19. Lee KT, Xie W, Teh KC (2005) Notational analysis of international badminton competitions. In: ISBS-conference proceedings archive
20. Leger LA, Lambert J (1982) A maximal multistage 20-m shuttle run test to predict $$\dot V $$ O2 max. Eur J Appl Physiol 49(1):1–12

21. Walklate BM, O'Brien BJ, Paton CD, Young W (2009) Supplementing regular training with short-duration sprint-agility training leads to a substantial increase in repeated sprint-agility performance with national level badminton players. J Strength Cond Res 23(5):1477–1481
22. Ooi CH, Tan A, Ahmad A, Kwong KW, Sompong R, Mohd Ghazali KA, Liew SL, Chai WJ, Thompson MW (2009) Physiological characteristics of elite and sub-elite badminton players. J Sports Sci 27(14):1591–1599
23. Aydoğmuş M, Arslanoğlu E, Özmen T (2015) Effect of badminton specific training versus badminton match on aerobic fitness. Online J Recreat Sport 4(2):12
24. Alcock A, Cable NT (2009) A comparison of singles and doubles badminton: heart rate response, player profiles and game characteristics. Int J Perform Anal Sport 9(2):228–237
25. Cornelissen VA, Verheyden B, Aubert AE, Fagard RH (2010) Effects of aerobic training intensity on resting, exercise and post-exercise blood pressure, heart rate and heart-rate variability. J Hum Hypertens 24(3):175–182
26. Li S, Zhang Z, Wan B, Wilde B, Shan G (2017) The relevance of body positioning and its training effect on badminton smash. J Sports Sci 35(4):310–316
27. Lee KT, Xie W, Teh KC (2005) Notational analysis of international badminton competitions. In: ISBS-conference proceedings archive
28. Torres-Luque G, Blanca-Torres JC, Cabello-Manrique D, Kondric M (2020) Statistical comparison of singles badminton matches at the London 2012 and Rio De Janeiro 2016 Olympic Games. J Hum Kinet 75(1):177–184
29. Ordóñez EG, Pérez MDCI, González CT (2016) Performance assessment in water polo using compositional data analysis. J Hum Kinet 54:143

Cues and Striding Performance in Skilled and Less-Skilled Riders in Three Types of Equine Gaits

Balqis Nabila, Hosni Hasan, Adam Linoby, Mohd Syrinaz Azli, and Shariman Ismadi Ismail

Abstract This study was carried out to observe the cues (leg and hand) and striding performance of the horse ridden by skilled and less-skilled riders in three types of equine gaits (walk, trot, and canter). This study used purposive sampling technique which includes 14 participants (7 less-skilled and 7 skilled riders). All riders were asked to ride the same horse between the distance consisted of a 12-m start-transition-finish point track. Two video cameras were placed 10 m from the sagittal plane of the horse's straight-line pathway. Results showed that there was a significant difference in the movement kinematics (knee angle) between skilled and less-skilled riders during the equine gaits. In addition, there were also significant correlations between the skilled rider's (elbow angle) and horse movement kinematics (head angle, stride length) with the horse speed. Therefore, it can be concluded that effective riding aids (rider's movement kinematics) will influence the better speed of the horse to achieve better performance.

Keywords Equestrian · Speed · Trot · Canter · Riding aids · Leg cues

1 Introduction

Equestrian is a horse riding sport in which men and women compete in the same class equally [1]. In equestrian, there are several events listed in the Olympic sports like dressage, jumping, and eventing [1]. According to The Equestrian Australia, 2005, dressage is where the horse and riders need to move harmonically as one

B. Nabila · H. Hasan (✉) · S. I. Ismail
Faculty of Sports Science and Recreation, Universiti Teknologi MARA, 40450 Shah Alam, Selangor, Malaysia
e-mail: hosnihasan@uitm.edu.my

M. S. Azli
Akademi Kecergasan Pertahanan, Universiti Pertahanan Nasional Malaysia, 57000 Sungai Besi, Kuala Lumpur, Malaysia

A. Linoby
Faculty of Sports Science and Recreation, Universiti Teknologi MARA, Cawangan Negeri Sembilan, Kampus Seremban, Seremban, Malaysia

© The Author(s), under exclusive license to Springer Nature Singapore Pte Ltd. 2023

S. F. Syed Omar et al. (eds.), *Innovation and Technology in Sports*, Lecture Notes in Bioengineering, https://doi.org/10.1007/978-981-99-0297-2_4

presenting the smoothness of the transition, the capability to maintain the rhythm, the understanding of the horse to move at different speeds of the same gaits according to the given cues and the suppleness of the horse body to bend and move sideways.

In equestrian, riding aids are the cues given by the riders to their horse to give command on what to do. As important communication tools, riding aids can be classified into natural aids (for example leg, hand, seat, and voice) and artificial aids (for example spurs and whip). Riders' communication with the horse is essential in order to command the horse to initiate the movement, change direction, maintain their rhythm and pace, and transition works. Leg aids are the main natural aids used towards the horse to get the desired movement from the horse, whereby the spurs which are artificial aids, are used by the riders only to remind the horse to move according to command when it does not listen to the natural aids [2]. It is because when riding, riders have to engage with the movement of the horse from stopping to slowing down or speeding by using aids from the legs first before using hands and seat aids [2].

Horse movements have four different gaits which are walk, trot, canter, and gallop [3]. The walk is the slowest gaits movement while the gallop is the fastest gaits speed movement. During this movement, riders need to have the capability to absorb the movement of the horse. The riders also have a few riding positions to suit the movement of the horse according to the need of the event the riders are participating in such as sitting, rising, and two-point (light seat). During dressage events, the priority position is sitting and rising which is only applicable during trotting. Rising is the movement of sitting on the saddle and standing on the stirrup according to the horse's movement.

Another exercise that is essential for a horse's performance is transition work. Transition is where the horse needs to shift its phase from slower to faster gaits and the other way around [4]. Riding for forwarding transitions is where the riders need to drive the horse forward with their legs and sit stably on the saddle, while downwards transitions required the riders also to sit stably on the saddle and use their legs to help the horse to step under itself as it slows down, this is to avoid the horse from losing its activity during the exercise [4]. Training the horse to stop properly is also important to acknowledge it to respond to the cues given by the riders.

Horseback riding does categorize riders into a few categories, which are Para-rider, expert, less-skilled, skilled, professional, amateur, experienced, and leisure riders [5]. Less-skilled riders are where the rider who is inexperienced in riding with no participation in any competition or some of them only enter in-house competitions. Skilled riders on the other hand have already competed at the national level in the equestrian discipline [5]. The capability of the less-skilled riders to give a clear leg cue is less effective compared to the skilled riders. The less-skilled riders also have difficulties maintaining the stability of their lower legs or keeping their legs quite on the body of the horse which in return disturbs the movement of the horse as the horse gets confused with the cues given.

Riders play an important role in influencing the performance of the horse. The riders need to have good chemistry with the horse in terms of body position, ability to stay balanced, and capability to give correct aids to their horse. To achieve that, riders need to train frequently in a proper manner. In dressage mainly, the judging score also includes the rider's seat and position, where the rider is observed in their upright body position, having a balanced seat, and able to follow the natural movement of the horse [6]. The riders also were judged by observing the effectiveness of aids given to the horse. This is where the rider are succeed to achieve certain movement need with proper submission, suppleness, and impulsion without obvious or noticeable effort [6]. The capability of the rider to stop the horse with all four legs in a square order will be given a high mark. To stop the horse, the rider needs to close both of their legs slightly on the horse's body before seating deeper in the saddle and squeezing back the rein.

In horse riding, the usage of leg aids as a driving aid is important to ensure the horse moves according to the rider's needs. It is because the riders need to come out with strategies for using their legs in order to support the movement of the horse [7]. Riders need to focus on their leg cues exerted on the horse. This is because riding aids start from cues given using the legs. This is where the horse receives the pressure from the rider's leg on its torso and the pressure is released once they receive the desired response from the horse [8]. Riders also need to make sure their riding aids given to the horse are effective to enhance the performance of the horse. Although riding aids are compulsory communications between the riders and the horse there are riders who are unable to give correct leg cues which affect the horse's performance and behavior [5]. There are still a few studies that investigate the leg cues by less-skilled and skilled riders and information on this issue is still limited. Further research on leg cues will be able to characterize the best leg cues in riding technique [8]. Therefore, the purpose of this study is to determine the relationship between leg cues and the striding performance of the horse among less-skilled and skilled riders.

2 Methodology

2.1 Participants and Horse

Fourteen riders (seven less-skilled and seven skilled riders) between the age of 18 to 48 years old were selected for this study. The less-skilled participants never ride competitively and had little riding experience at the recreational level while the skilled riders were recruited from local riding clubs and had at least 4 years of participation in the competitions and had formal training throughout the year [5, 9].

In this study, only one horse was used throughout the test by all participants to identify the same performance outcomes and duration of feedback from the horse towards the rider's leg cues [8, 10]. The horse gender is a gelding type with a height of 15.2 hands (154.43 cm) and a weight of around 480 kilograms. The horse had experience competing in dressage, jumping, cross country, and horseback archery. The horse already competing for 6 years in various competition levels such as the

National Horse Show, International Horse Show, and Eventing League. The selected study area was the Malaysian Armed Forces Equestrian Centre (MAFEC) in Sungai Besi, Kuala Lumpur. The safety and consent upon the horse's well-being before being ridden were given by the veterinary personnel from the Universiti Putra Malaysia (UPM). A written informed consent form also was provided for the riders, and the study was performed in accordance with the university ethics committee guidelines (Universiti Teknologi MARA).

2.2 Data Collection

Since all participants were using the same horse, the recording was recorded for one participant a day. The participant was given 15–20 min to warm up the horse prior to the recording session. There was a rest interval between the set of movements. This step was taken to avoid fatigue and learning effect of the horse that can affect the result [11]. Three types of equine gaits such as walk, trot, and canter were measured in this study. This experiment covers the stride length of the horse in relation to the cues of pressure intensity exerted by the rider's leg. This experiment measured the horse head's vertical line angle, the horse speed, and the maximum leg extension angle. A 10-meter distance for the testing was measured using the measuring tape (see Fig. 1 for the track pattern). Passive reflective markers were placed on the rider's joint (shoulder, elbow, wrist, hip, knee, and ankle) and also on the horse's knee joint and hoof (see Fig. 2). There also reflective markers were placed on the rider's knee and toe for frontal view recording. The reflective marker is used to ensure that the rider and horse joint can be easily spotted during data analysis using movement analysis software. The placement of the passive markers was based on a previous study by Balqis et al. [10]. Once all the calibration process was done, the rider's body angle (hip joint, knee joint and elbow joint), the horse head vertical line angle, the horse knee joint angle, the horse stride length, and speed are measured.

2.3 Data Analysis

The data of the movement was recorded using the Sony handy cam video camera at the frame rate of 25 frames per second (fps). Next, the data were analyzed using the Kinovea movement analysis software and which has been proven as a reliable, low cost and easy-to-use software for sports applications [12]. This is where the video from the handy cam is extracted into the software and used to find the body angles of the rider and the horse as well as to find the horse's speed and stride length. The sample size of this had been determined according to the previous studies [8] with 95% statistical power and an effect size of 0.25. All the data calculations and statistics were analyzed using the Statistical Package for Social Science version 20(SPSS). The obtained data were then analyzed using an Independent T-test (to compare movement

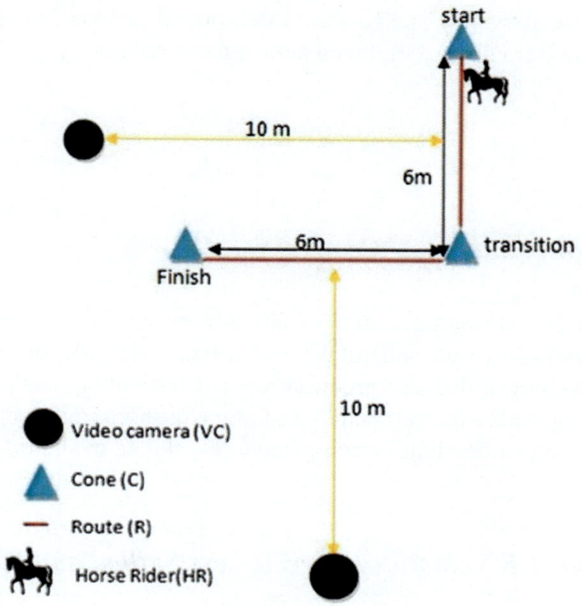

Fig. 1 Track pattern of riding by the participant

Fig. 2 Markers placement on the riders and horse body

kinematics between the less-skilled and skilled riders) and the Pearson correlation coefficient method (to find a correlation among the variables).

3 Result

3.1 Descriptive Data

The data in Table 1 shows the results obtained from this study. The widest elbow angle of the rider during walk and trot was by the less skilled with an angle of 137.62 and 142.57. The less-skilled also produced wider knee and hip angles compared to the skilled during walk, trot, and canter. In addition, the horse rode by the less skilled also recorded longer speed time as compared to the skilled in all horse gait.

3.2 Movement Kinematics Between Less Skilled and Skilled Riders

Results showed that there was no significant difference ($p>0.05$) between less-skilled and skilled riders on all kinematic variables during walk, and trot. However, there was a significant different between less skilled and skilled riders on knee angle ($t = 2.359$, $df = 12$, $p = 0.036$) during canter. For horse movement kinematics, there was

Table 1 Rider's and horse's body angle and speed

Variables	Group/gait	Walk	Trot	Canter
Rider's elbow angle (°)	Less-skilled	137.62 (16.31)	142.57 (9.78)	137.91 (15.00)
	Skilled	135.38 (13.58)	140.09 (12.26)	145.29 (8.40)
Rider's knee angle (°)	Less-skilled	113.91 (10.25)	112.71 (8.51)	111.05 (5.86)
	Skilled	109.05 (12.63)	109.62 (10.19)	103.14 (6.65)
Rider's hip angle (°)	Less-skilled	133.48 (7.35)	134.28 (6.47)	124.57 (10.73)
	Skilled	138.62 (10.94)	139.86 (8.79)	128.24 (7.55)
Horse's head angle (°)	Less-skilled	85.34 (4.16)	86.76 (10.75)	81.48 (10.08)
	Skilled	73.6 (4.61)	70.24 (5.29)	71.28 (4.57)
Horse's knee angle (°)	Less-skilled	167.05 (2.37)	165.38 (5.28)	164.95 (5.50)
	Skilled	164.90 (2.06)	165.28 (3.26)	168.19 (3.46)
Horse's stride length (m)	Less-skilled	1.46 (0.41)	1.78 (0.13)	2.06 (0.52)
	Skilled	1.51 (0.07)	1.79 (0.09)	2.20 (0.28)
Horse's speed (ms^{-1})	Less-skilled	1.48 (0.13)	2.57 (0.36)	4.63 (0.42)
	Skilled	1.62 (0.18)	2.41 (0.26)	3.95 (0.23)

a significant different ($t = 4.990$, $df = 12$, $p = 0.005$) on the head angle during walk, during trot ($t = 3.650$, df = as, $p = 0.003$) and during canter ($t = 2.436$, $df = 12$, $p = 0.031$). In addition, there was also a significant difference between the less-skilled and skilled riders on the speed of the horse during canter ($t = 3.776$, $df = 12$, $p = 0.003$).

3.3 Pearson Correlation Between the Rider's Body Angle and Horse Speed

Results showed that there were significant correlations between the rider's elbow angle and horse head angle with the horse speed, and the rider's hip angle with horse stride length among the skilled participants ($p < 0.05$). In addition, there was also a significant correlation between horse head angle with horse stride length among the less-skilled participants ($p < 0.05$).

4 Discussions

This study was conducted to investigate the leg cues and striding performance in skilled and less-skilled riders in three types of equine gaits. The findings of this study are the less-skilled rider body angle and knee angles are wide throughout the walking and trotting gaits compared to skilled riders. This is where the rider tends to push their leg forward instead of remaining their leg at the side of the horse's torso. During cantering, the less-skilled rider's body angle is narrower than the skilled rider's as the upward transition needs the rider to engage the horse's movement by maintaining slight pressure on the horse's torso to encourage the forward movement of the horse.

Walking is symmetrical gaits movement with a lateral sequence of hoof placements and there is no moment of suspension in this gait movement [13]. There are a few types of walking which are free walk on the long rein and active walk. The performance of the horse during walking depends on the aids given. For walking, there are two types of transition, which are upward transition which consists of a halt to walk, and downward transition which consist of walk to halt [4]. To start walking from a halt, the rider needs to put slight pressure on the horse's torso to encourage the horse to move forwards. The stride length and the speed of the horse are depending on the rider's hip joint posture which will influence the placement of the rider's heels on the horse's body. The heel placement can be behind or in front of the vertical line of the rider's body.

Trotting is a symmetrical running gait in which the movement of diagonal limb pairs is synchronized [13]. There are four types of trots which are, collected, working, medium and extended trots. There are three types of trotting movement rider's body adaptation, which are sitting trot, rising trot, and two-point position trot. For trotting, there are 2 types of upward transitions which are halt to trot and walk to trot, whereas there are 2 types of downward transitions which are trot to walk and trot to halt [4]. To post the trot movement, the rider needs to place their lower leg on the horse's torso and then apply pressure on it. Once the horse gives a response, the pressure applied will be released. In dressage, the type of trot that usually been used is a working trot. To post collected trot, the rider needs to support the horse movement using their lower leg then seat and their hands to ensure the horse head is in a vertical line. As these are the first two phases of the horse movement (walk and trot), our study shows the movement kinematics performed by skilled and less-skilled riders are similar. This is due to the low speed of the horse's movements, and the horse does not require a complex footfall pattern compared to the canter.

Canter is an asymmetrical leaping gait in which the leading limb is on the same side of the fore and the hind pairs. The footfall sequence is trailing hind leg, leading hind, and trailing forelimbs together then leading forelimb. Canter gaits are also characterized by having a rocking motion of the rider's body [13]. There are 3 types of upward transition, which are halt to canter, walk to canter, and trot to canter and there are 3 types of downwards transition, which are canter to trot, canter to walk, and canter to halt [4]. To pose this movement, the rider needs to ensure their horse is in a calm situation in whatever current gaits they are doing, and the rider needs to slide their outside leg behind the girth and apply pressure with both of their legs. Our study shows that the skilled riders produce better knee angle movement in controlling the horse as compared to the less-skilled riders, maybe due years of experience in horse riding. In addition, to ensure the horse maintains forward motion, the inside leg of the rider is needed to keep slight pressure. The rider's hand needs to follow the horse's head and neck motion to ensure that steady contact between the rider and horse is always maintained.

This study showed that the skilled participants produced better movement kinematics that led to faster horse movement as compared to the less-skilled participants. Despite that there was only one significant difference in the kinematic variables (knee angle), however, there was a trend of better body positioning (as measured in the range of motion in this study) for the skilled riders during all horse gaits. This sport is characterized by the winners by the time taken to complete the course, there were more significant relationships produced by the skilled participants on the riders and horse movement kinematics with the horse speed. With years of practice and experience, skilled riders not only move in phase with the horse, but they also can improve the consistency of the horse's movements [13]. In addition, findings by Olivier et al. [14] showed that skilled riders showed better postural stability during riding as compared to less-skilled counterparts.

5 Conclusions

In equestrians, riding aids or cues are essential as a communication tool between the rider and their horse. Positioning of the riding aids also needs to be observed according to the horse's response. The horse needs constant leg pressure on its torso to achieve the movement desired by the rider and there is also a horse that only needs a very light leg cue but needs more attention than the seat aids. In addition, riders also need to learn to listen and understand their horse's need to achieve the best performance accordingly. Our studies show that skilled riders produce better riding aids or cues to the horse as compared to the less-skilled riders in a more complex horse footfall pattern (canter). In addition, future studies may compare skilled and less-skilled movement kinematics during the walk, trot, and canter with the effect of their physical characteristics.

References

1. Katherine B (2018) What are the Olympic equestrian sports. In: The spruce PETS. Retrieved from https://www.thesprucepets.com/what-are-the-olympic-equestrian-sports-1886888
2. Steffen P, Beth B (2017) How to fine-tune your dressage aids with Steffen Peters. Retrieved from https://dressagetoday.com/instruction/how-to-fine-tune-your-dressage-aids-with-steffen-peters/
3. Muir S, Sly D, Houghton K (2003) The complete horse and rider: a practical handbook of riding and an illustrated guide to riding tack. Hermes House, London
4. Katherine B (2017) Learn about riding transitions. In: The spruce PETS. Retrieved from https://www.thesprucepets.com/learn-about-riding-transition-1885745
5. Williams J, Tabor G (2017) Rider impacts on equitation. Appl Anim Behav Sci 190:28–42. https://doi.org/10.1016/j.applanim.2017.02.019
6. The Equestrian Australia (2005) EA NOAS dressage judges course readings. Retrieved from https://www.equestrian.org.au/sites/default/files/EA%20NOAS%20Dressage%20Judges%20Course%20Readings.pdf
7. de Cocq P, Mooren M, Dortmans A, van Weeren PR, Timmerman M, Muller M, Van Leeuwen JL (2010) Saddle and leg forces during lateral movements in dressage. Equine Vet J Suppl(38):644–649. https://doi.org/10.1111/j.2042-3306.2010.00201.x
8. Hawson LA, Koprinska I, McLean AN, McGreevy PD (2013) Deciphering the cues from riders' legs. J Vet Behav Clin Appl Res 8(2):e9. https://doi.org/10.1016/j.jveb.2012.12.021
9. Hasan H, Davids K, Chow JY, Kerr G (2016) Compression and texture in socks enhance football kicking performance. Hum Mov Sci 48:102–111
10. Balqis N, Ismail SI, Sulaiman N, Adnan R (2016) The relationship between rider's body posture and horse speed during rising trot in equestrian sports. In: Proceedings of the 2nd international colloquium on sports science, exercise, engineering and technology 2015 (ICoSSEET 2015). Springer, Singapore, pp 89–96
11. Powers PNR, Harrison AJ (2007) Influences of a rider on the rotation of the horse–rider system during jumping. Equine Comparat Exerc Physiol 1(1):33–40. https://doi.org/10.1079/ECP20032
12. Adnan NMN, Ab Patar MNA, Lee H, Yamamoto SI, Jong-Young L, Mahmud J (2018) Biomechanical analysis using Kinovea for sports application. In: IOP conference series: materials science and engineering, vol 342, no 1. IOP Publishing, pp 012097

13. Clayton HM, Hobbs S-J (2017) The role of biomechanical analysis of horse and rider in equitation science. Appl Anim Behav Sci 190:123–132. https://doi.org/10.1016/j.applanim.2017.02.011
14. Olivier A, Viseu JP, Vignais N, Vuillerme N (2019) Balance control during stance-a comparison between horseback riding athletes and non-athletes. PLoS ONE 14(2):e0211834

A Cluster Analysis and Artificial Neural Network of Identifying Skateboarding Talents Based on Bio-fitness Indicators

Aina Munirah Ab Rasid, Muhammad Zuhaili Suhaimi, Anwar P. P. Abdul Majeed, Mohd Azraai Mohd Razman, Mohd Hasnun Arif Hassan, Nasree Najmi, Noor Azuan Abu Osman, and Rabiu Muazu Musa

Abstract This research aims to identify talented skateboarding athletes with reference to their bio-fitness indicators. A total of 45 skateboarders (23.09 ± 5.41 years) who were playing for recreational purposes were recruited for the study. Standard assessment of their bio-fitness as well as their skateboarding performances was performed. The bio-fitness investigated consisted of stork balance, star excursion balance test, vertical jump, standing broad jump, single-leg wall sits, plank and sit-up while the related-skill performances consisted of the observation on skateboarding tricks execution, namely Ollie, Nollie, Frontside 180, Pop-Shuvit and Kickflip. To achieve the objective of the study, a hierarchical agglomerative cluster analysis (HACA) was performed to cluster the athletes into groups in reference to the level of their bio-fitness markers. The clusters identified two groups of performance named High-Potential Skaters (HPS) and Low-Potential Skaters (LPS) following their skateboarding performance scores. An Artificial Neural Network (ANN) was conducted to ascertain the classified athletes into the clusters (HPS and LPS) based on

A. M. Ab Rasid · M. Z. Suhaimi · R. M. Musa (✉)
Centre for Fundamental and Continuing Education, Universiti Malaysia Terengganu, 21030 Kuala Nerus, Terengganu, Malaysia
e-mail: rabiu.muazu@umt.edu.my

A. P. P. Abdul Majeed
School of Robotics, XJTLU Entrepreneur College (Taicang), Xi'an Jiaotong-Liverpool University, Suzhou 215123, China

M. A. Mohd Razman
Innovative Manufacturing, Mechatronics and Sports Laboratory, Universiti Malaysia Pahang, 26600 Pekan, Pahang Darul Makmur, Malaysia

M. H. A. Hassan
Faculty of Mechanical and Automotive Engineering Technology, Universiti Malaysia Pahang, Kuantan, Malaysia

N. Najmi
Terengganu State Sports Council, Jalan Lapangan Terbang, 21300 Kuala Terengganu, Terengganu, Malaysia

N. A. Abu Osman
Faculty of Engineering, Department of Biomedical Engineering, Centre for Applied Biomechanics, University of Malaya, 50603 Kuala Lumpur, Malaysia

© The Author(s), under exclusive license to Springer Nature Singapore Pte Ltd. 2023

S. F. Syed Omar et al. (eds.), *Innovation and Technology in Sports*, Lecture Notes in Bioengineering, https://doi.org/10.1007/978-981-99-0297-2_5

the bio-fitness indicators evaluated along with the skateboarding tricks performance scores. The result demonstrated that ANN accomplished a high classification accuracy of 91.7% indicating excellent performance from the classifier in classifying the skateboarding athletes. Similarly, the area under the curve of the classifier was found to be 0.988 signifying further the validity of the model developed. Overall, these results suggest that the proposed technique was able to classify the skateboarding athletes reasonably well which will in turn possibly assist coaches to identify talents in this sport through the bio-fitness indicators examined.

Keywords Skateboarding · Talent identification · Bio-fitness · Machine learning · Hierarchical Agglomerative Clustering · Individual extreme sport

1 Introduction

Skateboarding is among the popular extreme sports, especially among youths who are highly engaged with the sport nowadays despite the injury risks involved in playing the sport. The popularity of the sport could be owing to its ability to harbour self-expression, and creativity, as well as its freestyle and rules' free nature [1]. Moreover, skateboarding was also recently featured in the 2020 Summer Olympic Tokyo as one of the sports competing not only to draw the attention of the youngsters as spectators but also to offer fair chances to the skateboarders out there to display their talents in such a productive way [2]. Accordingly, discovering talents to fulfil the role of skateboarding athletes is crucial primarily to represent the nation, seeing that the pressure to excel in sports internationally has never been greater especially after the world experience an epic lockdown about two years ago [3].

On one hand, being an athlete, having good bio-fitness is an essential criterion to improve one's physical performance, identify talents and reduce as many possible injuries [4]. Equally important, a high level of bio-fitness not only assist in improving the efficiency of learning sport skills but can also help the athlete in building high endurance as a key parameter of delaying exhaustion [5]. Lower body limbs related ability could be essential for skateboarding athletes considering that the athlete in the sport largely utilizes the lower parts of the body to execute tricks smoothly. Machine learning is a method that is capable of evaluating athletes' performances which has shown promising results in forecasting and classifying athletes' performances with respect to certain input parameters [6]. The machine learning technique is one of the many artificial intelligences that operate by producing an output algorithm predicted from the inputs which consist of large data [7]. Additionally, it can be branched into two parts that are supervised machine learning and unsupervised machine learning where the supervised machine learning caters to labelled data before being processed further for output prediction. Meanwhile, unsupervised machine learning, by contrast, is a model that works on an un-labelled dataset allowing users to implement extra problematical tasks [8]. Thus, considering the significant aspect of machine learning, this technique will be implemented for the categorization and classification purpose of skateboarding athletes.

Fundamentally, the study has considered skateboarding athletes as experimental subjects since many previous studies only reviewed the most common sport such as football, basketball and athletics [9–11]. Therefore, the purpose of this study is to identify their talents based on their bio-fitness and skateboarding performances with the assistance of technology of machine learning. This is non-trivial as it provides an effective way of evaluating and searching for talents based on few related performance markers which could save enormous resources, time, as well as the energy of talent Scouters.

2 Methodology

2.1 Participants

The participants (n = 45) of the current research were recruited among amateur skateboarders across the East Coast of Malaysia. The participants were playing for recreational purposes at different locations and were not associated with any authorized clubs within the government nor had they received any special training prior to the collection of data in this study. It is important to highlight that before the commencement of the study informed consent was obtained from the skater and ethical approval was received from the university human research ethic committee (UMT/JKEPHMK/2021/53).

2.2 Bio-fitness Assessments

The bio-fitness assessments consist of seven tests including the stork balance test, star excursion balance test (SEBT), vertical jump test [12], single-leg wall sit test [13], standing broad jump test [14], plank and sit-up test. The participants were given full explanations regarding the experiment and underwent health screening tests to ensure the participants were healthy with no serious injuries. In preparatory to the assessments, the participants were instructed to do some warming and stretching up for 3 min to avoid any injuries. Then, they were submitted to each of the tests where the tests were divided into sections. Additionally, in between each test, they were given 5 min water break and rest. The stork balance was to test the athletes' static balances [15] while SEBT, was to evaluate the athletes' dynamic balances [16]. Vertical jump, single-leg wall sits and standing broad jump were to examine the athletes' leg peak power, leg strength and explosive leg power respectively. Finally, plank and sit-up were to measure the strength of core muscle [12] and to test the abdominal muscle strength as well as endurance of the athletes respectively [17].

2.3 Skateboarding Tricks' Execution

Upon completing all the bio-fitness assessments, the athletes were given extra minutes to take a break before skateboarding tricks execution was examined. The five tricks employed in the current research are the most fundamental tricks that have been scrutinized and investigated in several other literatures previously [18–23]. The tricks included were Ollie, Nollie, Frontside 180, Pop-Shuvit and Kickflip [24]. The participants were given three trials to perform each trick as the researchers observed each trick's execution and marked successful if the athletes were able to perform the tricks and land on the skateboard successfully without staggering. Conversely, when an athlete is unable to land successfully on the skateboard or fails to execute the trick, the performance is deemed failed. These evaluation were done via the researcher's real-time observation with the aids of judgement of other skateboarders who were watching.

2.4 Hierarchical Agglomerative Clustering Analysis (HACA)

Hierarchical Agglomerative Clustering Analysis (HACA) is a type of unsupervised machine learning often employed to cluster the dataset obtained via the experiment. HACA clustered data into random groups intending to organize and simplify the set of data. This analysis was performed via Orange 3.29.3 software. At first, the proximity of individual points and other points was differentiated as an individual cluster. Then, the clusters that consisted of similar characteristics were amalgamated together into a single cluster. After that, the proximity of new clusters is then again calculated and merged with other clusters forming another new cluster. Finally, the steps of proximity calculation of new clusters were repeated until all clusters were merged forming only a single cluster. HACA can be visualized as a tree-like diagram that records the sequences of merges or splits (dendrogram) [25].

2.5 Artificial Neural Network (ANN)

An artificial neural network (ANN) is one of the supervised machine learning algorithms utilized for the purpose of classification. This algorithm operates by predicting the classes of the athletes based on the clusters obtained from HACA. The neural network organized neural units into multiple layers. In the first layers, the input was computed and processed until the output was produced. The output was then transmitted to the other layers to generate another output [26]. The steps will keep on repeating until all the athletes were classified. The analysis of classification was performed via Orange 3.29.3 software.

Table 1 Descriptive statistics of the measured variables

Variable	Obs	Min	Max	Mean	Std. dev.
Age	45	14.000	35.000	23.089	5.405
Stork balance (s)	45	5.360	64.760	23.082	10.870
Single-leg wall sit (s)	45	13.000	70.050	32.184	16.952
Vertical Jump (m)	45	0.600	1.940	0.955	0.245
SEBT normComposite (%)	45	59.165	91.290	73.636	7.272
Plank (s)	45	21.590	127.200	65.149	27.453
Standing broad jump (m)	45	1.400	2.850	2.246	0.306
Sit-up (max no sit up/min)	45	17.000	40.000	27.467	5.476
Tricks' performance (%)	45	0.000	1.000	0.686	0.238

3 Results and Discussion

3.1 Descriptive Statistics

The descriptive statistics of the measured variables were summarized in the following-tabulated table (Table 1). As can be observed from the table, the observations, the variables over which the skateboarders were examined, the minimum boundary, maximum boundary, the mean as well as the standard deviation are displayed.

3.2 Identification of Clusters Predicated on Bio-fitness and Skill-Related Assessments

The succeeding figure (Fig. 1) presents the category of each skateboarder in accordance with their performances projected by HACA. The clustering process was in reference to the level of similarity of all bio-fitness components investigated. It can be ascertained from the illustration that two clusters were identified: high-potential skaters (HPS) and low-potential skaters (LPS) following their skateboarding performance scores. From the analysis, 31 out of 45 participants were clustered in HPS while the remaining 14 participants were clustered in LPS.

3.3 Descriptive Statistics Differences of HPS and LPS

Table 2 displayed the differences descriptive statistics between HPS and LPS with the number of observation 31 which represents 31 skateboarders for HPS and 14

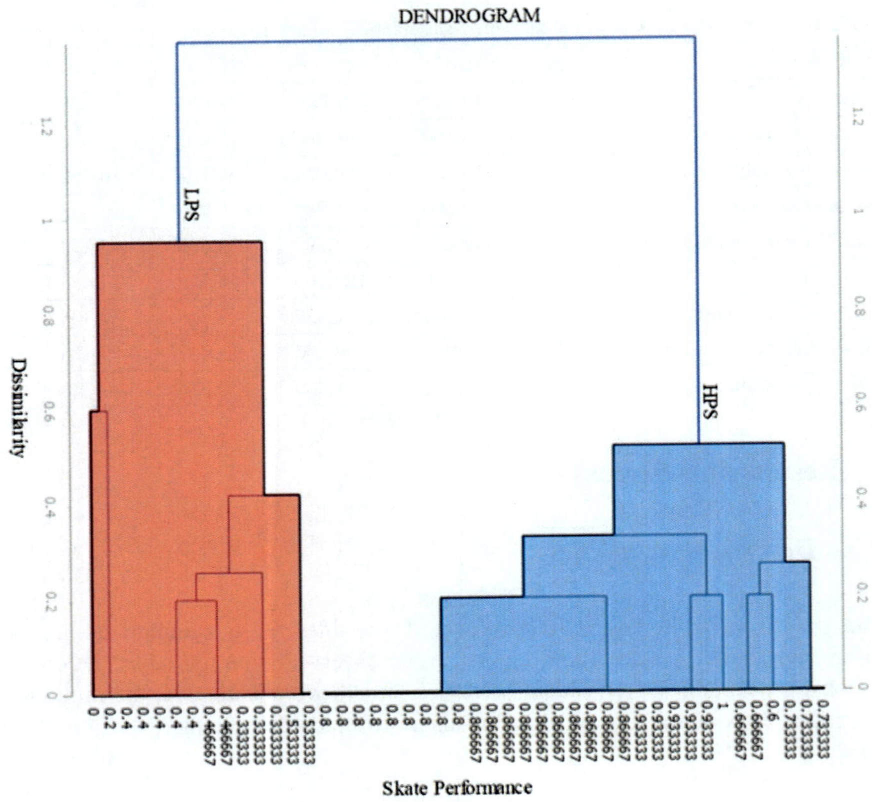

Fig. 1 Two performance groups defined by HACA via Dendrogram

represents 14 skateboarders from LPS. The minimum, maximum, mean and standard deviation values were also tabulated.

As observable in Table 2, there were significant difference between HPS and LPS. For stork balance, the mean score for HPS is higher at 27.283 s compared to LPS at 13.779. Whereas the maximum time for HPS to hold the stork balance position was 64.76 s which indicates excellent static balance compared to LPS with maximum time at 27.84 demonstrating average static balance [27]. For single-leg wall sit, the mean score for HPS (37.775 s) higher than LPS (19.802 s) showing average scores for HPS and poor scores for LPS [13]. The next variable is vertical jump, where HPS executed 0.993 m somewhat greater than LPS whom exhibits 0.872 m where both classes have excellent leg peak power. Nevertheless, if compared the minimum value of both classes, HPS is higher than LPS at 0.693 and 0.600 respectively that differentiate between very good and above average athletes [28]. Then, the HPS SEBT mean scores exhibits slightly higher value than LPS at 75.598% and 69.292% respectively which shows majority of them have good dynamic balances. For plank on the other hand, the mean score for HPS is 71.308 s higher than LPS at 51.511,

Table 2 The descriptive statistics of HPS and LPS

Class	Variables	Obs	Min	Max	Mean	Std. dev.
HPS	Stork balance (s)	31	15.91	64.76	27.283	10.019
LPS		14	5.36	27.84	13.779	5.744
HPS	Single-leg wall sit (s)	31	13	70.05	37.775	17.572
LPS		14	14.1	28.89	19.802	4.528
HPS	Vertical jump (m)	31	0.693	1.94	0.993	0.268
LPS		14	0.6	1.25	0.872	0.161
HPS	SEBT normComposite (%)	31	60.813	91.29	75.598	6.482
LPS		14	59.165	82.56	69.292	7.246
HPS	Plank (s)	31	22.31	127.2	71.308	28.909
LPS		14	21.59	83.4	51.511	18.204
HPS	Standing broad jump (m)	31	1.687	2.85	2.318	0.285
LPS		14	1.4	2.467	2.087	0.302
HPS	Sit-up (max no sit up/min)	31	20	40	28.613	5.13
LPS		14	17	35	24.929	5.54
HPS	Tricks' performance (%)	31	0.6	1	0.828	0.089
LPS		14	0	0.533	0.371	0.138

demonstrated the differences of above average and below average athletes [29]. Next, for both standing broad jump and sit-up, only slight differences were shown between HPS and LPS regardless the parameters (min, max and mean) showing average scores for the athletes [30]. Lastly, for the trick performances mean scores, HPS scores significantly higher than LPS at 82.8% and 37.1% correspondingly which discern the athletes between the classes.

3.4 Analysis of Machine Learning Model in Predicting HPS and LPS

The analysis through the machine learning model (ANN) in predicting HPS (Table 3) and LPS (Table 4) was associated with all bio-fitness markers evaluated along with the skateboarding tricks performance scores.

The Table 3 presents the analysis of the ANN model in forecasting HPS. From the table, it can be observed that the Area Under the Curve (AUC) is 0.988. The

Table 3 Evaluation of ANN in predicting HPS

Model	AUC	CA	F1	Precision	Recall
ANN	0.988	0.917	0.943	0.926	0.962

Table 4 Evaluation of ANN in predicting LPS

Model	AUC	CA	F1	Precision	Recall
ANN	0.988	0.917	0.824	0.889	0.800

AUC gives an aggregate of the performances of the model at a threshold value (0.5). In other words, the closer the value of AUC to 1.000, the perfect the classification model is [31]. Furthermore, the classification accuracy (CA) of the model indicates how well could the classifier classify HPS. In this case, ANN has 91.7% of CA suggesting excellent accuracy of classification. Additionally, the F1 or F Score is closely related to Precision and Recall, that is it is a weighted average score for both. Therefore, the Precision for ANN is 0.926 while for Recall, it has a value of 0.962. This demonstrates that the ANN model accounted for about 93% of positive prediction and above 95% of actual positive classes, indicating excellent prediction for HPS.

Table 4 shows the evaluation of the ANN model in predicting LPS. The results of AUC and CA of the classifier pointed out the same readings as HPS while F1, Precision and Recall were modified in evaluating LPS. These changes could be due to the modification of the classification threshold value (0.5) as AUC and CA are classification-threshold invariants, which means the changes in threshold do not affect both aforementioned scores [32]. As mentioned previously, F1 is linked to Precision and Recall as a weighted average score. Hence, F1 has approximately above 80% value while Precision and Recall both have 88.9% and 80% value respectively demonstrating excellent prediction for LPS.

4 Conclusion

In this study, we have effectively conducted an experiment to identify skateboarding talents in accordance with bio-fitness and skill-related performances with the aid of machine learning algorithms. The bio-fitness markers considered in the study represent the lower body strength of the athletes which could generally assist in the execution of skateboarding tricks. From the analysis, HACA has successfully clustered the athletes into two clusters namely HPS and LPS. This clustering analysis appeared useful in providing a reasonable grouping of the athletes into clusters. Hence in predicting the classes of these two clusters, ANN was employed to classify the athletes into their respective clusters. The ANN model demonstrated an excellent classification accuracy of 91.7%. In conclusion, the outcomes of this analysis could assist coaches, sports personnel and talent Scouters to practically identify talents given a few bio-fitness elements which are essential as the technique could assist the stakeholders in identifying talented skateboarders systematically thereby saving energy and resources, especially during talent identification programmes.

Acknowledgements The authors would like to acknowledge the Ministry of Education Malaysia for supporting this study under the Fundamental Research Grant Scheme (FRGS/1/2020/TK0/UMT/03/2). The authors thank the athletes and all the relevant stakeholders for their cooperation throughout the data collection process.

References

1. Reuters T (2021) Skateboarding legend Tony Hawk shreds new Olympic venue, calling it 'surreal. CBC Sports Canada
2. Borden I (2016) Ollies at the Olympics: why having skateboarding at Tokyo 2020 is a winning move, no Aug 2016
3. Johnston K, Wattie N, Schorer J, Baker J (2018) Talent identification in sport: a systematic review. Sport Med 48(1):97–109. https://doi.org/10.1007/s40279-017-0803-2
4. Henriques-Neto D, Minderico C, Peralta M, Marques A, Sardinha LB (2020) Test–retest reliability of physical fitness tests among young athletes: the FITescola® battery. Clin Physiol Funct Imaging 40(3):173–182. https://doi.org/10.1111/cpf.12624
5. García-Pinillos F, Cámara-Pérez JC, Soto-Hermoso VM, Latorre-Román PÁ (2017) A High Intensity Interval Training (HIIT)-based running plan improves athletic performance by improving muscle power. J Strength Cond Res 31(1). https://journals.lww.com/nsca-jscr/Fulltext/2017/01000/A_High_Intensity_Interval_Training__HIIT__Based.17.aspx
6. Kevin Andrews S, Lakshmi Narayanan K, Balasubadra K, Josephine MS (2021) Analysis on sports data match result prediction using machine learning libraries. J Phys Conf Ser 1964(4):1–10. https://doi.org/10.1088/1742-6596/1964/4/042085
7. Van Eetvelde H, Mendonça LD, Ley C, Seil R, Tischer T (2021) Machine learning methods in sport injury prediction and prevention: a systematic review. J Exp Orthop 8(1). https://doi.org/10.1186/s40634-021-00346-x
8. Rajoub B (2020) Supervised and unsupervised learning. Biomed Signal Process Artif Intell Healthc 51–89. https://doi.org/10.1016/b978-0-12-818946-7.00003-2. (April)
9. Taha Z et al (2018) The identification of high potential archers based on relative psychological coping skills variables: a support vector machine approach. IOP Conf Ser Mater Sci Eng 319(1):1–7. https://doi.org/10.1088/1757-899X/319/1/012027
10. Maliki ABHM et al (2018) A multilateral modelling of Youth Soccer Performance Index (YSPI). IOP Conf Ser Mater Sci Eng 342(1):1–10. https://doi.org/10.1088/1757-899X/342/1/012057
11. Razali N, Mustapha A, Yatim FA, Ab Aziz R (2017) Predicting player position for talent identification in association football. IOP Conf Ser Mater Sci Eng 226(1):1–9. https://doi.org/10.1088/1757-899X/226/1/012087
12. Gür S, Soyal M, Doğan Ö (2022) Investigation of the relationship between vertical jump and core performance on competition shooting performance in elite female basketball players. J Phys Educ Sport 22(4):948–954. https://doi.org/10.7752/jpes.2022.04120
13. Chimera NJ, Knoeller S, Cooper R, Kothe N, Smith C, Warren M (2017) Prediction of functional movement screen^{TM} performance from lower extremity range of motion and core tests. Int J Sports Phys Ther 12(2):173–181
14. Domínguez-Díez M et al (2021) Comparison of multidirectional jump performance and lower limb passive range of motion profile between soccer and basketball young players. PLoS One, 16(1):1–18. https://doi.org/10.1371/journal.pone.0245277. (January)
15. Miller D (2013) Measurement by the physical educator why and how, 7th edn. MCGRAW-HILL US HIGHER ED
16. Chang WD, Chen S, Tsou YA (2021) Effects of whole-body vibration and balance training on female athletes with chronic ankle instability. J Clin Med 10(11). https://doi.org/10.3390/jcm10112380

17. Chizewski A, Box A, Kesler R, Petruzzello SJ (2021) Fitness fights fires: exploring the relationship between physical fitness and firefighter ability. Int J Environ Res Public Health 18(22). https://doi.org/10.3390/ijerph182211733
18. Borden I (2016) Ollies at the Olympics: why having skateboarding at Tokyo 2020 is a winning move. Conserv https://theconversation.com/ollies-at-the-olympics-why-having-skateboarding-at-tokyo-2020-is-a-winning-move-63349
19. Frederick EC, Determan JJ, Whittlesey SN, Hamill J (2006) Biomechanics of skateboarding: kinetics of the Ollie. J Appl Biomech 22(1):33–40
20. Corrêa NK, de Lima JCM, Russomano T, dos Santos MA (2017) Development of a skateboarding trick classifier using accelerometry and machine learning. Res Biomed Eng 33(4):362–369. https://doi.org/10.1590/2446-4740.04717
21. Shapiee MNA et al (2020) The classification of skateboarding trick manoeuvres through the integration of image processing techniques and machine learning, vol 632. Springer Singapore
22. Rasid AMA et al (2021) Development of skill performance test for talent identification in amateur skateboarding sport. Adv Intell Syst Comput 1350:385–390. https://doi.org/10.1007/978-3-030-70917-4_37. (AISC)
23. Abdullah MA, Ibrahim MAR, Bin Shapiee MNA, Mohd Razman MA, Musa RM, Abdul Majeed APP (2020) The classification of skateboarding trick manoeuvres through the integration of IMU and machine learning. In: Intelligent manufacturing and mechatronics, pp 67–74
24. Shapiee MNA, Ibrahim MAR, Razman MAM, Abdullah MA, Musa RM, Abdul Majeed APP (2020) The classification of skateboarding tricks by means of the integration of transfer learning and machine learning models. In: Embracing industry 4.0, 2020, pp. 219–226.
25. Albiach JP, Mir-Jimenez M, Moreno VH, Moltó IN, Martínez-Gramage J (2021) The relationship between VO2 max, power management, and increased running speed: towards gait pattern recognition through clustering analysis. Sensors 21(7). https://doi.org/10.3390/s21072422
26. Claudino JG, de Capanema DO, de Souza TV, Serrão JC, Machado Pereira AC, Nassis GP (2019) Current approaches to the use of artificial intelligence for injury risk assessment and performance prediction in team sports: a systematic review. Sport Med Open 5(1). https://doi.org/10.1186/s40798-019-0202-3
27. Abdul Rahman K, Azaman A, Latip HF, Mat Dzahir MA, Balakrishnan M (2017) Comparison of tibialis anterior and gastrocnemius muscles activation on balance training devices and hoverboard. Malaysian J Fundam Appl Sci 13. https://doi.org/10.11113/mjfas.v13n4-2.820
28. Babbit D (2020) The Vertical Jump and it's variations. Stack
29. Strand SL, Hjelm J, Shoepe TC, Fajardo MA (2014) Norms for an isometric muscle endurance test. J Hum Kinet 40:93–102. https://doi.org/10.2478/hukin-2014-0011
30. Krishnan A, Sharma D, Bhatt M, Dixit A, Pradeep P (2017) Comparison between Standing Broad Jump test and Wingate test for assessing lower limb anaerobic power in elite sportsmen. Med J Armed Forces India 73(2):140–145. https://doi.org/10.1016/j.mjafi.2016.11.003. (Apr 2017)
31. Li S-F, Huang M-L, Li Y-Z (2022) Exploring and selecting features to predict the next outcomes of MLB games. Entropy 24(2):288. https://doi.org/10.3390/e24020288
32. Rahman MM, Rahman S, Rahman R, Hossain BMM, Shoyaib M (2017) DTCTH: a discriminative local pattern descriptor for image classification. Eurasip J Image Video Process 2017(1):1–25. https://doi.org/10.1186/s13640-017-0178-1

Social Network Analysis and Data Visualization of Football Performance Preceded to the Goal Scored

M. Syafiq M. Fauzi, K. Imran, and Zulkifli Mohamed

Abstract Social network analysis reveals a significant set of essential information on the behaviour of the players and teams: such as passing sequences between players in the attacking third position of attacking performance. Despite their usefulness, network metrics related to expected goal values for soccer analysis are minimal. Thus, the study compared network approaches that led to Chelsea FC's goal and playing style in the English Premier League in season 21/22. Moreover, 'expected goal values' (xG) show the probability of the goal scored, which can be related to the goals scored passing network. The study used centrality in network analysis such as degree prestige, degree centrality, and betweenness centrality to find the significant contributor of the player's position independently during the match that led to the goal scoring and did not consider the playing style of Chelsea FC. Furthermore, the xG values of shots in every game were visualized using Tableau software. A set of adjacency metrics were computed using the highlight videos of goals for every match, and the results of network analysis found that the most received the ball from their teammates were left wing-back and right wing-back, defensive midfielders and right wing-back have the highest degree centrality and attacking midfielder and left wing-back have the highest betweenness centrality. Furthermore, the statistical data from the visualization such as cross percentage, passing percentage, shots and xG percentage can be used to enhance the team performance.

Keywords Match analysis · Passing network · Expected goals

M. S. M. Fauzi · K. Imran (✉)
School of Mechanical Engineering, College of Engineering, Universiti Teknologi MARA, Shah Alam, Shah Alam, Malaysia
e-mail: imransainan@uitm.edu.my

Z. Mohamed (✉)
Sports Engineering Artificial Intelligence Centre (SEA-IC), UiTM, Shah Alam, Malaysia
e-mail: zulkifli127@uitm.edu.m

© The Author(s), under exclusive license to Springer Nature Singapore Pte Ltd. 2023

S. F. Syed Omar et al. (eds.), *Innovation and Technology in Sports*, Lecture Notes in Bioengineering, https://doi.org/10.1007/978-981-99-0297-2_6

1 Introduction

During attacking play, the passing sequence unveils the way of cooperation between the players and the team's playing style. Network analysis was a feasible method to identify the interaction during the match. The use of statistics in soccer has exploded in recent years due to gaining a competitive edge and achieving success in local games and international competitions [1]. Another study used network analysis to determine the prestigious position of Spain's positional roles during FIFA World Cup 2014 [2]. Recently, passing factors have been used to construct social networks meant to represent connections players have built. These have been done to understand better how an individual's performance can be enhanced by increasing the team's overall performance.

Passing networks between teammates can play a role in scoring goals and building better performance, which developed in the training sessions to be applied in the actual match [3]. Meanwhile, the opposition team also has their defending positions to block any possible attempt to score a goal [4]. Previous research indicates that passing quality leads to a better way of going forward and contributes to the cooperation of playing the passes as a team. Furthermore, the fluidity of transitions between players enhances the technicality of the passes between the players themselves [5]. The individual attributes of the passing sequence type also play a significant role in a completed pass. However, research on analyzing the passing sequence that leads to the goal being scored or conceded by a team by coming out with data that measure the effectiveness of the attacking formation in the attacking third of the field should be studied. This measurement was compared with the team's playing style. Another research indicates the importance of the pitch zones, which enhance the team performance by locating suitable positions to score goals through a passing network, freekick, or corners [6]. However, the probability of achieving a goal should be studied regarding the location of shot attempts and player position on the defensive side. This measurement should be in comparison between teams that have an increased value of goals scored.

Rather than that, a pitch zone also can increase the team performance on the attacking and defensive sides in the previous study [7]. Previous research indicates the importance of the pitch zones, which enhance the team performance in locating the suitable positions to score goals through a passing network, freekick or corners [8]. However, the probability of achieving a goal should be studied regarding the location of shot attempts and player position on the defensive side. This measurement focused on comparing teams with an increased value of scored goals.

Passing networks are dynamical systems in and of themselves, and fully identifying and quantifying how factors influence the evolution of a team's game are still unsolved issues [9]. Indeed, because the game cannot be free of arbitrary factors and the tremendous sophistication of its fundamental dynamics, modelling and forecasting a football match is a complex challenge [10]. Fortunately, Network Science can assist in differentiating interference from determinism since it is possible to measure the amount of randomness in the network's topology and the dynamics that occur in it. Quantifying players' movements and capturing broader trends in the game's overall flow were the primary uses for the vast quantity of data produced by fully automated and semi-automated player tracking systems [11]. With this information, coaches can concentrate their training strategies on the most crucial elements of players' fitness. Expected goals value has a challenge in obtaining the raw data such as players' position. In that case, a more sophisticated approach is needed to categorize the data corresponding to shot type and attribute a separate probability for each class. Some analyze various shot properties, resulting in a complicated mathematical model. xG was shown to be a valuable and reliable tool in professional football by previous research [12]. xG was influenced more by the combination of distance and angle than by distance alone. Despite the lack of direct practical application, this strategy could be integrated into training exercises during the attacking and defensive phases to help players better comprehend the game's requirements.

Utilizing web technologies, create and execute an information visualization with a distinct viewpoint [13]. They could work out the incorporated features in great depth by applying an iterative design technique to acquire the final product, paired with informal input. Thus, the user study concluded that while visualization aids hypothesis creation, some features, such as hovering, can slow down users. It thus shows the significance of conducting user research before launching a real-world project. Using visualizations in these specific methods can potentially be beneficial in scouting scenarios [14]. It's possible for a sports player to be interested in understanding the habits and behaviours of their opponent to devise their methods for competing against such characteristics.

The implementation of the playing style for Chelsea FC, which resulted in fewer goals scoring in the attacking third, has been analyzed in this study. Comparing the statistical data of the post-match for Chelsea FC, such as cross attempted and cross successful, shots and goals, and passing network, led to determining the prestigious positional roles between players. This approach compares teams' performance values [15]. Since the focused team, Chelsea FC uses their playing style during attacking, which can be correlated with the challenge to score a goal. In addition, a significant comparison by relating the expected goals value in the study to evaluate the team performance with the social network analysis.

2 Methodology

2.1 Data

This study collected and analyzed thirty-eight matches from Chelsea FC's English Premier League 2021/2022 datasets. The datasets collected on the FBref website (*Premier League Stats | FBref.Com*, n.d.) comprise player and fixture statistics for every match in season 2021/2022. The collected data have been separated into two files to ensure it does not mess up when visualizing. The fixture data contains statistics for every match in one season, such as opponents, goals, passes, corners, aerial duels, and shots in the match. The second data comprises the player statistics such as a corner, shot timeline, expected goals (xG), assist, the outcome of the shots, and pass for every match.

2.2 Observation

An aggregate of 142 adjacency metrics was generated considering the player interactions of passes that led to the goal and converted to 3 network metrics analysis. To develop adjacency measurements, the researcher uses an approach using video highlights of goals to determine the passing trajectory between players. Passes that were successfully received were recorded and used pass links as an example; from player 6 to player 9, considered the clear footage of the video highlights.

2.3 Player Positional Roles

Some criteria have been taken by labelling the node, which consists of 11 nodes with the player's positional role during the matches (Table 1). This methodology categorizes the player in the team's tactical formations done by the previous study [17]. Player positional roles were thorough for every match by considering multiple players for one positional role. For example, Chelsea plays with three centre-backs; therefore, an approach has been discussed by placing one positional role for two nodes, such as centre back 1 and centre back 2. Moreover, the adjacency matrix generated corresponded to the pass live that led to the goal and did not include set pieces, corners, and penalties. After evaluating the video highlight, overall adjacency metrics were generated. Since the video highlight did not show the overall passing between players, the value of the adjacency matrix was small.

2.4 Network Analysis

The social network analysis consists of the goal-scoring passing network by constructing the network matrix and diagram comprising nodes (player positional roles in every match) and edges (number of passing between players). For example, player 2 passes the ball to player 3 five times in a match considering the five passes were the edges. This matrix combines all the links in the adjacency matrices [18]. Once the overall adjacency data had been generated based on the passes between players for every match in season 21/22, it was imported into Social Network Visualizer (SocVizNet) to be evaluated. Given one weighted digraph G with n vertices, the total links index, *LD W*, of G can be computed as follows [17]:

Table 1 List of nodes reliable to the position roles

Node	Positions	Glossary
1	Goalkeeper	GK
2	Centre Back 1	CB 1
3	Centre Back 2	CB 2
4	Left Back/Left Wing Back	LB/LWB
5	Right Back/Right Wing Back	RB/RWB
6	Defensive Midfielder	DM
7	Centre Midfielder	CM
8	Attacking Midfielder/Centre Midfielder	AM/CM
9	Striker	ST
10	Left-Wing Forward	LWF
11	Right Wing Forward	RWF

$$L_D^W = \sum_{i=1}^{n} \sum_{j=1, j \neq 1}^{n} a_{ij} \quad (1)$$

where a_{ij} are components of a G's weighted adjacency matrix. In this study, the centrality levels of Chelsea FC players with positional roles were analyzed using three network analyses: degree prestige; (ii) degree centrality; and (iii) betweenness centrality.

2.4.1 Degree Prestige

Degree prestige indicates the incoming links that a player receives from other teammates. In these circumstances, players with a higher degree of prestige often receive the ball from their teammates during passing sequences. The method to calculate the degree prestige is as follows:

$$P'^{W}_{D-in}(n_i) = \frac{k_i^{w-in}}{\sum_{i=1}^{n} \sum_{j=1, j \neq i}^{n} a_{ij}} \quad (2)$$

where k_i^{w-in} is the index of the vertex and n_i and a_{ij} are the elements of the weighted adjacency matrix of G [19].

2.4.2 Degree Centrality

The degree centrality or out-degree centrality represents the overall outbound connections from a player to his teammates. Other than that, players with a higher degree of centrality contribute more with their passes to other teammates during attacking play. The normalized degree of centrality can be evaluated as:

$$C'^{W}_{D-out}(n_i) = \frac{k_i^{w-out}}{\sum_{i=1}^{n} \sum_{j=1, j\neq i}^{n} a_{ij}} \tag{3}$$

where k_i^{w-out} is the index of the vertex and n_i and a_{ij} are the elements of the weighted adjacency matrix of G [19].

2.4.3 Betweenness Centrality

The betweenness centrality of a node is measured by the number of times it lies on the path with the shortest distance to other nodes. It means players with a higher betweenness centrality are likely to be frequently located between their teammates, acting as a link or 'bridge' between them. The standardized betweenness centrality can be calculated as [20]:

$$C_B(n_i) = \sum_{j\neq k\neq i} g'^{i}_{jk} g_{jk} \tag{4}$$

where g'^{i}_{jk} is the number of most excellent passing linkages between players j and k by player i and g_{jk} is the number of most excellent passing linkages between players j and k. All the centrality matrix that has been mentioned above is used to identify the prominent positional roles of the player in the 56 goals scoring passing network for Chelsea FC.

2.5 *Visualization*

Data visualization shows data in a graphical or pictorial style understandably and efficiently [21]. Many sports tools have recently evolved to increase team and player performance. In this study, an application that has been used to visualize the data is Tableau software since it is easy to use and allows more creative analysis. All the data collected were visualized using this software, such as passing types (successful and unsuccessful), shots timeline including the expected goals value, number of goals scored, and shot types (goals, wayward, and block).

In this study, the collected data from FBref.com and understats.com were segmented into 3 folders: fixtures stats, shot timeline stats, player stats, and player stats for a single match through the season 21/22 for Chelsea FC. In advance, utilizing the type of collected data was essential for visualization in this study. As in Tableau, an excel file was used to connect the data to the data source. All the files were union together and filtered using the date of the match. The software interprets the data based on the column header as a field name. Choosing the appropriate charts to show

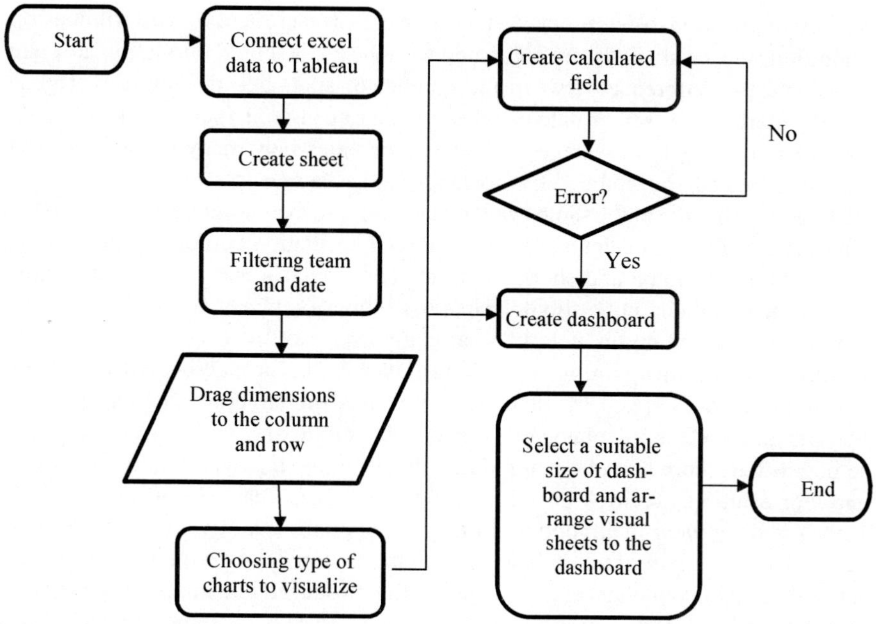

Fig. 1 Flowchart diagram for visualizations

the data using one of these visuals is one of the most critical aspects of the analysis process. An approach has been taken to visualize the data using Tableau software (Fig. 1).

Calculated fields let generate new data from existing data in the data source. When a user creates a calculated field, they are effectively adding a new field to the data source, and the values or members of that field are determined by a calculation used in this study. A transformation that can perform on the values in visualization is a table calculation. Table calculations are a particular kind of calculated field that perform computations using local data in Tableau. They do not consider any measures or dimensions filtered out of the visualization and are derived based on what is currently displayed.

2.5.1 Teams' Data Visualizations

The team dashboard interactively presents the team statistics using pie charts, stacked bar charts and line charts. In detail, teams' statistics such as crosses, shots, goals and xG. This approach is to visualize the team performance using such variables. Regarding Chelsea FC, they usually concentrate on the wide play by crossing into the 18-yard-box. It can be evaluated using the statistical data for cross percentage by constructing a calculated field in Tableau. However, the visual data for successful and attempted cross were computed using a running total calculation through 38 game weeks, which is used to determine the total sum of a measure across a table dimension. In addition, shots statistics were computed with the goals scored throughout a season to evaluate the difference between the number of shots and the number of goals scored. Therefore, a set of calculated fields has been constructed to identify the goals/shots on target percentage. This value indicates how many possibilities that have been created contribute to the goals scored in the season.

Moreover, the sum of the shots from match week 1 to match week 38 was computed using the running sum calculation in the Table Calculation comparing the running sum of shots attempted and shots on target. This analysis was to show the teams' performance on their shot attribute. The possibility of a player scoring a goal from a particular position on the field in a particular aspect of the event was assessed to determine whether an opportunity signified a chance. This probability is referred to as expected goals or xG [22]. The statistics were based on numerous factors before the shot was taken. xG is calculated on a scale from zero to one, with zero representing the inability to score and one being the probability that a player would score every time. For example, Team A has an xG of 1,9, whereas Team B has an xG of 0.7. Team A had a greater likelihood of scoring based on the quality of its attempts than team B. However, it does not indicate that Team A should have won. Expected goals measure the quality of chances and the possibility produced into goals. This approach determined the difference in xG values between Chelsea FC and their opponents for each game analyzed in this study. This picture depicted the attacking performance of Chelsea FC's team in comparison to that of their opponents in each match.

2.5.2 Fixtures Dashboard

The fixtures dashboard represents the match statistics in season 21/22 for Chelsea FC. This approach visualizes the match analysis, such as the shot timeline in 90 min, the player that played in the match and computing the number of passes and shots (players). For the shot timeline, data is required for constructing the path chart, such as shot types (goals, wayward, blocked and saved), players' names, and xG value. Running total calculations were used; the visual data presented the xG value in total between the home and away teams, creating chances for scoring the goals. In addition, the number of shots and passes for individual players was computed using a stacked bar chart. The visual data presented the contribution of individual players in terms of passes and shots in the match. Data for xG has been obtained from the understats.com website, where they represent the statistics of the xG based on the team post-match.

Furthermore, xG is the percentage based on the attributes of the shot and the location in which it is taken that would result in a goal. There are several factors to consider, including the shooter's position, the field's slope, distance from the goal, and part of the body using their head or foot. The category for passes such as short passes, a cross, a set-piece, or something else entirely. Some xG models can generate the same result by using approaches and variables equivalent to one another (StatsBomb).

$$RMSE = \sqrt{\frac{\sum_i^n xG_i - G_i^2}{n}} \quad (5)$$

where xG is the model's prediction, which might have a probability between 0 and 1 for the shot, indicates the index i and G are the actual result.

3 Results and Discussion

Figure 2 shows the total interactions between players depending on their positional role for every matches. The analysis results from the passing network for goal scoring (Table 2) in EPL season 21/22 show that node 5 (right wing-back), node 4 (left wing-back), and node 9 (striker) have the highest value of % D.P. (degree prestige). Such players frequently receive the ball from their teammates in creating a goal score action, whether from short or long passes. It proves that Chelsea FC focused on the 'wide defenders' position and the 'attacker playing position' based on short passes and crosses from the side of the field into the 18-yard box.

Furthermore, the results in this study indicate that node 6 (defensive midfielders), node 5 (right wing-back), node 8 (attacking midfielder), and node 4 (left wing-back) have the highest value for % D.C. and % B.C. The value of % D.C. denotes that the defensive midfielders and wing-back defenders begin to build up a passing sequence to help the offensive transition from the centre of the field spaces. Besides, the value of % B.C. shows that defensive midfielder and wing-back defenders were the most

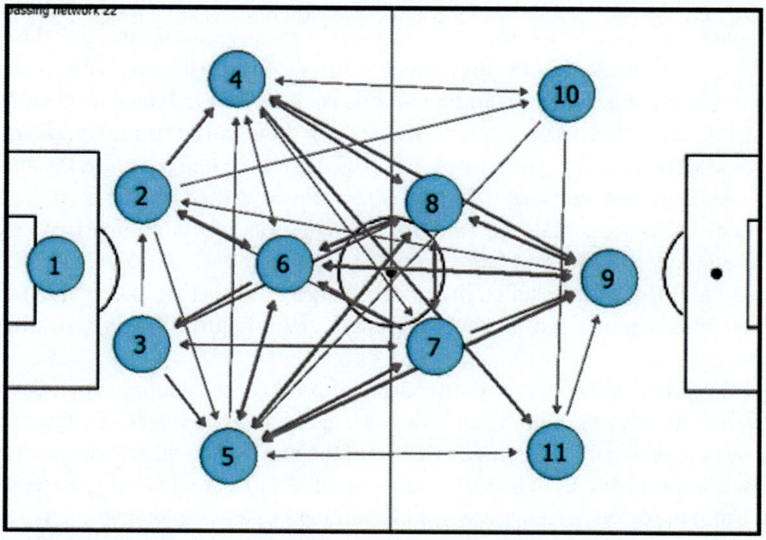

Fig. 2 Graph of player interaction preceded the goal scored

Table 2 Centrality values of positional roles in Chelsea FC

Node	Position	%DP	%DC	%BC
1	GK	0.00	0.00	0.00
2	CB 1	8.16	6.34	2.78
3	CB 2	4.08	4.93	2.78
4	LWB	14.29	9.86	38.19
5	RWB	16.33	18.3	41.44
6	DM	12.24	21.1	49.31
7	CM	8.16	10.5	0.00
8	AM	12.24	14.7	19.44
9	ST	14.29	9.86	1.39
10	LWF	4.08	2.82	0.00
11	RWF	6.12	1.41	0.00

(% D.P.—Degree Prestige, % D.C.—Degree Centrality, % B.C.—Betweenness Centrality)

linked players between players acting as a 'bridge'. As a result, a player's lower betweenness is likely to imply a better ability to keep the ball moving, with less reliance on single players [17].

In addition, the first visual design in this study is called the Teams' dashboard (Fig. 3), comprising shots, goals, and cross statistics for overall the season for Chelsea FC. Since Chelsea FC used to play a wide playing style in attacking transitions, resulting in 496 crosses attempted, and the cross-accuracy percentage found that 13.91% of overall crosses successfully delivered the ball to the players in the 18-yard box. In this case, they were positioning a player with excellent mobility and astute space in terms of dribbling and passing, which can penetrate through the central area [25], allowing Chelsea FC to create more goals during the attacking phase when facing the low block defense. Some players, such as Reece James, Mason Mount, Marcos Alonso, and Hakim Ziyech, often crossed the ball into the opponent's 18-yard box and found that the quality of the crossing or long ball passes in the attacking transition should be emphasized. It can also be related to their positional roles during the match. Other than that, the running sum of shots was a comparison between shots attempted, and shots on target using an area chart as Chelsea FC has 585 shots attempted and 199 shots resulted in a shot on target. Comparing and analyzing goals and shots throughout a season revealed that 37.7% of shots on target resulted in a goal.

Figure 4 presents a Fixtures dashboard comprising the running sum of expected goals value (xG) for both home and away teams in a single match. Different match can be seen by using the date filter. Although, the player's name and statistics during the match were included, indicating the passes and shots created using a stacked bar graph. From the dashboard, we can see the number of passes that have been produced for each player. Meanwhile, the line graph represents a shot timeline through 90 min of the match, indicating the running sum of xG for both teams. In the analyses, it

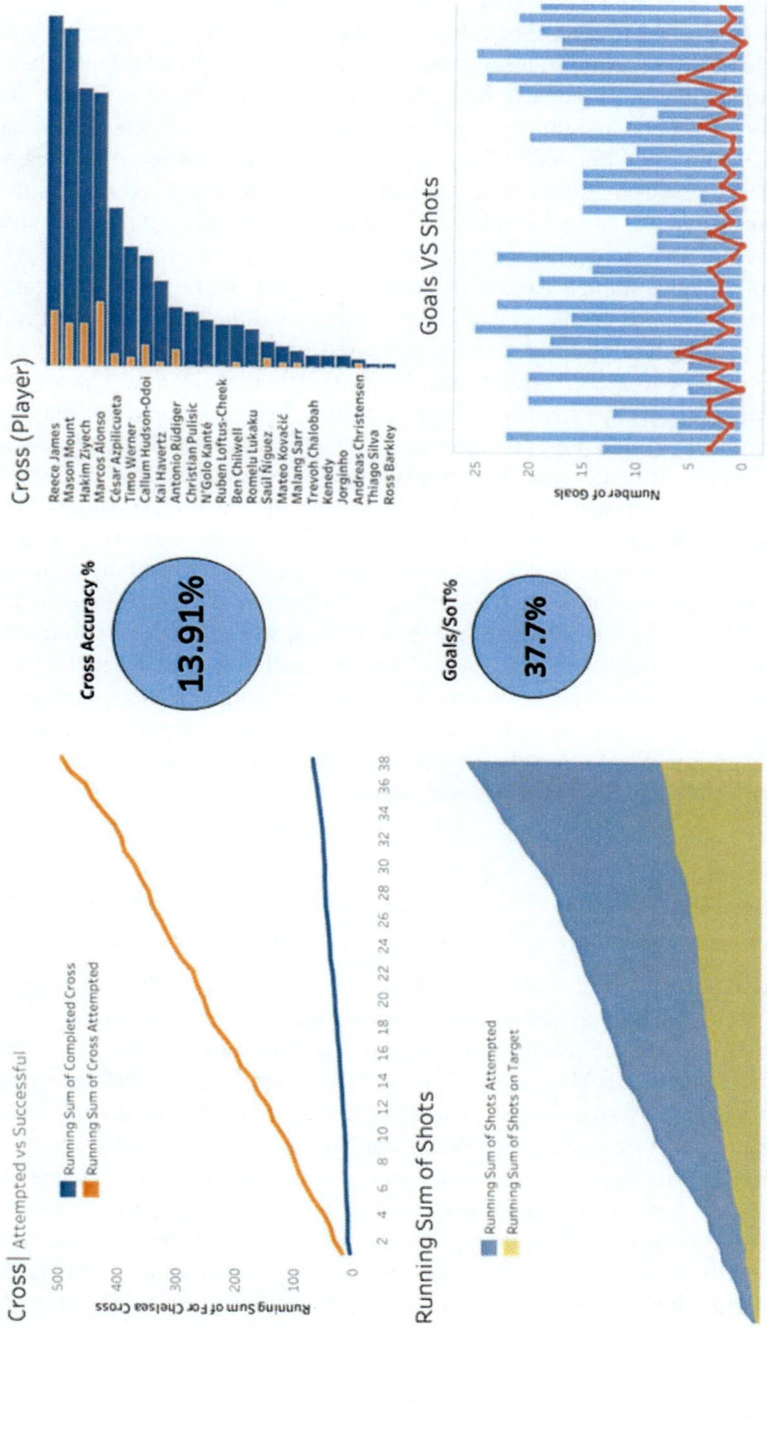

Fig. 3 Teams' dashboard comparing shots, crosses, and goals

was found that the difference between xG of Chelsea FC (67.2%) and xG against (35.8%) was 31.4% from game week 1 until game week 38 (Fig. 5).

Despite being used to determine values for specific shots, the xG statistic is more effective across larger time frames, such as a whole season. In addition to being applied to individuals, it may also be utilized for teams. Compared to simple data such as shots and possession, the expected goals statistic helps comprehend and quantify performance effectively. For instance, it might be inferred that a team's finishing has been strong if they have scored more goals than their xG implies they should have scored. xG can determine whether a player or team is setting themselves up well to score. Expected goals can influence a single game's performance analysis by the team's relative strengths and weaknesses. In analogy, Team A's xG is 3 against Team B's xG, which is 1 implies that Team A dominated the game. However, it is possible that after scoring three goals in the first period, Team B planned to take a backseat and try to preserve their lead. Even though Team A might have more opportunities, Team B had already left the field when Team A entered the match.

Another visual design presents the passing statistics based on the match's win, draw, and lose. A pie chart makes the visual design more comprehensive in showing the data. The passing statistics referred to the passing such as short, medium, long, ground, high (the height of the ball during passing), and cross attempted (Fig. 6). This dashboard shows the corresponding passing types based on the winning or losing game for the team in controlling the possession of the ball during the match. The visual data shows that the number of passes attempted was high when winning the match, indicating that the team might perform on creating goal chances. Previous research revealed that most goals were scored via short passing sequences, supporting their third prediction that most goals would be scored via relatively short passing interactions. Normalized attempts to goal ratios were more desirable for shorter passing sequences than longer passing sequences [26].

4 Conclusion

To conclude, this research focused on employing a network method to comprehend teammates' interactions better as they manifested in passing sequences. Consequently, network measurements were employed in this study to determine team patterns during their attacking moves that resulted in goals scored (Sarmento et al., 2020). Only open-play goals were examined in the case of goals scored, eliminating goals scored through free kicks, penalties, or corners. As a result, the initial stage was to categorize the strategic allocation of players from the observed team on the field. Then the connections were determined based on passes between player positions. It was feasible to discover the most significant values of % D.P. based on the analysis done on the goals scored. The players with the most outstanding % D.P. values were the right wing-back (node 5), left wing-back (node 4), and striker (node 9),

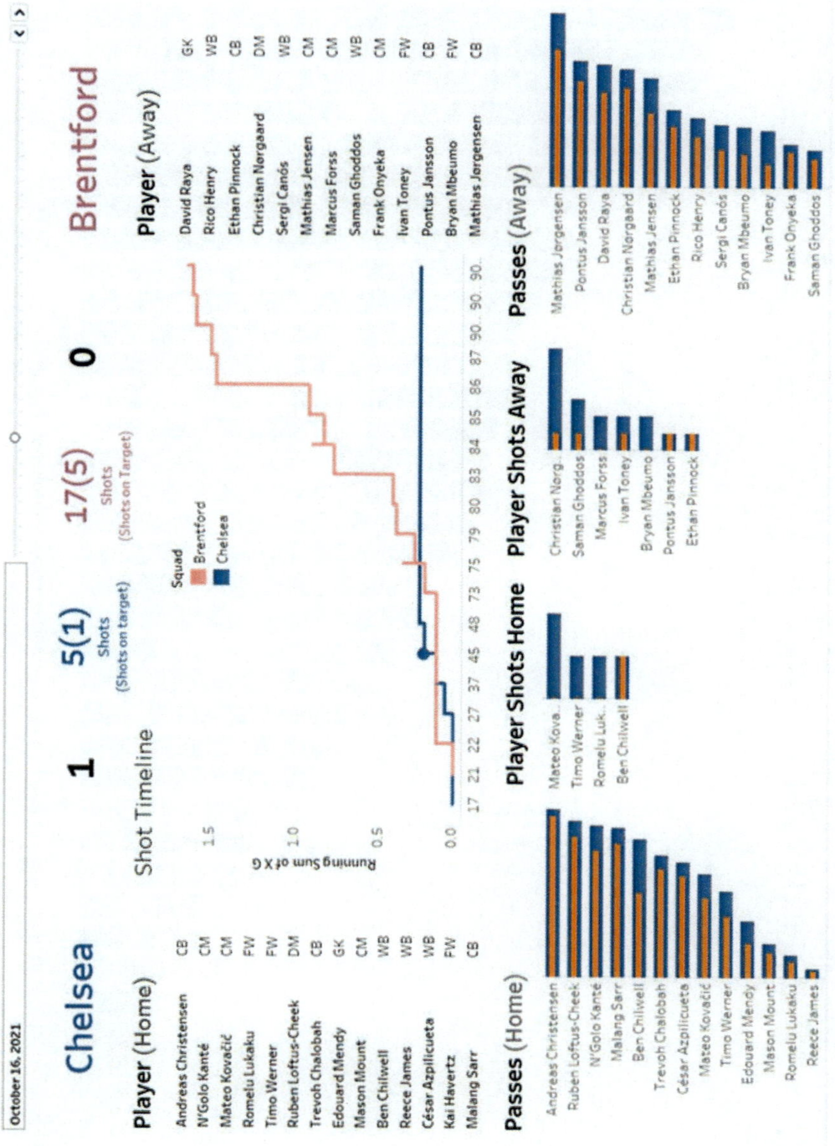

Fig. 4 Fixtures dashboard of post-match

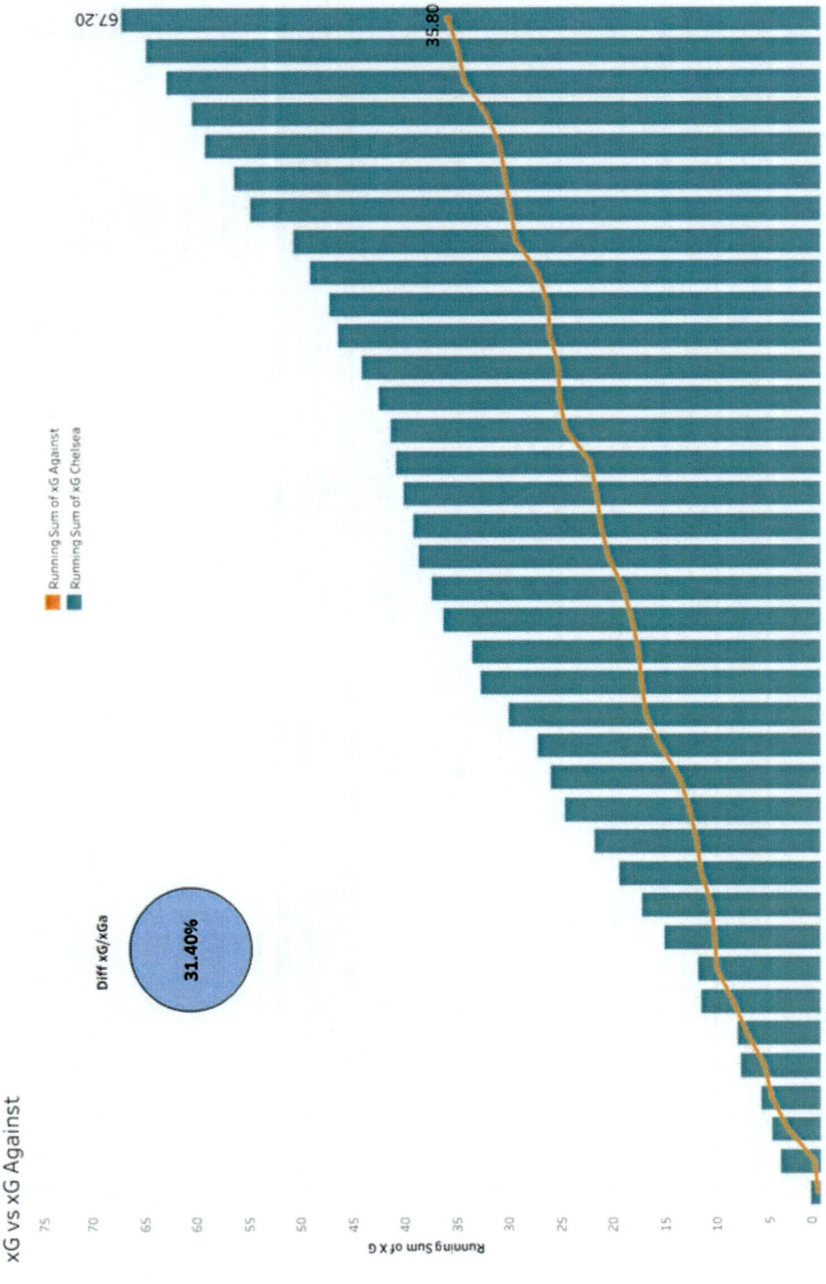

Fig. 5 Comparison of running sum of xG between Chelsea FC and their opponents through a season

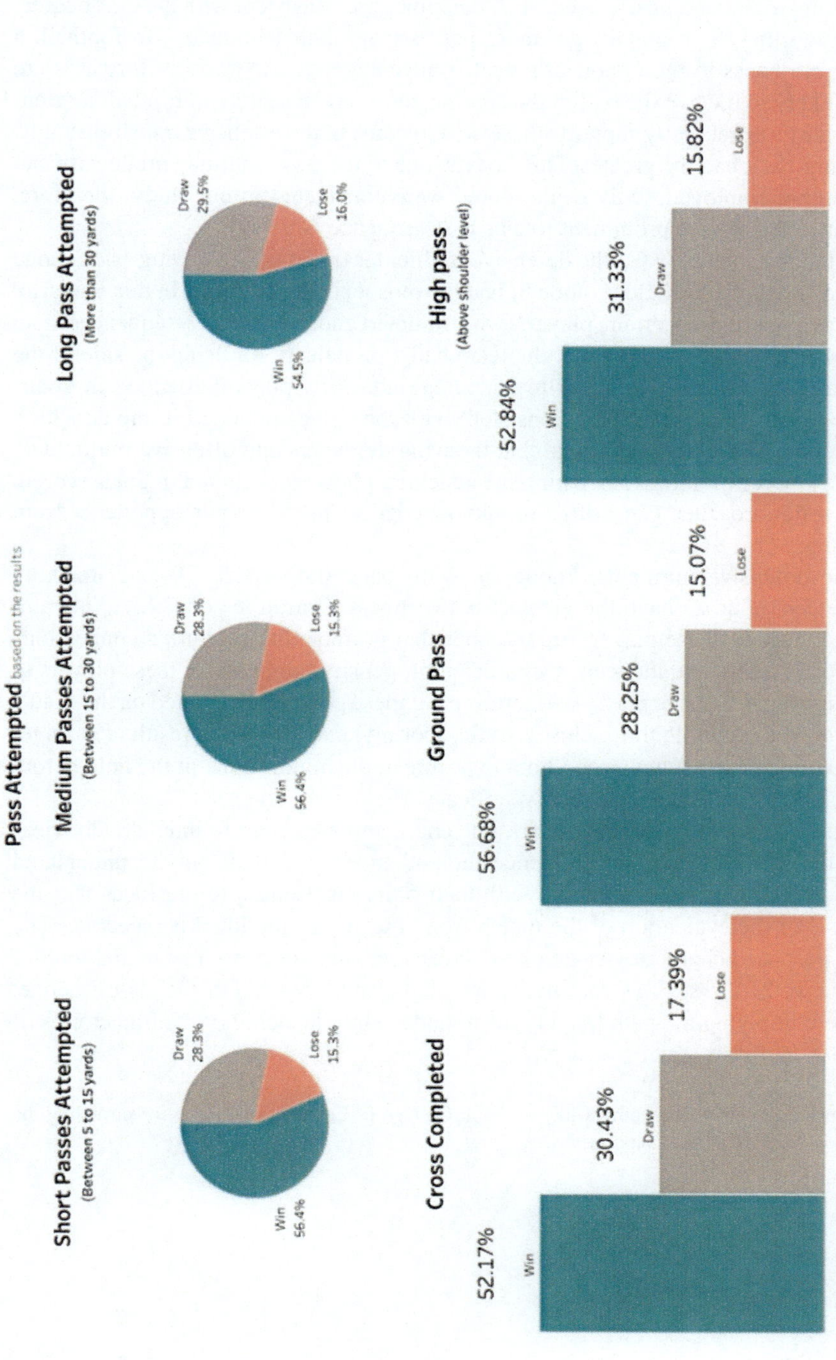

Fig. 6 Pass dashboard for attempted short, medium, long, cross, ground, and high passes

according to the findings. Indeed, such outcomes are consistent with the D.P. degree, indicating that these players got more passes from their teammates. In football, a closer closeness to the opponent's goal is advantageous. As a result, forwards are more likely to receive the ball in the scoring zone. The research on football's offensive transition generally supports these conclusions. In the offensive transition study, the wing-back had the greatest D.P., maybe due to the team's unique propensity and the sample employed. Only goals scored were considered in our study; therefore, forwards had a more prominent role as pass sequence targets.

In the instance of D.C., the defensive midfielder (node 6), right wing-back (node 5), and attacking midfielder (node 8) had the most excellent ratings. In this scenario, D.C. denotes those certain players who initiated more passes or sequences than others. Such numbers show that the team had a propensity for using the side of the field and the central midfielder to encourage attacking plays that ended in goals. These results, once again, are consistent with the values obtained in the attacking transitions. These findings reveal that the wing defenders and offensive midfielders are the key contributors to boosting the attacking plays that resulted in goals scored. As a result, coaches can utilize this knowledge to prevent their opponents from attacking.

The defensive midfielder (node 6), wing back (node 4 & 5), and attacking midfielder (node 8) have the greatest betweenness centralities (B.C.). As a result of these values, these players are the ones that promote the network's connections the most. As a result, the team's significant propensity might be for these players to participate in offensive plays. Furthermore, all these players are located on the field's sides or in the middle third, closest to the scoring zone [28]. As a result, a stronger inclination to use the wings and limit exposure to the middle zone of the field before the last pass contributes to offensive efficacy.

Finally, using a stacked bar chart, line chart, and area chart to interactively visualize the statistics data can influence the audience or football fans to understand the team's statistics and come up with their opinions. Central to our ideas was the analysis of the evaluation of the match from a team or individual perspective. The study proposed charts to visualize the player attributes for every match and total of the season 21/22, such as passing, shots, goals, and crosses. For the shot timeline, we were implying the path line glyphs to understand the detail of the trajectories of the shots during a match.

Acknowledgements The authors like to thank College of Engineering UiTM for providing the facilities and the financial support.

References

1. Malqui JLS, Romero NML, Garcia R, Alemdar H, Comba JLD (2019) How do soccer teams coordinate consecutive passes? A visual analytics system for analysing the complexity of passing sequences using soccer flow motifs. Comput Graph 84:122–133. https://doi.org/10.1016/j.cag.2019.08.010
2. Ievoli R, Palazzo L, Ragozini G (2021) On the use of passing network indicators to predict football outcomes [Formula presented]. Knowl-Based Syst 222. https://doi.org/10.1016/j.knosys.2021.106997
3. de Silva et al V (2018) Player tracking data analytics as a tool for physical performance management in football: a case study from Chelsea football club academy. Sports 6(4). https://doi.org/10.3390/SPORTS6040130
4. Ichinose G, Tsuchiya T, Watanabe S (2021) Robustness of football passing networks against continuous node and link removals. Chaos, Solitons Fractals 147:110973. https://doi.org/10.1016/J.CHAOS.2021.110973
5. Buldú JM, Busquets J, Echegoyen I, Seirul. lo F (2019) Defining a historic football team: using network science to analyze Guardiola's F.C. Barcelona. Sci Rep 9(1). https://doi.org/10.1038/s41598-019-49969-2
6. Stein M et al, From game events to team tactics: visual analysis of dangerous situations in multi-match data
7. Mclean S, Salmon PM, Gorman AD, Stevens NJ, Solomon C (2018) A social network analysis of the goal scoring passing networks of the 2016 European Football Championships. Hum Mov Sci 57:400–408. https://doi.org/10.1016/J.HUMOV.2017.10.001
8. Stein M et al (2016) From game events to team tactics: visual analysis of dangerous situations in multi-match data. In: TISHW 2016 - 1st international conference on technology and innovation in sports, health and wellbeing, proceedings. https://doi.org/10.1109/TISHW.2016.7847777
9. Narizuka T, Yamamoto K, Yamazaki Y (2014) Statistical properties of position-dependent ball-passing networks in football games. Phys A 412:157–168. https://doi.org/10.1016/j.physa.2014.06.037
10. Caicedo-Parada S, Lago-Peñas C, Ortega-Toro E (2020) Passing networks and tactical action in football: A systematic review. Int J Environ Res Public Health 17(18):1–19. https://doi.org/10.3390/ijerph17186649
11. Hanson H, Harland A, Holmes C, Lucas T (2012) Method for understanding football ball motions using video-based notational analysis. Proc Eng 34:164–169. https://doi.org/10.1016/j.proeng.2012.04.029
12. Rathke A (2017) An examination of expected goals and shot efficiency in soccer. J Human Sport Exerc 12(Proc2). https://doi.org/10.14198/jhse.2017.12.proc2.05
13. Bartels P, de Buyser T, van Ussel J, A football data visualization: the Belgian first division. http://bit.ly/1cVJLuK
14. Perin C, Vuillemot R, Stolper CD, Stasko JT, Wood J, Carpendale S (2018) State of the art of sports data visualization. Comput Graph Forum 37(3):663–686. https://doi.org/10.1111/cgf.13447
15. Grund TU (2012) Network structure and team performance: the case of English Premier League soccer teams. Soc Netw 34(4):682–690. https://doi.org/10.1016/J.SOCNET.2012.08.004
16. Premier League Stats | FBref.com. https://fbref.com/en/comps/9/Premier-League-Stats (Accessed 04 July 2022)
17. Clemente FM, Martins FML (2017) Network structure of UEFA champions league teams: association with classical notational variables and variance between different levels of success. Int J Comput Sci Sport 16(1):39–50. https://doi.org/10.1515/ijcss-2017-0003
18. Korte F, Lames M, Link D, Groll J (2019) Play-by-play network analysis in football. Front Psychol 10. https://doi.org/10.3389/fpsyg.2019.01738
19. Clemente FM, Sarmento H, Aquino R (2020) Player position relationships with centrality in the passing network of world cup soccer teams: win/loss match comparisons. Chaos, Solitons Fractals 133:109625. https://doi.org/10.1016/J.CHAOS.2020.109625

20. Winship C, Wasserman S, Faust K (1996) Social network analysis: methods and applications 91(435):JSTOR. https://doi.org/10.2307/2291756
21. Musa SM, Akujuobi C, Sadiku MNO, Shadare AE, Akujuobi CM, Perry RG (2016) Data visualization. Int J Eng Res Adv Technol 12:2454–6135. https://www.researchgate.net/publication/311597028
22. Lamberts M (2021) Data metrics explained: expected goals (xG) – Her Football Hub. https://herfootballhub.com/data-metrics-explained-expected-goals-xg/. (Accessed 25 June 2022)
23. StatsBomb, Evaluating Mohamed Salah - StatsBomb|Data Champions. https://statsbomb.com/articles/soccer/evaluating-mohamed-salah/. (Accessed 25 June 2022)
24. Clemente FM, Manuel F, Martins L, Mendes RS (2016) Analysis of scored and conceded goals by a football team throughout a season: a network analysis. Kinesiology 48:103–114. https://doi.org/10.26582/k.48.1.5
25. Rein R, Raabe D, Memmert D (2017) 'Which pass is better?' Novel approaches to assess passing effectiveness in elite soccer. Hum Mov Sci 55:172–181. https://doi.org/10.1016/j.humov.2017.07.010
26. Wright C, Atkins S, Polman R, Jones B, Sargeson L (2011) Factors associated with goals and goal scoring opportunities in professional soccer. Int J Perform Anal Sport 11(3):439–449. https://doi.org/10.1080/24748668.2011.11868563
27. Sarmento H, Clemente FM, Gonçalves E, Harper LD, Dias D, Figueiredo A (2020) Analysis of the offensive process of AS Monaco professional soccer team: a mixed-method approach. Chaos, Solitons Fractals 133:109676. https://doi.org/10.1016/J.CHAOS.2020.109676
28. Arriaza-Ardiles E, Martín-González JM, Zuniga MD, Sánchez-Flores J, de Saa Y, García-Manso JM (2018) Applying graphs and complex networks to football metric interpretation. Hum Mov Sci 57:236–243. https://doi.org/10.1016/j.humov.2017.08.022

Data Visualization of Football Using Degree of Centrality

M. Syukri Mazlan, K. Imran Sainan, and Zulkifli Mohamed

Abstract Previous research indicated that passing networks can increase the performances of players in a football team. This can be achieved with the aid of data visualization and analysis using post-match data. This paper provides a taxonomy of sports data in football visualization and summarizes the data from three aspects of data types, main tasks, visualization techniques, and visual analysis with the use of Tableau software. The objective of this paper is to identify the playing pattern for Liverpool FC during Jurgen Klopp's era. To identify the playing pattern, this paper will display the diagram of the passing networks from the goals created in the match. Besides, networks and graph theory using Social Network Visualizer is to investigate social structures from the passes data that created goals from an open play. It describes networked systems in terms of nodes and the links between them. The playing pattern may thus be determined by examining the degree of centrality, degree of prestige, and betweenness centrality from nodes and linkages. This paper introduces a visual analysis of competitive football, using the social network from passes to construct degree centrality, and finally discusses the playing pattern for Liverpool FC. For this paper, collecting and flexibly presenting large and complex data is the main concern to increase the understanding of the analysis. In summary, it was feasible to draw the conclusion that network metrics can give sport analysts knowledge that is complimentary to traditional notational analysis by offering a novel visualisation and comprehension of team members' behaviour as well as by characterising particular play patterns.

Keywords Data visualization · Passing network · Social network analysis · Playing pattern

M. S. Mazlan (✉)
Universiti Teknologi MARA (UiTM), Shah Alam, Malaysia
e-mail: syukri159@gmail.com

K. I. Sainan · Z. Mohamed
Sports Engineering Artificial Intelligence Centre (SEA-IC), UiTM, Shah Alam, Malaysia

© The Author(s), under exclusive license to Springer Nature Singapore Pte Ltd. 2023

S. F. Syed Omar et al. (eds.), *Innovation and Technology in Sports*, Lecture Notes in Bioengineering, https://doi.org/10.1007/978-981-99-0297-2_7

1 Introduction

It has seen significant growth in recent years, aided by new technological advancements and new multidisciplinary techniques, which mostly employ methodologies from the social sciences and mathematics to conduct this study. Passing statistics, possessions and locations are the main variables to complete the project. In this paper, network analysis is the main aspect that needs to be considered. As applied to football, network analysis can provide a unique and significant contribution to the categorization of play patterns and the identification of specific interplays between teammates [1].

To relate the playing pattern with the passing network requires post-match data and network analysis to tackle the objectives in this paper. Passing networks are developed from the observation of ball exchange between players, in which network nodes (or vertices) represent football players and links (or edges) represent the number of passes made between any two players on a team [1].

Passing networks are dynamic systems made up of flexible and interacting components that enable the quantification, and development of the game through time. In the network structure, techniques from graph theory, scientific experiments, and big data are combined [2].

This study investigates the network interaction analysis related to playing patterns and defensive actions. A dataset of at least 50,000 passes between players and post-match data analysis for the team for 76 total matches implements the investigation of the network analysis.

It offers a quick way to identify trends or strong and weak ties between players and their spots in the lineup, and it can help demonstrate a player's abilities, a team's strategies, and relationships between positions [3].

However, a single application of a network graph enables coaches and analysts to quickly detect some team characteristics, so assisting them in characterizing the connectivity between teammates. As was previously noted, most of the network analysis in sports concentrated on observing the attack-related processes [4]. Several limitations of the earlier research are determined after a review of the group performance literature. The methodology section discusses the analytical approach and the visualization of data with the available approach that is utilized, followed by the introduction of the variables and measurements for defensive actions and network structure. The section on methods is followed by the results section. The discussion of the results will be concluded at the end of this paper.

2 Literature

A study states that individual contributions, particularly the intermediary function of players, have not been discovered in football based on a separate study of passing networks [4]. Only aggregated passing data from many periods or the full match was used in the studies. As stated in [5], playing patterns can increase team performances. It can come up with various kinds of styles and can be determined with the aid of network analysis.

Additionally, betweenness and proximity measurements are typically applied to the whole passing contact between players throughout a game to analyze the intermediary function of players to connect their teammates as bridging players by distributing the ball [6].

Recently, social networks representing the ties formed by teammates have been created using passing variables to better understand how team performance can be improved [1].

Furthermore, another study states that the ability of the team to self-organize following the behaviors of the teammates is the primary basis for the stochastically flipped dynamics of the players during the game. demonstrated how a data-driven method may offer significant potential to precisely predict the success of the team [7].

The method simply offers a discrete description of players' behaviors, although relevant information can be derived from the mentioned performance indicators and passing networks [8]. The results of this research are most likely based on performance-related outcomes, hence the knowledge about those performances' causes is quite rudimentary because it only provides a partial view of the team's behavior. To understand why and how players and teams control their performance in place and time to accomplish a common goal, notational analysis is therefore insufficient.

The authors did show that a complex systems perspective on football passing data has the potential to identify hidden patterns and behaviors. Within this context, network centrality algorithm computations could be utilized to identify the team members who have the most impact on performance and comprehend the information flow [9]. However, playing patterns are not consistent throughout the match. Thus, it can be understood by the visualization of passing networks [6]. All of the problems in the field of passing networks and network analysis that have occurred over the last few years will be demonstrated by this system need. In this case, the information and the source may aid in the identification of similar information that may be used in project research. Even though scoring status is crucial to this passing network analysis, no research has looked at how scoring status affects defensive actions in a single team.

An information visualization's design is given certain general rules. The most significant finding is that human skills and limits should be taken into consideration when designing information representations. This was the subject of the multimedia course's opening lecture as well [10].

2.1 Limitations

The major findings cannot be generalized or compared to other teams because only one team was studied. Future research must address this problem by including a wider range of data on other teams, even those with varying ultimate scores [4].

The crucial issue in the study is positioning and playing pattern of a team could be very hard to determine. The positioning of players on the field has recently become one of the most important indicators of team performance [7]. This method determines whether players' paired displacement trajectories exhibit a regular and predictable pattern, which may reveal information about their tactical behaviour [1].

When there are transitions, counterattacks, or goalscoring attacks throughout a game, the passing patterns change. The pattern and, more crucially, the team's current scoring situation can both have an impact on a player's style of play and the players' decisions made during the passing process [5].

3 Methodology

3.1 Sample Data Collection

A total of 76 Liverpool FC matches have been recorded. This includes post-match data from the 18/19 and 20/21 English Premier League seasons. The choice of selected seasons was made because of the differences in their achievements. The dataset that was retrieved for this experiment goes through data processing steps to make sure that only the relevant features are used to evaluate team performance statistics and shorten the experiment's run duration [11].

3.2 Observation

Video recordings of available broadcasting official channels were used to conduct observations. Completed passes in goal-scoring passing sequences were recorded. Passing direction was used as an example; passing from player 7 to player 8 is deemed distinct from passing from player 8 to player 7 [3].

3.3 Player Position Analysis

A total of 11 nodes were set according to the player position. The strategic distribution of players in the squad was discovered for the team-members connection study, to identify each player's position on the field. Every position holds a different role in the field. This influences the flow of passing in the team related to their tactical formation in every match. The formation consists of four defenders which are two central defenders, one right and left-back defender. Three midfielders who are defensive, central, and attacking midfielders. Lastly, the three forward positions are left, right and central forward. This was the team's strategic exposition for all the matches that were observed.

3.4 Social Network Analysis

Two nodes corresponding to these two players are connected by an undirected edge when a player passes the ball to another player in the game. Many edges are allowed if multiple passes between the same nodes are made. The passing network quantifies the collection of all passes made by one team during a game. A value of one (1) was given for each pass between nodes, while a code of 0 was given if there were no passes between teams zero (0). Each player's tactical position was codified based on the tactical line-up. A techno-tactical assignment to positional roles was adopted [12]. To be more accurate, only passes made by teammates are considered and when a player is replaced by a reserve, the node for the new player gets the same number as the node for the replacement player [7]. Football match analyses have used network analysis to identify interteam communication, goal scoring passing networks, and overall match passing features [13].

All data were imported to the software Social Network Visualizer (SocNetV) to perform the network analysis. The graphs might well be constructed using the data that has already been created for teammates' interactions and regional connectivity. Aside from graph construction, a set of metrics that mathematically defined the graph's connection could be computed. Degree centrality, degree prestige, and betweenness centrality were computed and calculated as the results of this research.

3.5 Degree Centrality

Degree centrality assigns a numerical to each node based only on the number of links it holds [14]. This indicates the higher degree of centrality are the players that are involved more in the team's passing distribution [15].

As stated in another study, in a conventional SNA, the nodes do not communicate with one another. For instance, in a passing SNA, players cannot pass to one another. As a result, a statistic was needed to determine which zones were used for the player's passing [16].

$$D(i) = \frac{1}{n-1}\sum_{j=1}^{n} A_{ij} \tag{1}$$

where A_{ij} is the *ij*-the element of the adjacency matrix *A* and n is the number of vertices in the graph.

3.6 Degree Prestige

Degree prestige shows it was clear that the players with the highest score were the ones whose teammates opted to pass the ball more frequently [17].

$$P(n_i) = \frac{\sum_{j=1}^{n} X_{ij}}{n-1} \tag{2}$$

where n_i and X_{ij} are weighted adjacency matrix of the element.

3.7 Betweenness Centrality

The role of a player in providing information to flow from one section of the network to the other is captured by betweenness centrality [1].

$$B(i) = \frac{1}{(n-1)(n-2)} \sum_{u,v=1,u \neq i \neq v}^{n} \frac{\partial_{uv}(i)}{\partial_{uv}} \tag{3}$$

Where ∂_{uv} number of paths from node u to v.

3.8 Team Passing Networks

All potential approaches for evaluating data that depict interactions between a group of units (players) to examine patterns and community structures are included in network theory. In this paper, the nodes represent the players from each position. It is linked by some variables which are the passing relationship between players and weights are the number of passes made in each match [18].

$$\tilde{\omega}_{ij} = \frac{P_{ij}}{\sum_{i=1}^{n} \sum_{j=1}^{n} P_{ij}} \tag{4}$$

where emax = n(n − 1)

3.9 *Visualisation*

There are numerous ways to represent data, and it is frequently possible to combine them to achieve a superior outcome. An interactive presentation, in which extra information can be displayed by interacting with each data input, is one technique to accomplish this.

The type of chart chosen relies on how the data is presented, and what conclusions are being drawn from. There are restrictions on the number of variables, datasets, and data types that each chart can depict. Each chart type also has advantages, and some are more effective than others at illustrating how different pieces of information relate to one another.

In this paper, datasets of Liverpool FC players in seasons 18/19 and 20/21 were displayed using Tableau software. The most important element of this research is deciding which type to utilize and how to build it [19]. During the development, box plots, bubble charts, pie charts, tables, line charts, and several types of bar charts were all utilized [20].

For the entire season, data was gathered from FBref.com [21]. Next, the step is connecting data to the data source and importing all datasets into an excel file.

To be visualized, it will be loaded into the Tableau software. The programming language will take care of selecting the prepared data to be visualized, which will be in excel format.

Equations are made to generate analysis and data value for the graphs. As for the percentage of clearances, Total clearances for each player are calculated and computed in a total of 100%. Goal differences for both seasons were calculated and analyzed using stacked bar graphs.

Some of the approaches that have been completed to represent the data are stacked bar graphs and pie charts. As stated in the figure below, the graphs were made to visualize and analyze team performances.

As stated in the figure below, the goal difference for both seasons was calculated. This approach would be an innovative way to help coaches to analyze the statistics for each player.

Team statistics, goals scored and against, defender comparison for both seasons relating to defensive actions, and post-match data analysis for the UEFA Champions League Final for 18/19 were the four aspects of the results.

It may not be possible to put all the variables from a dataset into a single chart but separating them into two and tying them together might be an alternative. Depending on the interactions in the first chart, the second chart will behave differently. The type of chart chosen relies on how the data is presented, what conclusions are being drawn from it, and the trainer's comments. There are restrictions on the number of variables, datasets, and data types that each chart can represent. Each chart type also has advantages, and some are more effective than others at illustrating how different pieces of information relate to one another (Fig. 1).

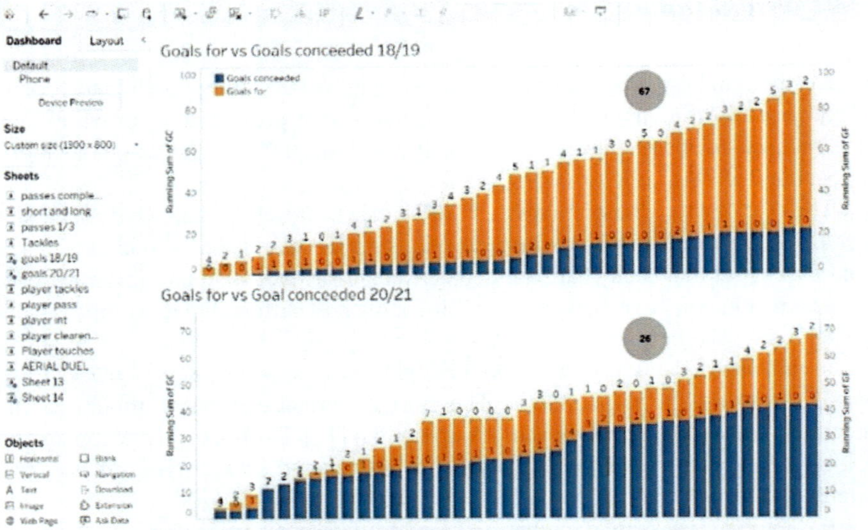

Fig. 1 Dashboard for goals statistics in both seasons

Data is summarised and abstracted in data visualisation. Key components of the data are spatial factors including position, size, and shape. The original dataset should be transformed, reduced in size, and projected on a screen using a visualisation system (Table 1).

Table 1 List of nodes number according to their position

Nodes	Position
1	Goalkeeper (GK)
2	Right back (RB)
3	Left back (LB)
4	Central defender (CB)
5	Central defender (CB)
6	Defensve midfielder (DM)
7	Right wing forward (RWF)
8	Central midfielder (CM)
9	Central forward (CF)
10	Attacking midfieder (AM)
11	Left wing forward (LWF)

4 Results and Discussions: A Case Study of Liverpool FC Statistics on Seasons 18/19 and 20/21

4.1 Social Network Analysis

Playing patterns for Liverpool FC in season 18/19 and 20/21 shows numerous kinds of play. The playing pattern can be determined by the match analysis at the end of the match. As we can observe, the achievements of Liverpool FC in season 18/19 are impressive as they won the UEFA Champions League during the season.

The main factor of comparison between these periods is the presence of Virgil Van Dijk as their centreback. It has been such a big loss for the team as he was out because of an anterior crucial ligament injury (ACL). The team was forced to play with 18 different pairings of centrebacks during the 19/20 season. The chemistry, style of play, and intensity of the game are yet to be different. Liverpool FC managed to win the UEFA Champions League cup during the 18/19 season while they ended the 20/21 season with no trophy achieved.

The results were computed and evaluated to determine the team's playing style. The theory implies that Virgil Van Dijk's presence influenced team performance based on his performances. Depending on the defensive line and command, a team can play in a variety of ways. It has been demonstrated that in season 20/21, Liverpool used a deeper defensive line to keep the opposition from scoring. In that season, the midfield's compactness decreased slightly, and the opponent has been exposed to greater space which leads to conceding goals.

The figure above shows the result analysis for season 18/19. The outer circle indicates that the players were not highly involved in creating goals for the team. Player no 3 (LB) is located at the center of the circle. This corresponds to the results stated in the table with a score of 16.292135%. (Table 2).

The left-back (3) has the highest degree centrality score of 16.292135%, as can be seen in the diagram above. This demonstrates that player 3 is a vital part of the team's attacking strategy. The team will most likely start and establish the attacking play on the left side (Figs. 2 and 3).

Meanwhile, players 7 and 11 are the teammates who want to provide passes and transfer the ball for degree prestige (both 19.101124%). They are the team's right and left-wing forwards, respectively. Using both of the wings forwards, the squad created numerous chances and scored. The adoption of a 'false 9' tactical formation is advantageous because the central forward will penetrate deep inside the middle, creating opportunities for the wing forward to score.

With a score of 43.666667% for betweenness centrality, player 6 is a defensive midfielder who was significantly involved relative to offensive play. The player usually feeds the ball to the attacking third from the defensive third and disperses it to begin the attack. This player act as a bridge to move the ball either in offensive or defensive play (Table 3).

Table 2 Result analysis for season 18/19

Nodes	%DC'	%DP'	%BC'
1	1.685393	0	0
2	11.797753	8.426966	23.888889
3	16.292135	10.11236	31.462963
4	5.05618	1.685393	12.166667
5	2.808989	0.561798	1.111111
6	4.494382	2.247191	43.666667
7	12.359551	19.101124	3.925926
8	12.921348	13.483146	4.907407
9	11.797753	17.41573	9.703704
10	10.11236	7.865169	5.37037
11	10.674157	19.101124	0

(%DC'—Degree centrality, %DP'—degree prestige, %BC'—Betweenness centrality)

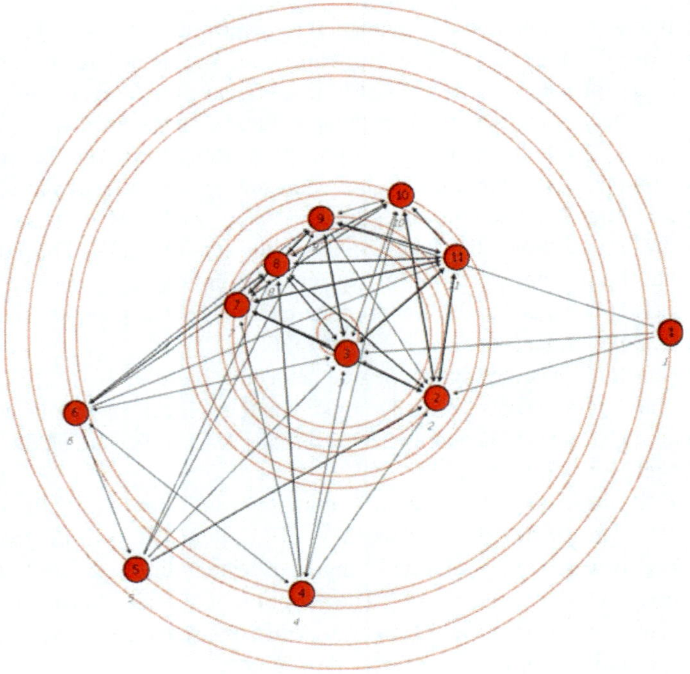

Fig. 2 Radial degree centrality graph for season 18/19

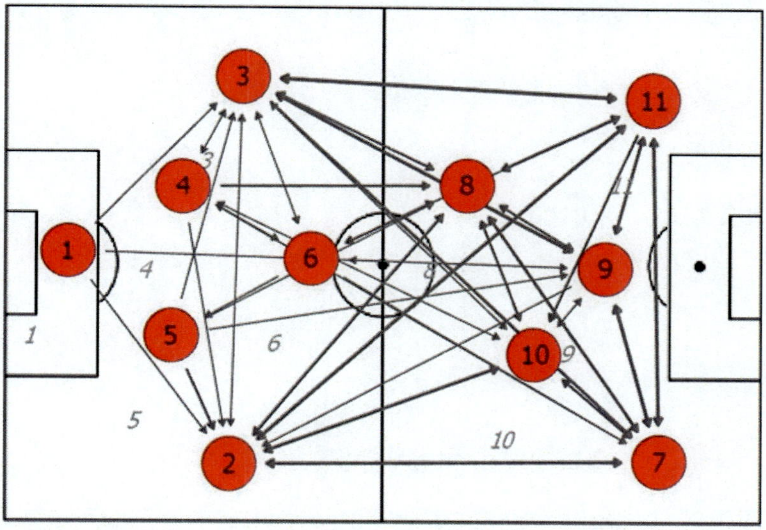

Fig. 3 Passing network for season 18/19

Table 3 Result analysis for season 20/21

Nodes	%DC'	%DP'	%BC'
1	0	0	0
2	20.833333	5.208333	0.555556
3	11.458333	6.25	4.444444
4	2.083333	0	0
5	1.041667	3.125	1.111111
6	6.25	4.166667	0
7	8.333333	18.75	22.222222
8	10.416667	10.416667	3.888889
9	10.416667	22.916667	28.333333
10	13.541667	13.541667	10
11	15.625	15.625	21.666667

(%DC'—Degree centrality, %DP''—degree prestige, %BC''—Betweenness centrality)

The squad, on the other hand, was keen to focus more on the right-hand side in season 20/21. The highest degree centrality score (20.833333%) from player 2 demonstrates this. Less accurate long balls will be sent into their pathways without Virgil Van Dijk (VVD). The right-back, who was linked up with the right-wing striker and central midfielder, was heavily involved in their attacking play (Figs. 4 and 5).

Furthermore, player 9 who is their center striker has the highest degree of prestige and betweenness centrality score (DP- 22.916667%, BC- 28.333333%). The team

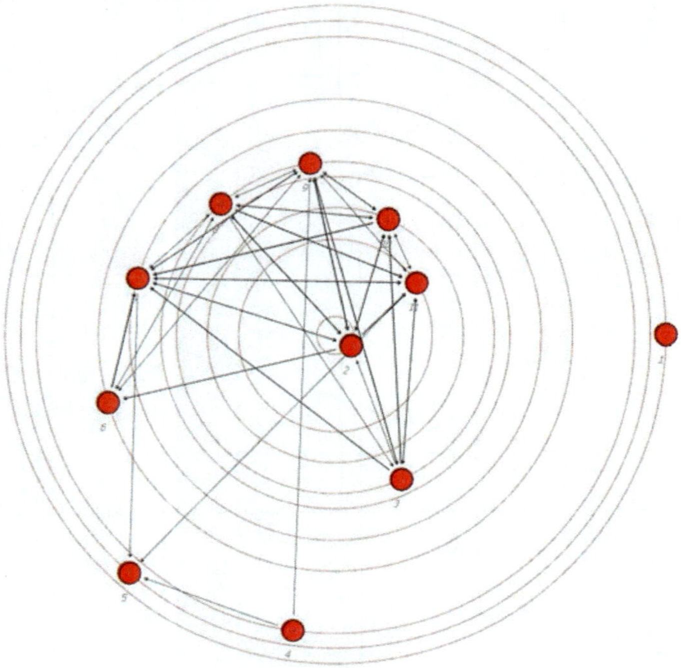

Fig. 4 Radial degree centrality graph for season 20/21

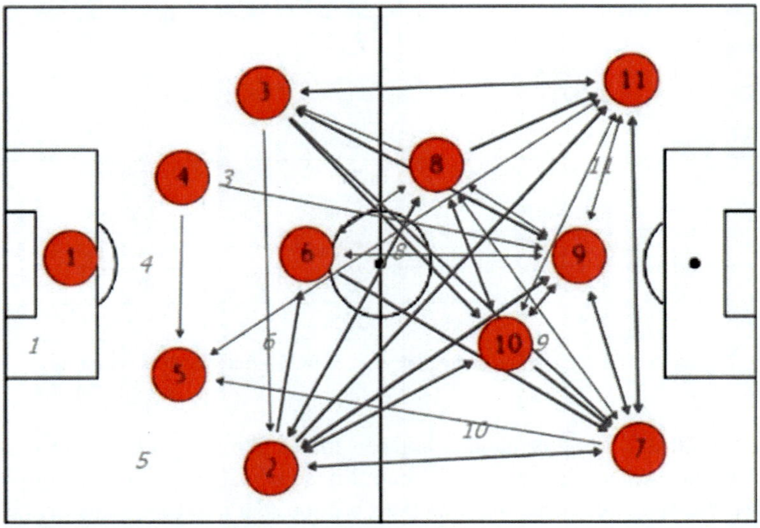

Fig. 5 Passing network for season 20/21

has changed their playing pattern as they were forced to play a deeper defensive line. The team's offensive play gives space and freedom to the central striker. He will feed the ball from the center of the field to be dispersed in the wide space.

Liverpool uses long diagonal passes more frequently this season. When an opportunity to play in their front-three arises during build-up phases, such as when playing up against a high-line, these passes typically attempt to provide it. The adoption of 'false 9' tactical in the game is not effective as in the previous seasons. The striker (9) was needed to feed the ball and disperse them to the wide play. Liverpool has occasionally switched to a 4–2-3–1 system in place of Fabinho (CDM) in midfield, giving the striker (9) more room to express himself in the Liverpool lineup. Fabinho, a distinctively defensively-focused player for Liverpool under the manager's high-pressing style, seldom leaves the center as he controls the midfield and anchors the defense. This form of play is preferred by Liverpool because it enables one of their midfielders to participate actively without upsetting the balance of the squad.

A 4–3-3 usually enables the addition of a defensive midfielder as a midfield defensive backup. Three narrow midfielders who protected the defenders and moved across the field in synchronization were used in the 4–3-3 formation. In order to spread out the attack, the three forwards play in a wider position [22].

Figure 4 shows that the right-back (2) is located at the center of the circle. Liverpool's capacity to maintain possession of the ball and play out from the back is one of their greatest strengths. The Reds' pass percentage of 86.2 percent places them second in the league.

Due to the present midfielder having to swap roles to centreback, the team will most likely rely on their central striker to contribute to offensive play. This makes it more difficult for them to score goals during the season. Instead, their right defender has been forced to play long passes from deeper in the field, where he has been less productive and daring in linking up down the right and moving into positions where these types of passes are more likely to result in a goal [23].

The playing pattern for the team has changed to survive for the whole season. It can be seen as the players were forced to play in their unnatural position are the main contribution to team performance.

4.2 Visualization

In the 18/19 season, the total number of passes in the defensive third was lower than in the 19/20 season. The high defensive line utilized for the tactical in 18/19 is to blame for this.

As shown in Fig. 8, the number of tackles made by the defenders may be connected to this. According to the statistics, 617 tackles were made by the players in season 18/19 due to intense pressing and a strong defensive line. Players played deep within their half in season 20/21, and the intensity of pressing was reduced. To avoid making mistakes in their half, fewer tackles are made by the defenders (Table 4).

Table 4 Type of passes completed (%)

	Pass completed%	Short pass completed%	Long pass completed%
18/19	83.6	89.4	67.4
20/21	84.3	90.1	66.8

Modern football teams' tactical developments have led to a concentration on ball possession as well as defensive positioning in zones rather than man-to-man, which has led to a re-possession style of play where teams have been increasingly successful at reclaiming the ball from the opposition. This shows a key component of good performance is the ability to strategically defend and manage the opposition while a side has the ball [24] (Fig. 6).

As shown in Fig. 7, Liverpool FC has scored 89 goals in 18/19, the most in a single season in club history. Their aggressive play may have been the deciding factor in their exceptional performance. Furthermore, the presence of VVD over the season has demonstrated the strength of defending. During the season, he played every match in a total of 38 game weeks. They have only allowed 22 goals against them, the fewest of any Premier League team other than Manchester City (23). In season 20/21 however, they only managed to score 68 goals while conceding 42. They had a poor season due to the loss of VVD, their center-back. Fabinho and Henderson have played a lot of center-back this season. This has had a knock-on effect, affecting their usual midfield balance as well as the urgency and intensity with which their fullbacks usually drive forward.

A team's defensive measures are the most important factor in keeping a clean sheet in every game. When compared to the other center-backs on the club, VVD had the greatest tackling rate in the defensive third. The region's 28 tackles have proven to be the most in their own space, preventing opponents from scoring. Furthermore, Fabinho's instinct as a pure defensive midfielder was demonstrated by his 51 tackles. However, as he tried not to commit any fouls in the penalty area, it was insufficient to prevent the opposition from scoring goals. VVD also managed to make 217 clearances out of 537 clearances, which is the most among them (40.41%) (Fig. 8).

With VVD on the team, the team is more likely to feel safe and secure. He completed 3407 touches, over half of which were in their defensive third. This could be linked to his long ball, which shattered the opposition's defence. Throughout the season, he made 728 long balls with an 82 percent completion rate (Fig. 8). Every team requires a player who can deliver high-quality long passes to keep the offensive play flowing in the opposing half.

VVD's physique is excellent, and his height of 1.93 m helps him win aerial duels against opponents. During the season, he won 176 aerial duels and knocked out any other defenders in the position (80%). He also surpassed all other defenders in terms of interceptions and shot blocks, with 52 and 22 respectively. This indicates that strikers from the opposition had a difficult time scoring against Liverpool.

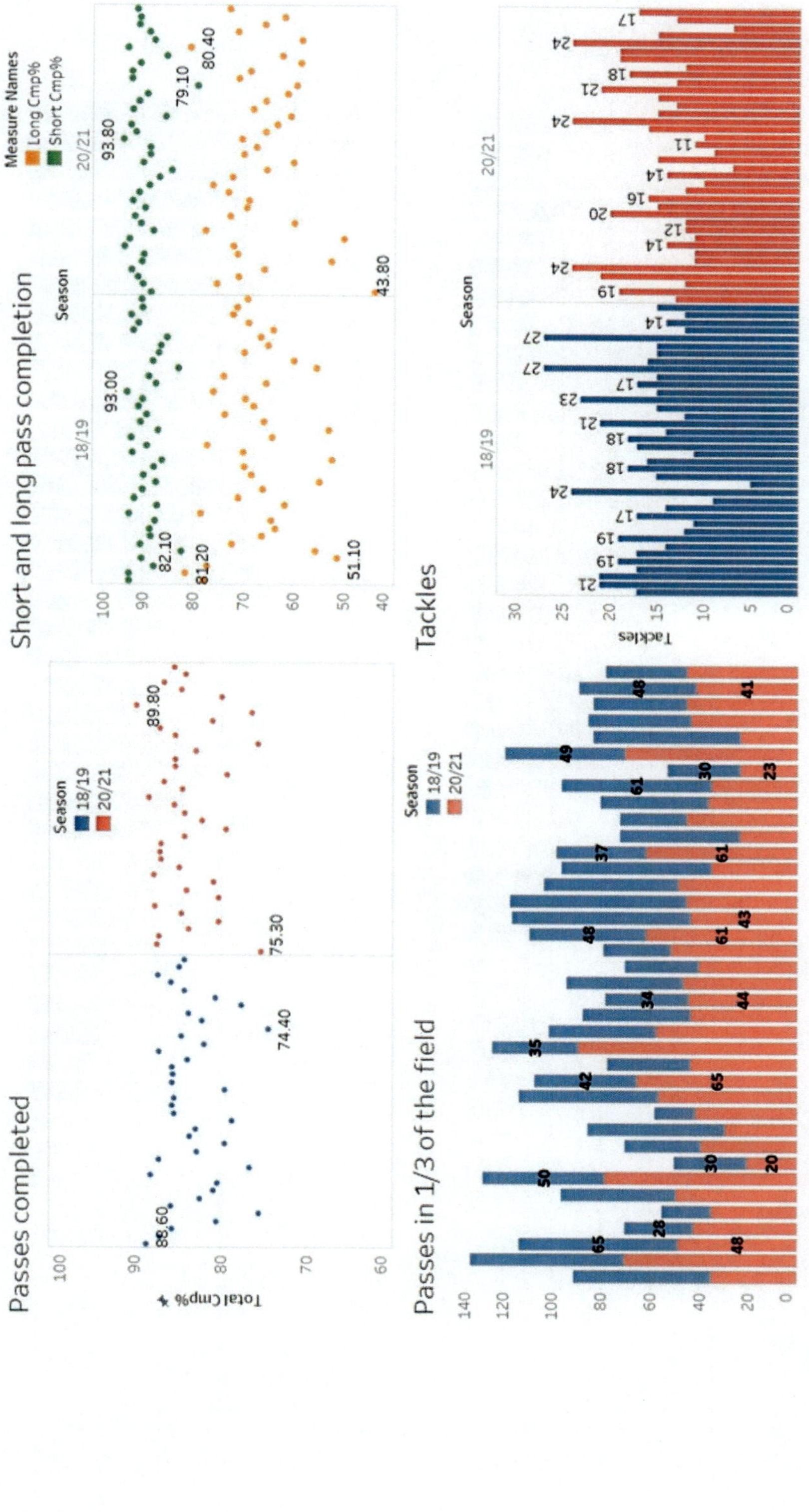

Fig. 6 Team statistics comparison for both season

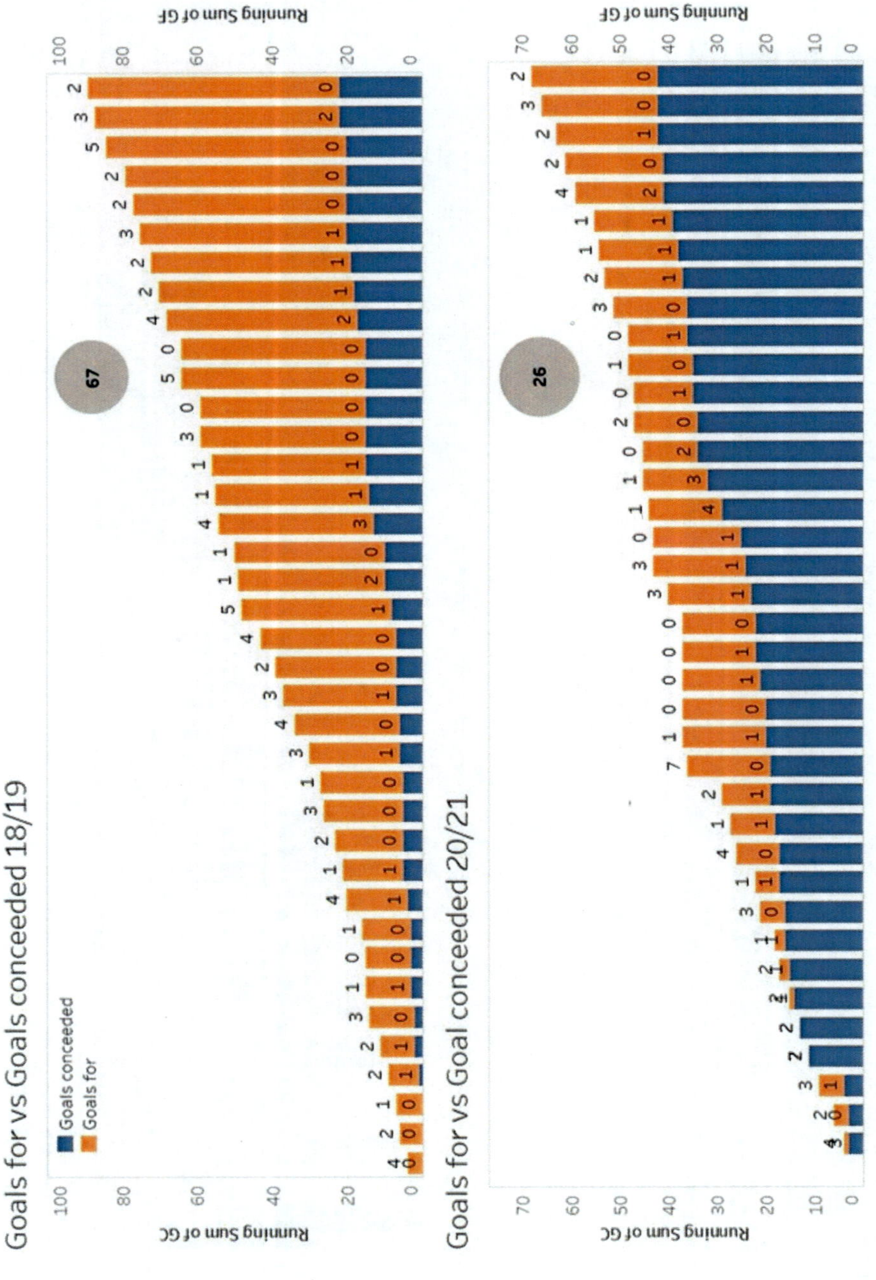

Fig. 7 Comparison of goals scored and against for both season

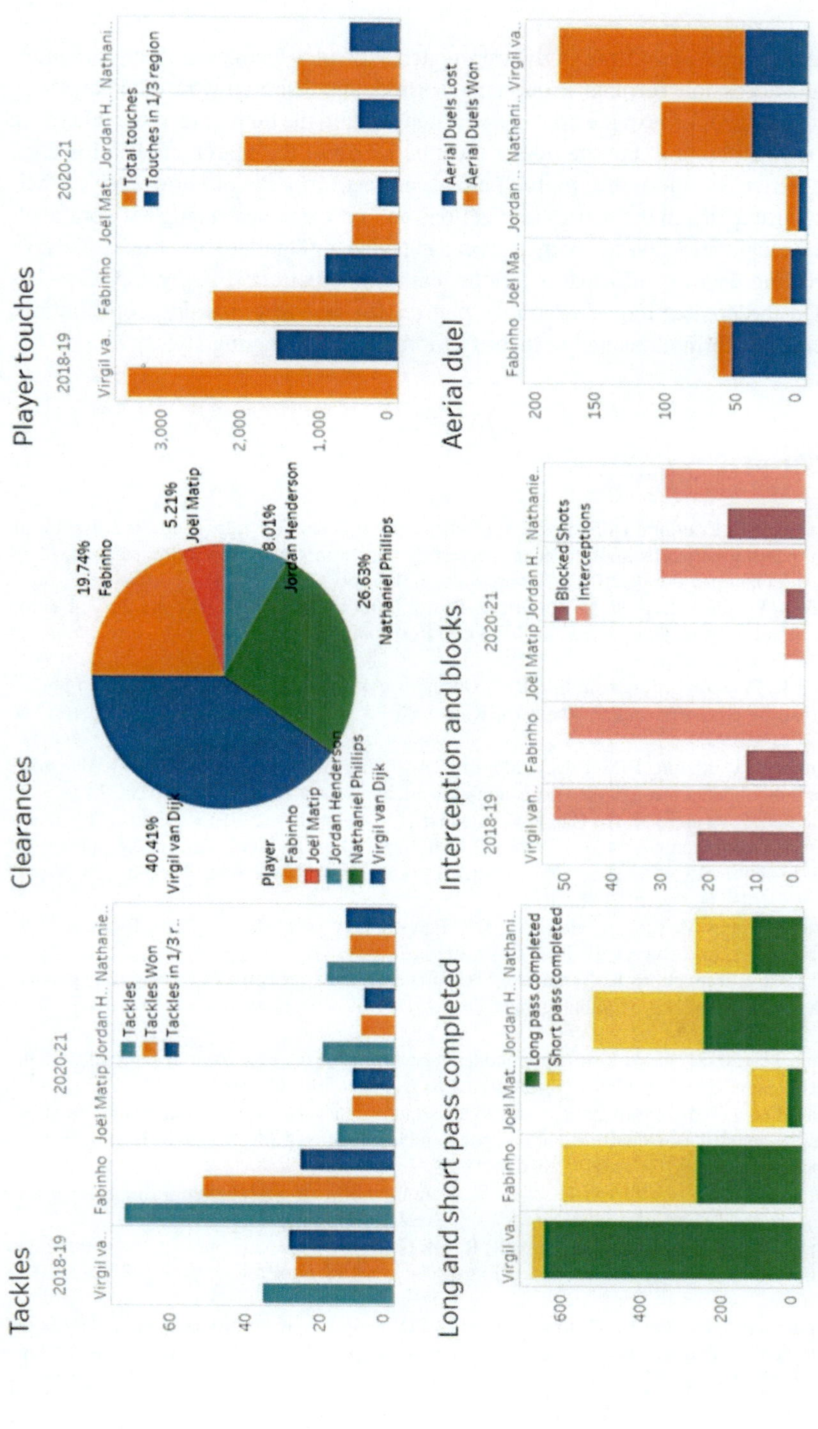

Fig. 8 Comparison of center-backs in both seasons

5 Conclusions

For readers to understand, the study was created to display team performance in interactive and accessible formats. Processing, sorting, and displaying the data have been shown to analyze the provided data easier than reading the data in its raw table form. Tables are still helpful, but the reader must be aware of the search criteria for them to be effective. Furthermore, football clubs can learn more about how each of their players participates in the team's interactions and performances using network analysis and visualization. Network metrics can enhance the standard notational analysis. By providing detailed information on the team's play pattern. Finally, the study has expanded the application of centrality metrics for assessing passing contributions across various formations and to identify football clubs' playing styles.

References

1. Gonçalves B, Coutinho D, Santos S, Lago-Penas C, Jiménez S, Sampaio J (2017) Exploring team passing networks and player movement dynamics in youth association football. PLoS One 12(1). https://doi.org/10.1371/journal.pone.0171156
2. Caicedo-Parada S, Lago-Peñas C, Ortega-Toro E (2020) Passing networks and tactical action in football: a systematic review. Int J Environ Res Public Health 17(18):1–19. https://doi.org/10.3390/ijerph17186649
3. Ievoli R, Palazzo L, Ragozini G (2021) On the use of passing network indicators to predict football outcomes[Formula presented]. Knowl Based Syst 222. https://doi.org/10.1016/j.knosys.2021.106997
4. Clemente FM, Manuel F, Martins L, Mendes RS (2016) Analysis of scored and conceded goals by a football team throughout a season: a network analysis. Kinesiology 48:103–114
5. Malqui JLS, Romero NML, Garcia R, Alendar H, Comba JLD (2019) How do soccer teams coordinate consecutive passes? A visual analytics system for analysing the complexity of passing sequences using soccer flow motifs. Comput Graph (Pergamon) 84:122–133. https://doi.org/10.1016/j.cag.2019.08.010
6. Korte F, Lames M, Link D, Groll J (2019) Play-by-play network analysis in football. Front Psychol 10. https://doi.org/10.3389/fpsyg.2019.01738
7. Narizuka T, Yamamoto K, Yamazaki Y (2014) Statistical properties of position-dependent ball-passing networks in football games. Phys A 412:157–168. https://doi.org/10.1016/j.physa.2014.06.037
8. Grund TU (2012) Network structure and team performance the case of english premier league soccer teams. Soc Netw. https://doi.org/10.1016/j.socnet.2012.08.004
9. Stein M et al (2016) From game events to team tactics: visual analysis of dangerous situations in multi-match data, Department of computer and information science. University of Konstanz. https://doi.org/10.1109/TISHW.2016.7847777
10. Bartels P, De Buyser T, Van Ussel J (2020) A football data visualization: the Belgian first division. Total Football Analysis. http://bit.ly/1cVJLuK
11. Rahman MHAA, Mustapha A, Fauzi R, Razali N (2018) Bayesian approach to classification of football match outcome. Int J Integr Eng 10(6):155–158. https://doi.org/10.30880/ijie.2018.10.06.022
12. Clemente FM, Mendes R, Martins FM (2014) (PDF) applying centrality metrics to identify the prominent football players. Instituto de Telecomunicações, Delegação da Covilhã. https://doi.org/10.13140/2.1.3171.2968

13. McLean S, Salmon PM, Gorman AD, Wickham J, Berber E, Solomon C (2018) The effect of playing formation on the passing network characteristics of a professional football team. Human Movement 2018(5):14–22. https://doi.org/10.5114/hm.2018.79416
14. Buldú JM et al (2018) Using network science to analyse football passing networks: dynamics, space, time, and the multilayer nature of the game. Front Psychol 9. Frontiers Media S.A. https://doi.org/10.3389/fpsyg.2018.01900
15. Clemente F, Sarmento H, Aquino R (2020) Player position relationships with centrality in the passing network of world cup soccer teams win or loss match comparisons. Chaos Solitons Fractals. https://doi.org/10.1016/j.chaos.2020.109625
16. Mclean S, Salmon PM, Gorman AD, Stevens NJ, Solomon C (2017) A social network analysis of the goal scoring passing networks of the 2016 European Football Championships. Hum Mov Sci. http://dx,doi.org/10.010/j.humov.2017.10.001
17. Mclean S, Salmon PM, Gorman AD, Stevens NJ, Solomon C (2022) The influence of social network prestige on in-service teachers' learning outcomes in online peer assessment. Comput Educ Open 3:100087. https://doi.org/10.1016/j.caeo.2022.100087
18. Clemente FM, Martins FML (2017) Network structure of UEFA Champions League teams: association with classical notational variables and variance between different levels of success. Int J Comput Sci Sport 16(1):39–50. https://doi.org/10.1515/ijcss-2017-0003
19. Buja A, Swayne DF, Littman ML, Dean N, Hofmann H, Chen L (2012) Data visualization with multidimensional scaling. Comput Graph Stat 17(2):444–472. https://doi.org/10.1198/106186008X318440
20. Ju W (2020) Football visualisation: capturing chaos and cultivating context the integration of physical-tactical actions view project return to play in professional football view project, football medicine and performance. www.fmpa.co.uk
21. 2021–2022 Liverpool Stats, Premier League I FBref.com, FBref.com, 2021. https://fbref.com/en/squads/822bd0ba/Liverpool-Stats (Accessed 26 June 2022)
22. Koay N (2022) Tactical analysis of Liverpool FC's 4–3–3 formation - HowTheyPlay," HowTheyPlay, May 17, 2022. https://howtheyplay.com/team-sports/Tactical-Analysis-on-Liverpool-Fcs-4-3-3-formation (Accessed 04 July 2022)
23. Desmond R (2021) Where Liverpool have gone wrong this season – tactical analysis – The MastermindSite," The Mastermindsite. https://themastermindsite.com/2021/02/07/where-liverpool-have-gone-wrong-this-season-tactical-analysis/ (Accessed 27 June 2022)
24. Mitchell F, Christopher M, Bruce L, Dwyer DB (2021) The effect of team formation on defensive performance in Australian football. J Sci Med Sport. 10.1016/1sars,2021.09,002

The Effectiveness of Five Minutes Calisthenic Exercise on Depression, Anxiety and Stress Levels Among Teenagers

Rosli Hamid, Syed Kamaruzaman Syed Ali, Ahmad Bisyri Husin Musawi Maliki, Megat Ahmad Kamaludin Megat Daud, and Ahmad Nadzmi

Abstract The study aims to identify the effectiveness of five minutes Calisthenics exercise on the depression, anxiety and stress levels among teenagers. A total of 180 16 years old students in Kota Setar district secondary schools were examined using quasi experimental method. Depression, Anxiety and Stress Scale (DASS) instrument was used for pre and post tests for eight weeks. The data were analyzed using the *Multivariate Analysis of Variance (MANOVA) and Multivariate Analysis of Covariance (MANCOVA)* to see the relationship between Calisthenics exercises and mental health levels of depression, anxiety and stress in pre and post-tests of treatment and control groups. In overall, multivariate test results with wilks' Lambda showed that there were significant effects on the three dependent variables $[F(6, 524) = 2.20, p < 0.05)$ for post-test and pre-test $[F(6,262) = 13.95, p < 0.05)$. The findings showed that students who practice five minutes Calisthenics exercise can reduce depression, anxiety and stress levels. In conclusion, the study showed that there is a significant relationship between five minutes Calisthenics exercise and depression, anxiety and stress levels. Female gender had higher depression, anxiety and stress levels in both tests.

Keywords Calisthenics exercise · Depression · Anxiety · Stress · Teenagers

R. Hamid · S. K. Syed Ali (✉) · M. A. K. Megat Daud
Faculty of Education, University of Malaya, Jalan Profesor Diraja Ungku Aziz, 50603 Kuala Lumpur, Malaysia
e-mail: syed@um.edu.my

A. B. H. M. Maliki · A. Nadzmi
Defence Fitness Academy, National Defence University of Malaysia, Kem Sungai Besi, 57000 Kuala Lumpur, Malaysia

© The Author(s), under exclusive license to Springer Nature Singapore Pte Ltd. 2023

S. F. Syed Omar et al. (eds.), *Innovation and Technology in Sports*, Lecture Notes in Bioengineering, https://doi.org/10.1007/978-981-99-0297-2_8

1 Introduction

Mental health is very important in every phase of human life. Mental health can affect everyday life, and more importantly, the future of a person, especially children and adolescents. Protecting mental health of children and adolescents is a very important factor in helping better individual development in the future. Approximately 450 million people worldwide are diagnosed with certain mental disorders such as depression, schizophrenia and bipolar disorder [20]. This situation will become more serious due to current predictions indicate that by 2030 depression will be the leading cause of disease burden globally [13].

The same goes to the situation in Malaysia. Mental health problems have increased by 15.6% or 400,227 people compared to 2009. By 2020, depression symptoms are expected to rank second in terms of burden of illness after heart disease (Ministry of Health Malaysia (MOH), (2006)). The report of Ministry of Health (2011) showed that mental illness in the country is increasing, as more individuals seek treatment for problems ranging from mild mental disorders to severe schizophrenia. According to the statistic by MOH (2017), there were 379,010 individuals sought treatment as psychiatric outpatients in government hospitals compared to only 324,344 in 2007. In addition, the MOH (2006) showed that more teenagers suffer from mental health problems. The survey showed that 19.5% were of the age group 70–74 and 14.4% were of the youngest group of 16–19 years old.

In addition, mental health problems among Malaysians are at an alarming rate. According to the statistics released by the National Health and Morbidity Study, about 30% or 4.2 million Malaysians over 16 years old suffer from mental health problems. Of the 6,540 screened students, 17.1% had severe anxiety symptoms, 5.2% had severe depression and 4.8% suffered from severe stress. Out of the total students screened, 2,345, i.e. 36% had participated in mental health skills training intervention.

In Malaysia, teenagers had a 10.0% (95% CI 9.2–10.8) prevalence of suicidal thoughts over the preceding year, and 6.9% (95% CI 6.2–7.7) of adolescents attempted suicide at least once during that period, according to the National Health and Morbidity Survey of 2017. In Malaysia, mental illness is a major contributor to disability and health loss, accounting for 8.6% of all years with a disability adjusted for (DALYs) [6]. An economic research determined that mental health issues in the workplace were anticipated to cost the Malaysian economy Malaysian ringgit 14.46 billion (£2.67 billion) in 2018. This rise in costs is correlated with the rising frequency of mental illnesses in Malaysia [6]. These findings show that mental illness have costs many sides effect to the Malaysian especially teenagers that can lead to serious problem such as suicide, disability and health loss and also this will effect on Malaysian economy as the money flows with unnecessary amount on single aspect without concerning other aspects of economy.

Many studies have shown that there is a relationship between physical activity and better mental health, lower risk for psychological stress and also prevention of mental health problems [2, 4]. This study was conducted to identify the effectiveness of five minutes Calisthenics exercise on depression, anxiety and stress levels based on gender among teenagers. Although many of the above studies prove the positive impact of Calisthenics exercise on the effectiveness of treating mental illness but in Malaysia, there is too less studies on the relationship between Calisthenics exercise and mental health. Hence, a study is needed to look into the effectiveness of Calisthenics exercise among Malaysians especially among school students. This study intends to investigate the effectiveness of five minutes Calisthenics exercise on the depression, anxiety and stress levels of teenagers.

1.1 Research Objectives

The implementation of this study was based on the following objectives:

1. To identify the significant difference between depression, anxiety and stress levels between treatment group and control group for pre-test by gender.
2. To identify the significant difference between depression, anxiety and stress levels between treatment group and control group for post-test by gender.

1.2 Research Questions

1. Is there a significant difference in depression, anxiety and stress levels between the treatment group and the control group for pre-test by gender?
2. Is there a significant difference in depression, anxiety and stress levels between the treatment group and the control group for post-test by gender?

2 Research Methodology

The study used a quasi-experimental design (equivalent group pre-post-test) involving pre and post-tests [26]. The sample consisted of 180 students aged 16 years from three secondary daily schools in the district of Kota Setar, Kedah. The selection of school was done randomly based on location, level of academic achievement and number of students based on the recommendation of Kota Setar District Education Office. For each selected school, 30 people in one class were recruited as treatment group while another 30 in one class were used as control group with intact sampling technique. Intact sampling technique was selected to avoid problems or interruptions in classroom management that have been set by school administrators. The control group and treatment group were determined by simple random sampling technique.

2.1 Instrument of Study

In this study, the researcher used an instrument adapted from the Depression, Anxiety, Stress Scale 42 (DASS 42) instrument to assess students' mental health levels of depression, anxiety and stress. The selection of DASS 42 is because the instrument was designed not only to measure the emotional level but to further process for understanding, comprehending and measuring emotion status significantly that is often described as stress. DASS can be well-used by groups or individuals for research purposes [25].

Each subject involved were given pre-test. They will respond to Depression, Anxiety, Stress Scale 42 (DASS 42) instrument. For the treatment group, they went through five minutes Qigong exercise every morning before the starting of teaching and learning activities and after the recess time. The process was guided by the Qigong exercise video provided by the researcher which is presented in front of the class using the Liquid Crystal Display (LCD) projector. This exercise video was shown by the subject teacher who enters the class during the first period of class and during the first period after recess throughout the week for eight weeks. Meanwhile, the control group underwent conventional eight weeks classes.

2.2 Validity and Reliability of DASS 42 Instrument

The past studies showed that the validity and reliability of the DASS 42 instrument were at high Cronbach's Alpha value. Among these were studies by Musa et al. [32], who reported that the validity and reliability for depression 0.94, anxiety 0.90 and stress 0.87 while Edimansyah's [12] study reported depression 0.91, anxiety 0.88 and stress 0.89.

For this study, the researchers conducted a pilot study and the Cronbach Alpha value was depression 0.82, anxiety 0.90 and stress 0.93.

Permission to conduct the study was obtained from the Education Planning and Research Division, Malaysia Ministry of Education, then from the State Education Department, the District Education Department and the school principals. Parental consent was obtained first because all subjects are underage. Information regarding the implementation of the study was provided to all students and form teachers. After the explanation was given, all subjects were given a pre-test.

After the pre-test, the treatment group carried out Qigong exercise for five minutes before starting the first lesson every morning and before starting the first lesson after the recess time for eight consecutive weeks. The control group was not given any intervention and attended classes as usual. After eight weeks of intervention, post-test was administered to both groups and data were collected for further action.

3 Findings of the Research

Statistical Package for Social Science (SPSS) 22.0 software was used to assist in the analysis of data collected from pre-post student questionnaires. Multivariate Analysis of Variance (MANOVA) was used to assess the effectiveness of Qigong exercise on the mental health levels of depression, anxiety and stress in pre and post-tests of treatment group and control group by gender among 16 years old students in the district of Kota Setar, Kedah. The findings of the study are as follows.

Table 1 shows that there was a significant main effect of gender independent variable [$F(3,266) = 17.88$, $p < 0.05$)] on all three dependent variables, namely depression, anxiety and stress in pre-test.

The researcher used the Levene's Test of Equality of Error Variances to see the variance similarity for each of the dependent variable categories studied. When the mean value obtained is less than 0.05 ($p < 0.05$), then it indicates that there are significant differences in the dependent variables. On the other hand, if the significant value obtained is greater than 0.05 ($p > 0.05$), then it shows no significant difference in the dependent variables tested [1].

Table 2 shows the value of Levene's test for the three dependent variables studied. The Levene test was used to test whether the variance of the dependent variable across categories in the independent variable is the same. Test results show that the three dependent variables yielded significant results at $p > 0.05$, anxiety $p = 0.700$, depression $p = 0.125$ and stress $p = 0.872$. Indirectly, these results indicated that all of the dependent variables are equal to the variance and do not violate the assumption of the MANOVA test (Table 3).

The results of the analysis in Table 3 of tests of between-subjects effects shows that in overall, there was a main effect of gender and group categories on the three dependent variables. The main effects of each dependent variable can be identified in Table 3 Tests of Between-Subjects Effects enabled a more detailed examination of the significant differences between genders in each dependent variable. The results

Table 1 MANOVA analysis of gender independent variable on depression, anxiety and stress levels dependent variables for pre-multivariate test

		Value	F	Hypothesis df	Error df	Sig.
Gender	*Wilks' Lambda*	0.855	14.984^{b}	3.000	266.000	0.000

Table 2 Levene's test of equality of error variance

	F value	Df1	Df2	Sig.
Depression	0.149	1	268	0.700
Anxiety	2.368	1	268	0.125
Stress	0.026	1	268	0.872

Table 3 Tests of between-subjects effects

	Dependent variable	Type III sum of squares	df	Mean square	F value	Sig.
Gender	Depression	30.670	1	30.670	40.886	0.000
	Anxiety	4.281	1	4.281	6.113	0.014
	Stress	2.315	1	2.315	4.799	0.029

a. R Squared = 0.132 (Adjusted R Squared = 0.129)
b. R Squared = 0.022 (Adjusted R Squared = 0.019)
c. R Squared = 0.018 (Adjusted R Squared = 0.014)

of the analysis showed that gender was a factor in depression [F (1,268) = 40.9, $p < 0.05$], anxiety [F (1,268) = 6.11, $p < 0.05$] and stress [F (1,268) = 4.79, $p > 0.05$].

The R^2 values below Table 3 indicated that gender contributed only 0.132 or 13.2% change on the dependent variable of depression. For the dependent variable of anxiety, gender contributed only 0.022 or 2.2%, and gender contributed only 0.018 or 1.8% to the dependent variable of stress.

In overall, the results of the multivariate wilks'Lambda test showed that there was a gender effect on all three dependent variables [F (3,266) = 17.88, $p < 0.05$)]. Based on the results of this analysis, it shows that gender is a factor for depression, anxiety and stress in students' mental health.

The analysis of the MANOVA test revealed that there was a significant difference between genders in the dependent variable of depression [F (1,268) = 2.64, $p < 0.05$] and anxiety [F (1,268) = 40.9, $p < 0.05$] and stress [F (1,268)) = 4.79, $p > 0.05$]. These results indicate that the gender of the students in the study affected depression, anxiety and stress.

By referring to the mean values of each dependent variable on the gender of the male and female students, it was found that female students experienced more depression, anxiety and stress. Depression (female mean score = 19,719; male = 19,044), anxiety (female mean score = 17.319; male = 17,067) and stress (female mean score = 22,444; male = 22,259) were compared with the male group.

However, gender factors contributed only 13.2% to depression, 2.2% to anxiety and 1.8% to stress.

The results of the Wilks' Lambda multivariate test showed that there were significant main effects of group and gender variables on all three dependent variables in the study F (6,524) = 2.20, $p < 0.05$). Therefore, the researcher reported that gender and group are factors in all three aspects of mental health, namely depression, anxiety and stress.

3.1 The Anxiety, Depression and Stress Levels Between the Treatment Group and the Control Group for Post-test by Gender

The table above shows the linear relationship between groups and anxiety for post-test. Based on the value of Deviation from Linearity Sig. is 0.521 greater than 0.05. It can therefore be concluded that there is a significant linear relationship between groups and anxiety for post-test (Tables 4 and 5).

The table above shows the linear relationship between groups and depression for post-test. The value of Deviation from Linearity Sig., 0.481 is greater than 0.05. It can therefore be concluded that there is a significant linear relationship between groups and depression for post-test. Once all assumptions in the MANCOVA analysis are followed, then only the MANCOVA analysis can be done. Here are the steps for the MANCOVA test.

The Table 6 shows the number of samples used in this study. There are two gender groups, male and female. The sample involved in this study was 270 people, which is 135 male and 135 female. Since the sample size used is over thirty people, so the study data is considered normal.

Table 7 above shows the mean values, standard deviations and sub-sample sizes of the three dependent variables across the two gender categories for post-test. The aspect of depression shows higher mean values among female students compared

Table 4 Linear tests between gender and anxiety for post test

			Sum of squares	df	Mean square	F	Sig.
Gender* anxiety	Between groups	(Combined)	2.365	4	0.591	2.406	0.050
		Linearity	1.810	1	1.810	7.362	0.007
		Deviationfrom linearity	0.555	3	0.185	0.753	0.521
	Within groups		65.135	265	0.246		
	Total		67.500	269			

Table 5 Linear testing between gender and pressure for post test

			Sum of squares	df	Mean square	F	Sig.
Gender* anxiety	Between groups	(Combined)	4.498	4	1.125	4.730	0.001
		Linearity	3.909	1	3.909	16.444	0.000
		Deviation from linearity	0.589	3	0.196	0.825	0.481
	Within groups		63.002	265	0.238		
	Total		67.500	269			

Table 6 Between-subject factors

		Label of value	N
Gender	1	Male	135
	2	Female	135

Table 7 Overall descriptive statistics of anxiety, depression and stress between gender of treatment group and control group for pre and post test

	Gender	Mean	Standard Deviation	N
Depression	Male	18.65	0.840	135
	Female	19.47	0.862	135
	Total	19.06	0.943	270
Anxiety	Male	16.93	0.816	135
	Female	17.19	0.796	135
	Total	17.06	0.816	270
Stress	Male	22.10	0.721	135
	Female	22.44	0.687	135
	Total	22.27	0.725	270

to male students (mean: female = 19.47, male = 18.65). For post-test of anxiety, it shows higher mean values among female students compared to male students (mean: female = 16.96, male = 1670) and finally the post test for stress shows female mean score over male students (mean = 22.44, male = 22.10).

The Box's M test was also performed to see the homogeneity of variance–covariance matrices. Table 8 below shows the results of the Box's M test to determine the homogeneity of variance–covariance matrices.

Table 8 above shows the results of the Box's M test was not significant at $p > 0.05$, $p = 0.076$. This indicates that the data meet the covariance similarities of the MANCOVA test, which in the sample variance study of the three dependent variables across theindependent variables is similar in population. Hence, this data does not violate the assumption and the above hypothesis can be tested using the MANCOVA test. Subsequently, the Levene's Test of Equality of Error Variances was run to see the variance similarity for each dependent variable category studied. When the significant value obtained is less than 0.05 ($p < 0.05$), then it indicates that there are significant differences in the dependent variables. On the other hand, if the significant value obtained is greater than 0.05 ($p > 0.05$), then the value shows no variance difference in the dependent variables tested [14].

Table 8 Box's test of equality of covariance Matrices[a]

Box's M	F Value	Df1	Df2	Sig.
31.623	1.470	21	264,168.555	0.076

Table 9 above shows the value of Levene's test for the three dependent variables studied. The Levene test was used to test whether the variance for the dependent variables across categories in the independent variable is the same. The test results show that the three dependent variables yielded insignificant results on post-test of depression $p = 0.267$, anxiety $p = 0.883$ and stress $p = 0.226$. Indirectly, this result indicates that all of these variables have equal variance and do not violate the assumptions of the MANCOVA test.

The results for the Wilks' lambda test in Table 10 show that there was a significant gender independent gradient effect [F (6,262) = 13.95, $p < 0.05$). Based on this decision, the researcher rejected the null hypothesis and concluded that gender was a factor in depression, anxiety and stress among students (Table 11).

The results in the table above indicate that there was a significant gender effect on all three dependent variables in the post test. The results show that there was a significant effect of gender on the three dependent variables in the study, namely depression [F (1,267) = 65.4, $p < 0.05$], anxiety [F (1,267) = 7.62, $p < 0.05$] and stress [F (1,267) = 16.85, $p < 0.05$]. Whereas the R^2 values below the table indicate that the gender independent variables for post-test, depression, accounted for 0.142 or 14.2%, 0.026 or 2.6% for anxiety and 0.025 or 2.5% for stress.

Table 9 Results of Levene's test of equality of error variance

	F value	Df1	Df2	Sig.
Depression	1.237	1	268	0.267
Anxiety	0.022	1	268	0.883
Stress	1.473	1	268	0.226

Table 10 Multivariate test

Effect		Value	F
Intercept	Pillai's Trace	0.997	12,821.409*
	Wilks' Lambda	0.003	12,821.409*
	Hotelling's Trace	293.620	12,821.409*
	Roy's Largest Root	293.620	2821.409*
Group	Pillai's Trace	0.105	5.147*
	Wilks' Lambda	0.895	5.147*
	Hotelling's Trace	0.118	5.147*
	Roy's Largest Root	0.118	5.147*
Gender	Pillai's Trace	0.242	13.949*
	Wilks' Lambda	0.758	13.949*
	Hotelling's Trace	0.319	13.949*
	Roy's Largest Root	0.319	13.949*

Table 11 Tests of between-subjects effects

	Dependent variable	Type III sum	df	Mean square	F value	Sig.
Gender 65.390.000	Depression	44.815		1		44.815
	Anxiety'	4.800	1	4.800	7.619	0.006
	Stress	8.181	11	8.181	16.854	0.000

a. R Squared = 0.142 (Adjusted R Squared = 0.136)
b. R Squared = 0.026 (Adjusted R Squared = 0.019)
c. R Squared = 0.025 (Adjusted R Squared = 0.017)

Table 12 Estimated marginal means

Dependent variable	Gender	Min error	Std	95% confidence interval	
				Lower bound	Upper bound
Post depression	Male	18.652*	0.071	18.512	18.792
	Female	19.467*	0.071	19.326	19.607
Post anxiety	Male	16.926*	0.068	16.791	17.060
	Female	17.193*	0.068	17.058	17.327
Post stress	Male	22.096*	0.060	21.978	22.214
	Female	22.444*	0.060	22.326	22.563

Table 12 shows the mean values of the gender independent variable across each dependent variable for post-test. Referring to the table above, it was found that female gender showed a higher mean value for depression dependent variable than male (mean score: female = 19.467[a], male = 18.652[a]). For dependent variable of anxiety, female gender showed a high mean score (mean score: female = 17.193[a], male = 16.926[a]). Meanwhile, the dependent variable of stress showed mean value (mean score: female = 22.444[a], male = 22.096[a]).

The results of the Wilks' Lambda multivariate test showed that in overall, there was a significant gender independent variable [$F(6,262) = 13.95$, $p < 0.05$) among the three dependent variables: depression, anxiety and stress for students in the Kota Setar district for the application of exercise. Subsequent analysis showed that there was a significant gender effect on the three dependent variables in the post test that showed a significant effect of gender on the three dependent variables in the study, namely depression [$F(1,267) = 65.4$, $p < 0.05$], anxiety [$F(1,267) = 7.62$, $p < 0.05$] and stress [$F(1,267) = 16.85$, $p < 0.05$]. Meanwhile, the R^2 values below the table indicate that the gender independent variable for depression post-test was 0.142 or 14.2%, 0.026 or 2.6% for anxiety and 0.025 or 2.5% for stress.

However, there were also effects of group control variables on all three dependent variables [$F(6,262) = 5.15$, $p < 0.05$). The results show that in overall, by controlling the group variable, gender is a factor in mental health levels of depression, anxiety and stress among secondary school 16 years old students.

The results of the analysis also indicate that the gender independent variables for depression post-test was 0.142 or 14.2%, 0.026 or 2.6% for anxiety and 0.025 or 2.5% for stress.

4 Discussion of the Research

The findings show that there is a significant difference in the formation of depression, anxiety and stress in the treatment group and control group for pre and post-test. The results of the pre-test analysis showed that there was a high level of anxiety, depression and stress among the two groups and does not show significant difference.

The findings of this study show that there is a difference between mental health and gender and is in line with the findings of Cohen [7], who suggested that Calisthenic exercise has a positive effect on stress treatment. This is because the practice of Calisthenic exercises has an emphasis on breathing. When the body and mind are at ease, the physical and mental functions of a person is much better with Calisthenic movements such as standing, doing correct movements, clearing thoughts and deep breathing. All of these physical activities help one to achieve good well-breing and reduce mental and physical stress. This situation proves that the practice of Calisthenic exercise is useful in relieving symptoms of depression [21].

The findings of this study show that there is a significant difference between the genders in terms of mental illnesses of depression, anxiety and stress. The results of the pre-test analysis in this study indicate that female gender has higher levels of anxiety, depression and stress than men. This finding is in line with the survey and psychiatrist reports that hospitalization rates for depression are more frequent among women compared to men (Lewinsohn et al. 1998). According to the report of Otten et al. [29], more women have mental disorders compared to men. According to the President of the Malaysian Psychiatric Association (MPA), Dr. Abdul Kadir Abu Bakar, mental illness was more prevalent among females at 12.1% compared to male at 10.4%.

Post-test findings show that there are significant differences between male and female genders in terms of depression, anxiety and stress. Although depression, anxiety and stress levels decrease compared to pre-test, female gender still showed higher levels compared to male gender. Looking at the level of depression, there are studies showing that female gender is higher compared to male gender, including the findings of [3], in their study to explore the relationship between depression history, gender differences and stress by assessing stress activity. The study also showed that female groups from both groups reported greater negative effects on post-test compared to men. Similarly, the results of the study conducted by [35] showed that female gender had a higher depression rate per sample in overall compared to male gender. This finding is further reinforced by the findings of [34], who showed that women have higher rates of depression than men. Although, the findings of the studies conducted by Özlem et al. [30] and Piccinelli and Wilkinson [31] show that gender differences in depression are still not fully determined.

The findings of Calisthenic exercise show that exercising can reduce depression [9, 15, 17, 18, 28, 33] and stress [10, 11, 19]. In fact, the positive effects of exercise on the individual psychological aspects have been proven by scientific research [5]. The effects of exercising clearly explain how exercises reduce psychological symptoms related to stress and anxiety [8, 27]. However, the concept of physiological and psychological combinations has been adapted and proposed in explaining the relationship interaction between mood changes and exercise.

The researcher concludes that teenagers 16 years old student should be given five-minute Calisthenic exercise to reduce mental health disorders of depression, anxiety and stress levels so that they can focus on lessons taught by the teachers and reduce the number of mentally illed patients in our country.

5 Conclusion

This study has proven the effectiveness of the Calisthenic exercise on the level of depression, anxiety and stress among teenagers. 5 min of Calisthenic exercise during each school day reduced the depression, anxiety and stress levels among students. Although female students showed higher depression, anxiety and stress levels than male students in pre and post-tests, both showed declination. As the practice of this exercise can reduce the level of depression, anxiety and stress among students, it is advisable for the school to implement it on a daily basis as this exercise does not involve any cost. Healthy body starts with a healthy mind and then students can focus on the lessons in schools.

References

1. Hashim A (2014) Panduan analisa data secara efisien. Dubook Press, Selangor
2. Astell-Burt T, Feng X (2019) Association of urban green space with mental health and general health among adults in Australia. JAMA Netw Open 2(7):e198209–e198209. https://doi.org/10.1001/JAMANETWORKOPEN.2019.8209
3. Bagley SL, Weaver TL, Buchanan TW (2011). Sex differences in physiological and affective responses to stress in remitted depression I. https://doi.org/10.1016/j.physbeh.2011.03.004
4. Bell SL, Audrey S, Gunnell D, Cooper A, Campbell R (2019) The relationship between physical activity, mental wellbeing and symptoms of mental health disorder in adolescents: a cohort study. Int J Behav Nutr Phys Act 16(1):1–12. https://doi.org/10.1186/S12966-019-0901-7/TABLES/3
5. Calders P, Elmahgoub S, De Mettelinge TR, Vandenbroeck C, Dewandele I, Rombaut L, Vandevelde A, Cambier D (2011) Effect of combined exercise training on physical and metabolic fitness in adults with intellectual disability: A controlled trial. Clin Rehabil 25(12):1097–1108. https://doi.org/10.1177/0269215511407221
6. Chua SN (2020) The economic cost of mental disorders in Malaysia. Lancet Psych 7(4):e23
7. Cohen KS (1999) The way of qigong: the art and science of chinese energy healing. Random House, Canada

8. Conradsson M, Littbrand H, Lindelöf N, Gustafson Y, Rosendahl E (2010) Effects of a high-intensity functional exercise programme on depressive symptoms and psychological well-being among older people living in residential care facilities: a cluster-randomized controlled trial. Aging Ment Health 14(5):565–576. https://doi.org/10.1080/13607860903483078
9. De Moor MHM, Beem AL, Stubbe JH, Boomsma DI, De Geus EJC (2006) Regular exercise, anxiety, depression and personality: a population-based study. Prev Med 42(4):273–279. https://doi.org/10.1016/J.YPMED.2005.12.002
10. Dugan SA, Bromberger JT, Segawa E, Avery E, Sternfeld B (2014). Association between physical activity and depressive symptoms: midlife women in SWAN. https://doi.org/10.1249/MSS.0000000000000407
11. Dunn AL, Trivedi MH, Kampert JB, Clark CG, Chambliss HO (2005) Exercise treatment for depression: Efficacy and dose response. Am J Prev Med 28(1):1–8. https://doi.org/10.1016/J.AMEPRE.2004.09.003
12. Edimansyah RN (2008) Effects of short duration stress management training on self-perceived depression, anxiety and stress in male automotive assembly workers: A quasi-experimental study. J Occupat Med Toxicol
13. Funk M (2016) Global burden of mental disorders and the need for a comprehensive, coordinated response from health and social sectors at the country level. https://aen.es/wp-content/uploads/2012/01/12.1_WHO_EBoard-Global_Burden_Resolut.pdf
14. George D, Mallery P (2018) General linear models: MANOVA and MANCOVA. In IBM SPSS Statistics 25 Step by Step. Routledge, pp 314–328
15. Harris AHS, Cronkite R, Moos R (2006) Physical activity, exercise coping, and depression in a 10-year cohort study of depressed patients. J Affect Disord 93(1–3):79–85. https://doi.org/10.1016/J.JAD.2006.02.013
16. Institute for Public Health (2017) National health and morbidity survey 2017: adolescent health survey 2017. Ministry of Health Malaysia, Malaysia
17. Jacka FN, Pasco JA, Williams LJ, Leslie ER, Dodd S, Nicholson GC, Kotowicz MA, Berk M (2011) Lower levels of physical activity in childhood associated with adult depression. J Sci Med Sport 14(3):222–226. https://doi.org/10.1016/J.JSAMS.2010.10.458
18. Jerstad SJ, Boutelle KN, Ness KK, Stice E (2010) Prospective reciprocal relations between physical activity and depression in female adolescents. J Consult Clin Psychol 78(2):268–272. https://doi.org/10.1037/A0018793
19. Jonsdottir IH, Rödjer L, Hadzibajramovic E, Börjesson M, Ahlborg G (2010) A prospective study of leisure-time physical activity and mental health in Swedish health care workers and social insurance officers. Prev Med 51(5):373–377. https://doi.org/10.1016/J.YPMED.2010.07.019
20. Kastrup MC, Ramos AB, Med D (2007) Global mental health. Danish Med Bulet 54(1):42–45. http://www.mentalhealth.com
21. Tang KC (1994) Tang KC. Qigong therapy—its effectiveness and regulation. Am J Chinese Med 235–242
22. Kementerian Kesihatan Malaysia (2011) Statistik Kesihatan dan Morbiditi Kebangsaan. Putrajaya
23. Kementerian Kesihatan Malaysia (KKM) (2006) Statistik Kesihatan dan Morbiditi Kebangsaan .,. Putrajaya.
24. Kementerian Kesihatan Malaysia (KKM) (2017) Statistik Kesihatan Mental
25. Lovibond SH, Lovibond PF (1995) Manual for the depression anxiety stress scales, 2nd edn. Psychology Foundation, Sydney
26. Marsden E, Torgerson CJ (2012) Single group, pre- and post-test research designs: some methodological concerns. Oxf Rev Educ 38(5):583–616. https://doi.org/10.1080/03054985.2012.731208
27. Matthews MM, Hsu FC, Walkup MP, Barry LC, Patel KV, Blair SN (2011) Depressive symptoms and physical performance in the lifestyle interventions and independence for elders pilot study. J Am Geriatr Soc 59(3):495–500. https://doi.org/10.1111/J.1532-5415.2011.03319.X

28. Mikkelsen SS, Tolstrup JS, Flachs EM, Mortensen EL, Schnohr P, Flensborg-Madsen T (2010) A cohort study of leisure time physical activity and depression. Prev Med 51(6):471–475. https://doi.org/10.1016/J.YPMED.2010.09.008
29. Otten D, Tibubos AN, Schomerus G, Brähler E, Binder H, Kruse J, Ladwig KH, Wild PS, Grabe HJ, Beutel ME (2021) Similarities and differences of mental health in women and men: a systematic review of findings in three large german cohorts. Front Public Health 9:40. https://doi.org/10.3389/FPUBH.2021.553071/BIBTEX
30. Özlem O, Haydari N, Sözeri C, Mattoni A, Treré E, Gencer ZT, Yanıkkaya B (2014) N聚氨酯/乙烯-辛烯共聚物共混体系的结构、性能及表征o Title. Telematics Inform 19(1):27–40. https://doi.org/10.1177/1742766510373715
31. Piccinelli M, Wilkinson G (2000) Gender differences in depression. Critical review. British J Psychiatry 177:486–492. https://doi.org/10.1192/BJP.177.6.486
32. Ramli Musa, Rosnani S Aidil, A. F. A. (2012). Psychometric Profile of Malaysian version of the Depressive, Anxiety and Stress Scale 42-item (DASS-42). MJP Online Early
33. Richards J, Jiang X, Kelly P, Chau J, Bauman A, Ding D (2015) Don't worry, be happy: Cross-sectional associations between physical activity and happiness in 15 European countries. BMC Public Health 15(1):1–8. https://doi.org/10.1186/S12889-015-1391-4/TABLES/3
34. Rosenfield S (1980) Sex differences in depression: Do women always have higher rates? J Health Soc Behav 21(1):33–42. https://doi.org/10.2307/2136692
35. Wade TJ, Cairney J, Pevalin DJ (2002) Emergence of gender differences in depression during adolescence: national panel results from three countries. J Am Acad Child Adolesc Psychiatry 41(2):190–198. https://doi.org/10.1097/00004583-200202000-00013

Sports Advisors' Perspectives and Satisfaction Level on Transformational Leadership of Senior Assistants of Co-curriculum

Dhavasini A. P. Arumugham, Syed Kamaruzaman Syed Ali, Nguang Ung Siong, Ahmad Bisyri Husin Musawi Maliki, and Ahmad Nadzmi

Abstract This study is conducted to study Sports Advisors' perspectives and level of satisfaction on Transformational leadership of Senior Assistants of Co-curriculum in primary schools in Selangor. This study uses survey causal-comparative with cluster random sampling involving 420 Sports Advisors (352 males and 68 females) in daily primary schools in Selangor state. Multifactor Leadership Questionnaire is used as research instrument to collect data on perspectives and levels of satisfaction from all the participants based on gender, location of schools and educational credentials. Statistic descriptive analysis shows very high mean values on perspectives and levels of satisfaction of Sports Advisors on Transformational leadership of Senior Assistants of Co-curriculum. Independent t-Test is used to identify differences between gender, location of schools and educational credentials in this study. Results show significant differences on perspectives and levels of satisfaction of Sports Advisors based on location of schools and educational credentials. However, there is no significant difference in gender on perspectives and levels of satisfaction of Sports Advisors on Transformational leadership of Senior Assistants of Co-curriculum. Therefore, these results clarify that Senior Assistants of Co-curriculum need to consider the difference in school locations and educational credentials of Sport Advisors to apply Transformational leadership in sport management in primary schools.

Keywords Perspectives · Satisfaction Level · Transformational Leadership · Senior Assistants of Co-Curriculum

---

D. A. P. Arumugham · S. K. Syed Ali (✉) · N. U. Siong
Faculty of Education, University of Malaya, Jalan Profesor Diraja Ungku Aziz, 50603 Kuala Lumpur, Malaysia
e-mail: syed@um.edu.my

A. B. H. M. Maliki (✉) · A. Nadzmi
Defence Fitness Academy, National Defence University of Malaysia, Kem Sungai Besi, 57000 Kuala Lumpur, Malaysia
e-mail: bisyri@upnm.edu.my

© The Author(s), under exclusive license to Springer Nature Singapore Pte Ltd. 2023

S. F. Syed Omar et al. (eds.), *Innovation and Technology in Sports*, Lecture Notes in Bioengineering, https://doi.org/10.1007/978-981-99-0297-2_9

1 Introduction

Sports management in schools is essential to meet the requirements of sports such as implementation of the One Pupil One Sports Policy (1M1S) which requires every student to participate in at least one sports activity in school. Through effective sports management in schools, the talent of the pupils can be discovered and developed through active participation in sports activities. Therefore, Sports management in primary schools is important to develop the pupils' potential and produce pupils who are physically, emotionally and spiritually balanced to achieve Malaysia Philosophy of Education. Besides, sports management is also important to fulfill Malaysian Education Development Plan 2013–2025 so that pupils can be educated and developed holistically by participating actively in club, sports or uniformed unit body. In addition, sports management will also influence the effectiveness in achieving the new curiculum learning objecitives in line with introduction of Primary school Standard-Based Curriculum (KSSR) in 2011.

In achieving systematic sports management in schools, Senior Assistant of Co-Curriculum need to use suitable leadership style which can guide and lead the sports advisiors when conducting sports in school to achieve school's vision and mission [40] such as to achieve the objectives of One Pupil One Sport Policy (1M1S) and Malaysian Education Development Plan. Thus, leadership in education has become more important and challenging at the same time for Senior Assistant of Co-Curriculum to accomplish and achieve sports excellence as well as to support Physical Education curriculum in schools.

Leadership style of school leader is crucial to ensure Ministry of Education can succeed in improving the quality of education in Malaysia and achieving national education goals. Therefore, Senior Assistant of Co-Curriculum is the most important leader to manage the implementation of sports activities in schools to achieve the vision and mission of Ministry of Education. So, they need to guide their Sports Advisors to carry out sports activities effectively as planned in schools. However, the goals and aspirations of the national education are difficult to achieve when teachers are less committed to sports development because school administrators are more likely to focus on administrative tasks and tend to ignore the leadership style in schools [6]. When there is growing imbalance between the management style and leadership style of the Senior Assistant of Co-curriculum in school on sports management, this will affect the effectiveness of sports developments and sports achievements in schools [20]. This is because sports management is closely related to sports achievements and outcome of sports activities in schools. In addition, success in sports activities are also a continuation and reinforcement of curriculum learning programmes in schools. Thus, a complete education in primary schools should emphasize on both academic learning and sports activities through the effective leadership style used by Senior Assistant of Co-Curriculum in sports management. Therefore, suitable leadership style of Senior Assistant of Co-Curriculum is essential to make sure successful implementation of sports activities for pupils in early stage.

However, leadership style of Senior Assistant of Co-Curriculum especially in sports management is also a factor that has contributed to the limitation of school-level sports performances. This is because leadership in sports is still not given as much attention by school leaders as in academic. In line with the development of the education system in Malaysia, Transformational leadership is a new leadership style that is suitable for school administrators such as Senior Assistant of Co-Curriculum to fulfill the needs of leadership in education and to achieve success in education [44]. Therefore, Transformational leadership style should be prioritized by Senior Assistant of Co-Curriculum to stimulate and motivate Sports Advisors in schools through charismatic, inspirational motivation, intellectual stimulation and individual consideration [7, 11]. Besides, Transformational leadership style also practices charismatic value to achieve its goals [44]. Therefore, Transformational leadership is more comprehensive and practical in terms of the structure, culture and achievement from the perspective of leadership.

In school, Sports Advisors are the people who support Senior Assistant of Co-Curriculum to implement sports activities and achieve schools' vision, mission, objectives and goals. So, Sports Advisors can manage the implementation of school sport plans systematically through suitable leadership style of the Senior Assistant of Co-Curriculum [20]. Therefore, Transformational leadership style can be practiced by Senior Assistant of Co-Curriculum to form a good collaboration with Sports Advisors [29]. This is because Transformational leadership is a well-organized leadership style that can greatly influence Sports Advisors to provide full support to Senior Assistant of Co-Curriculum in fulfilling their responsibilities in school. Therefore, the perspectives of Sports Advisors on leadership are important to ensure the school sports plans are implemented smoothly as they are the executors in the management and implementation of sports activities in schools [7, 11]. Furthermore, the demographic factors such as gender, school location and educational credentials also affect the leadership styles used by school administrators in managing sports activities in schools [31]. So, demographic data is useful to help Senior Assistants of Co-Curriculum to practice or apply the Transformational leadership effectively in schools based on the demographic factors to guide Sports Advisor in sports management to achieve school sports goals.

Besides, satisfaction level of Sports Advisors also need to be taken into consideration because level of satisfaction is a key element for Sports Advisors to ensure that sports activities are implemented effectively in school [17, 34, 35]. Therefore, Senior Assistants of Co-Curriculum need to pay attention to the Sports Advisors' job satisfaction so that they can achieve self-actualization [34] while performing sports activities as planned in schools. This is because the satisfaction level of the Sports Advisors on leadership style are important to ensure the effectiveness in implemention of school sports programmes to achieve the vision and mission. Thus, well-organized Transformational leadership can also motivate Sports Advisors to collaborate with Senior Assistants of Co-Curriculum to fulfill their responsibilities in the schools. So, Transformational leadership is suitable to be implemented by school administrator such as Senior Assistants of Co-Curriculum to achieve organizational goals by stimulating the needs of others [18]. This is because four dimensions such as charismatic, inspirational motivation, intellectual stimulation and individual consideration [7, 11] emphasized in Transformational leadership are useful to guide school Sports Advisors to implement their tasks in school. Furthermore, the study about perspectives and satisfaction level of Sports Advisors on Transformational leadership also helps Senior Assistants of Co-curriculum to understand the Sports Advisors' perspectives and satisfaction level on Transformational leadership practiced by Senior Assistants of Co-Curriculum as well as to identify the effects of demographic factors on their perspectives and satisfaction level on Transformational leadership. Through Transformational leadership, Sports Advisors can be effectively guided by Senior Assistants of Co-curriculum to perform the task more enthusiastically to contribute to the sports in school.

1.1 Research Questions

This study aims to provide answer to the following questions:

1. What is the perspective and level of satisfaction of Sports Advisor on Transformational leadership of Senior Assistant of Co-curriculum in primary schools in Selangor?
2. Is there any significant difference in the perspective of Sports Advisor on Transformational leadership of Senior Assistant of Co-curriculum based on demographic factors such as (a) gender (b) location of schools (c) educational credentials?
3. Is there any significant difference in the satisfaction level of Sports Advisor on Transformational leadership of Senior Assistant of Co-curriculum based on demographic factors such as (a) gender (b) location of schools (c) educational credentials?

1.2 Teachers' Perspectives and Satisfaction Level on Leadership Style

The studies of Transformational leadership and its relationship with teachers' job satisfication conducted by Abdullah [1], Saad [41], Amin [2], Jaafar [21] and Ghazali [13] showed strong Transformational leadership practiced by principals on four dimesions Transformational leadership based on the perceptions of respondents. Therefore, Transformational leadership existed among school administrators could lead to the changes and effectiveness of management in their organizations.

Meanwhile, the studies conducted by Abdullah [1], Saad [41] showed strong relationship between Transformational leadership practices by principals and teachers' job satisfaction whereby Transformational leadership practised by principals gave positive impacts on teachers' job satisfaction. Besides, a study by Mansor and Esa

[33] showed high mean values for job satisfaction of teachers on Democratic leadership style, Autocratic leadership style and Laissez faire leadership style practiced by headmasters in three different primary schools. In addition, the results showed non significant difference between those leadership styles on teachers' job satisfaction.

1.3 Teachers' Perspectives and Satisfaction Level on Leadership Style Based on Gender

A study by Ramli and Hamid [39] consisting 203 respondents from 7 secondary schools in Kajang, Selangor showed non-significant difference on teachers' perceptions on female principals' leadership styles based on their demographic factors. The results also found out that teachers' perceptions for school principals' leadership styles are low in Kajang, Selangor. A study also conducted by Mandell and Pherwani [32] to study gender difference in Transformational leadership style and their relationship to emotional competence among organizational managers. However, the results showed there was significant difference between gender (male and female) whereby female respondents scored higher on emotional competence than males. Thus, gender difference is a obstacle factor that can give an impact on leadership in administration.

In the studies by Hassan and Wahab [16], Wiyono [45], Gupta and Gehlawat [14] indicated non-significant difference between male and female teachers' levels of satification based on gender difference on leadership style in education. However, the study conducted by Kaur and Sidana [24] showed the levels of satification of male teachers on principals' transformation leadership were higher than female teachers in school.

1.4 Teachers' Perspectives and Satisfaction Level on Leadership Style Based on Location of Schools

A survey study was conducted by Tahir [43] that comprised 1705 teachers from 35 secondary schools in Johor to examine principals' leadership styles. The results found out that there was significant difference on principals' leadership style based on type of schools in different location. The studies by Abdullah [1] also showed that culture or type of school in different location will influence the principals' leadership styles in schools. However, a study by Muhammad [36] did not find any significant difference for teachers' perceptions on principals' leadership based on location of schools.

A study was conducted by Kadir [23] to evaluate relationship between teachers' job satisfaction and motivation in schools based on schools location. Data analysis showed that there was no significant difference on teacher's job satisfaction based on location of schools ($t = 0.494, p > 0.05$). Meanwhile, there was a survey study conducted by Sharif, Omar dan Mondus comprising 352 teachers (182 urban schools and 170 rural schools) in Sandakan to find out whether there was significant difference on organizational learning practices and levels of satisfaction between urban and rural teachers. The findings showed that teachers' satisfaction levels in rural schools were higher than teachers in urban schools in Sandakan, Sabah. Meanwhile, the results also proved that there was significant difference on job satisfaction among urban and rural school teachers. Besides, the study by Lim [30] also showed significant difference on job satisfaction among urban and rural school teachers due to of environmental factors.

1.5 Teachers' Perspectives and Satisfaction Level on Leadership Style Based on Educational Credentials

A survey study by Ramli and Hamid [39] consisted of 1476 respondents showed significant difference on teachers' perceptions of principals' leadership styles based on educational credentials. The results found out that there was signifcant difference on teachers' perceptions on leadership practiced by principals based on educational credentials. Besides, the study conducted by Kadir [22], Deal dan Peterson [10] also showed that teachers with bachelor's degree of education and without bachelor's degree of education (Diploma of education) had different perceptions on leadership practiced in schools.

Futhermore, the study by Emad [12] found out that the teachers with bachelor's degree in education and diploma in education indicated significant difference in job satisfaction whereby teachers with bachelor's degree of education were more motivated than teachers with diploma of education in schools. Meanwhile, a study by Gupta dan Gehlawat also showed the graduate teachers had higher levels of job satisfaction in performing the assigned task than the non-graduate teachers.

However, a study by Wiyono [45] indicated non significant difference on teachers' levels of satisfication based on educational credentials. Futhermore, teachers with or without bachelor's degrees in education showed the same level of satisfication in schools. Thus, teachers' competence and knowledge were the main factors that contributed to the different levels of satisfaction in schools [45]. This is in line with the study conducted by Catania and Randall [5] who also showed that there was no significant difference between the educational credentials and level of job satisfaction.

2 Methodology

2.1 Methodology and Sampling Method

This study used quantitative research design with causal-comparative survey. The purpose of this study was to identify the perspectives and satisfication level of Sports Advisors on Transformational leadership of Senior Assistants of Co-Curriculum in primary schools in Selangor. So, this study used descriptive and comparative causal method to collect data because this is a suitable method to collect data on large number of respondents and reduce the limitation in distributing the questionnaires to respondents [9, 19]. Population of this study are Sports Advisors in primary schools in Selangor state registered under Ministry of Education Malaysia. Cluster random sampling technique is used to choose the respondents for the study because it helps to manage the population easily in many areas and reduce the cost to collect data from samples and also from different location of Selangor in this study [4, 37, 19].

The respondents of this study were 420 Sports Advisors in ordinary daily primary schools in Selangor state. There are 10 District Education Offices in Selangor and each District Education Offices is considered as one group and was given numbers from 1 to 10. Then, 3 District Education Offices (Klang District, Kuala Selangor District and Petaling Perdana District) were selected randomly from 10 District Education Offices in Selangor. After that, 250 primary schools in Klang District (92 schools), Kuala Selangor District (72 schools) and Petaling Perdana District (86 schools) were given numbers from 1 to 250. From 250 primary schools, 5 primary schools were randomly selected from each District Education Office (15 primary schools) with each school represented by 28 Sports Advisors in this study. From 420 respondents selected in this study, there were 352 male respondents and 68 female respondents as well as 334 respondents of city schools and 86 respondents of rural schools. Besides, 326 respondents have bachelor's degrees of education and 94 respondents have diploma of education.

2.2 Validity and Reliablitity

Instruments are important tools in research to collect data [19]. Therefore, Multi-factor Leadership Questionnaire (MLQ5x) adapted by Habib Ismail [20] is used as instrument to measure Transformational leadership and to study about the perspectives as wells as satisfaction level on Transformational leadership practiced by Senior Assistants of Co-Curriculum in this study. This questionnaire has 3 sections in which, section A consists demography data such as gender, location of schools and educational credentials. Then, section B has 16 items to measure charismatic, inspirational motivation, intellectual stimulation and individual consideration [20] and section C consists 6 items to measure the levels of satisfaction of Sports Advisors [3].

Validity of instruments are important to ensure that it can measure what should be measured in research [19]. Thus, validity of questionnaire was focused in this study in order to make sure all items in questionnaire were able to measure data that need to be measured. So, validity of criterion-related evidence of the research instrument was processed by consulting three expert panelists to assess the validity of research instrument. Three senior lecturers from University of Malaya were selected due to extensive experiences and they have been involving in lots of management research in University of Malaya. Three of the expert panelists have confirmed and approved that the questionnaire was suitable to be used in this study.

A pilot study was conducted to obtain the reliability of all the items in questionnaire using test and retest reliability. Instruments showing high correlation in test and retest reliability have proven that the instrument used was consistent and have achieved high reliability [9]. A total of 35 primary school teachers were identified as subjects of the pilot study and have obtained reliability value of 0.840 for individual consideration items, 0.897 for charismatic items, 0.904 for inspirational motivation items and 0.954 for intellectual stimulation items as well as 0.946 for levels of satisfication items. Thus, reliability values of all items in questionnaire have reached 0.70 and indicated that the instrument was suitable to be used in this study [8].

3 Results

3.1 What is the Perspective and Level of Satisfaction of Sports Advisors on Transformational Leadership of Senior Assistant of Co-Curriculum in Primary Schools in Selangor?

Table 1 showed the mean scores for perspectives of Sports Advisors on Transformational leadership of Senior Assistants of Co-curriculum based on dimensions of Transformational leadership like individual consideration, charismatic, inspirational motivation and intellectual stimulation. The highest mean value in dimension of transformational leadership was Charismatic (M = 4.48; SD = 0.550) as shown on Table 1. Data analysis for all items in Charismatic found out that mean values of all items for Charismatic were very high which is, between 4.45 and 4.50. Besides, the results also showed very high mean value for individual consideration (M = 4.36; SD = 0.495) and intellectual stimulation (M = 4.30; SD = 0.530). Mean values of all items for individual consideration obtained were between 4.22 and 4.55 and from 4.22 to 4.46 for intellectual stimulation. According to Table 1, mean value of inspirational motivation was slightly lower than the other dimesions of transformational leadership (M = 4.28; SD = 0.513). However, all items in inspirational motivation still obtained mean values between 4.11 (high) and 4.49 (very high) for the perspectives of Sports Advisors on Senior Assistants of Co-curriculum. This findings showed perspectives of Sports Advisors on Senior Assistants of Co-curriculum were very high.

According to Table 2, mean values of six satisfaction items of Sports Advisors were very high on Transformational leadership practised by Senior Assistants of Co-curriculum in which, mean values were between 4.41 and 4.58. Overall, levels of satisfaction of the Sports Advisors were also very high (M = 4.47; SD = 0.490) in this study.

Normality Test Data

Skewness and Kurtosis normality test was used to examine if the data from gender, location of school and educational credentials were normally distributed or not in this study. So, Skewness and Kurtosis normality test was conducted on data collected

Table 1 Mean values for perspectives of sports advisor on transformational leadership of senior assistants of co-curriculum

Items	Mean	SD	Interpretation
Individual consideration items			
Guide me to develop my potential	4.22	0.640	Very high
Appreciate my ability	4.45	0.590	Very high
Treat me as a human being rather than a machine	4.55	0.578	Very high
Take time to guide me	4.23	0.638	Very high
Overall for Individual Consideration	**4.36**	**0.495**	**Very high**
Charismatic items			
Motivate teachers to achieve school vision	4.45	0.582	Very high
High expectations on teachers' abilities	4.46	0.583	Very high
Confident in decision making	4.50	0.572	Very high
Have a clear personal vision	4.49	0.572	Very high
Overall for Charismatic	**4.48**	**0.550**	**Very high**
Inspirational motivation items			
Provide usable facilities	4.35	0.582	Very high
Involved in performing tasks	4.15	0.621	High
Draf a shared vision	4.11	0.657	High
Motivate teachers to perform tasks	4.49	0.604	Very high
Overall for Inspirational Motivation	**4.28**	**0.513**	**Very high**
Intellectual stimulation items			
Open-minded to receive opinions	4.27	0.592	Very high
Provide different alternatives to problem solving	4.25	0.600	Very high
Explain the importance to set a clear goal	4.46	0.626	Very high
Re-evaluate critical responses for its relevance to issues	4.22	0.593	Very high
Overall for Intellectual Stimulation	**4.30**	**0.530**	**Very high**

Table 2 Mean values for levels of satisfication of sports advisors

Satisfaction items	Mean	SD	Interpretation
Tasks given in sports	4.47	0.600	Very high
Plan an annual plan for sports activities	4.44	0.577	Very high
Manage pupils in sports activities	4.41	0.582	Very high
Manage participants for sports activities	4.40	0.592	Very high
Evaluate student's sports achievement	4.50	0.568	Very high
Provide annual reports on sports in schools	4.58	0.532	Very high
Overall for Levels of Satisfication	**4.47**	**0.490**	**Very high**

Table 3 Skewness and Kurtosis normality test for perspectives of sports advisors on transformational leadership of senior assistants of co-curriculum based on gender, location of schools, educational credentials

Demographic factors	Skewness	Kurtosis
Gender	0.203	−1.459
Location of schools	0.413	−0.535
Educational credentials	−0.421	−0.091

from gender, location of school, educational credentials and the results were as shown below:

Based on Table 3, Skewness and Kurtosis normality values for gender were 0.203 (Skewness) and −1.459 (Kurtosis) while for the location of schools showed Skewness (0.413) and Kurtosis (−0.535). Besides, the educational credentials showed Skewness (−0.421) and Kurtosis (−0.091) for perspectives of Sports Advisors on Transformational leadership of Senior Assistants of Co-curriculum. Thus, Skewness and Kurtosis values for gender, location of school and educational credentials for perspectives of Sports Advisors were normally distributed as the Skewness and Kurtosis range was between −2.00 and 2.00.

Analysis for Perspectives of Sports Advisors on Transformational Leadership of Senior Assistants of Co-curriculum based on Gender

According to Table 4, Levene's test for equality of variance for gender showed that the significance value greater than 0.05 ($F = 0.586, p > 0.05$). The data did not violate the equality of variance assumption in this study. Meanwhile, Independent t-Test results indicated non significant difference on gender with $t(418) = 0.059$, $p > 0.05$. So, there was no significant difference between males' (M = 4.364; SD = 0.458) and females' (M = 4.36; SD = 0.503) perspectives on transformational leadership of Senior Assistants of Co-curriculum.

Table 4 Independent T-Test for pperspectives of sports advisors on transformational leadership of senior assistants of co-curriculum based on gender

Gender	N	Mean	SD	Levene's test for equality of variance		T-test for equality of means		
				F	*Sig*	*t*	*df*	*Sig*
Male	68	4.364	0.458	0.586	0.445	0.059	418	0.953
Female	352	4.360	0.503					

Table 5 Independent T-test for perspectives of sports advisors on transformational leadership of senior assistants of co-curriculum based on location of schools

Location of schools	N	Mean	SD	Levene's test for equality of variance		t-test for equality of means		
				F	*Sig*	*t*	*df*	*Sig*
Urban	334	4.28	0.447	5.112	0.24	−6.501	418	0.000
Rural	86	4.62	0.357					

Analysis for Perspectives of Sports Advisors on Transformational Leadership of Senior Assistants of Co-curriculum based on Location of Schools

Table 5 showed the significance value of Levene's test was greater than 0.05 ($F = 5.112, p > 0.05$). So, the data of location of schools did not violate equality of variance assumption in this study. Meanwhile, Independent t-Test results indicated significant difference on location of schools with t(418) = −6.501, $p < 0.05$. So, there was significant difference on perspectives of Sports Advisors between urban schools (M = 4.28; SD = 0.447) and rural schools (M = 4.62; SD = 0.357).

Analysis for Perspectives of Sports Advisors on Transformational Leadership of Senior Assistants of Co-curriculum based on Educational Credentials

According to Table 6, Levene's test for equality of variance for Educational Credentials showed the significance value greater than 0.05 ($F = 2.813, p > 0.05$). The data did not violate the equality of variance assumption in this study. Meanwhile, Independent t-Test showed that there was significant difference between Bachelor's degree of education (M = 4.38; SD = 0.430) and Diploma of education (M = 4.21; SD = 0.528) with t(418) = 2.813, $p < 0.05$ on Transformational leadership of Senior Assistants of Co-curriculum.

Table 6 Independent T-test for perspectives of sports advisors on transformational leadership of senior assistants of co-curriculum based on educational credentials

Educational credentials	N	Mean	SD	Levene's test for equality of variance		t-test for equality of means		
				F	*Sig*	*t*	*df*	*Sig*
Bachelor's degree of education	352	4.38	0.430	5.672	0.18	2.813	418	0.005
Diploma of education	68	4.21	0.528					

3.2 Is There Any Significant Difference in the Satisfaction Level of Sports Advisor on Transformational Leadership of Senior Assistants of Co-curriculum Based on Demographic Factors Such as (a) Gender (b) Location of Schools (c) Educational Credentials?

As shown in Table 7, Skewness and Kurtosis normality values for gender showed Skewness (−0.538) and Kurtosis (−0.161) while the location of schools showed Skewness (1.469) and Kurtosis (0.157). Meanwhile, Skewness and Kurtosis values for the Educational credentials were 1.842 (Skewness) and 1.401 (Kurtosis) for satisfaction level of Sports Advisors on Transformational leadership of Senior Assistants of Co-curriculum. Therefore, Skewness and Kurtosis values for gender, location of school and educational credentials for satisfaction level of Sports Advisors were normally distributed as the Skewness and Kurtosis range was between −2.00 and 2.00.

Analysis for Satisfaction Level of Sports Advisors on Transformational Leadership of Senior Assistants of Co-curriculum based on Gender.

Table 8 showed the significance value of Levene's test was greater than 0.05 ($F = 0.812$, $p > 0.05$). So, the data did not violate the equality of variance assumption in this study. Meanwhile, Independent t-Test showed that there was no significant difference on gender with $t(418) = -1.032, p > 0.05$. Thus, there was no significant difference between males' (M = 4.409; SD = 0.475) and females' (M = 4.476; SD

Table 7 Skewness and Kurtosis normality test for satisfaction level of sports advisors on transformational leadership of senior assistants of co-curriculum based on gender, location of schools, educational credentials

Demographic factors	Skewness	Kurtosis
Gender	−0.538	−0.161
Location of schools	1.469	0.157
Educational credentials	1.842	1.401

Table 8 Independent T-test for satisfaction level of sports advisors on transformational leadership of senior assistants of co-curriculum based on gender

Gender	N	Mean	SD	Levene's test for equality of variance		t-test for equality of means		
				F	*Sig*	*t*	*df*	*Sig*
Male	68	4.409	0.475	0.812	0.368	−1.032	418	0.303
Female	352	4.476	0.493					

= 0.493) levels of satisfaction on Transformational leadership of Senior Assistants of Co-curriculum.

Analysis for Satisfaction Level of Sports Advisors on Transformational Leadership of Senior Assistants of Co-curriculum based on Location of Schools

According to Table 9, Levene's test for equality of variance for location of schools showed the significance value greater than 0.05 ($F = 8.729, p > 0.05$). So, the data from location of schools did not violate equality of variance assumption in this study. Meanwhile, Independent t-Test results indicated significant difference on location of schools with t(418) = −5.523, $p < 0.05$. So, there was significant difference on satisfaction level of Sports Advisors between urban schools (M = 4.40; SD = 0.489) and rural schools (M = 4.72; SD = 0.409) on Transformational leadership of Senior Assistants of Co-curriculum.

Analysis for Satisfaction Level of Sports Advisors on Transformational Leadership of Senior Assistants of Co-curriculum based on Educational Credentials.

Levene's test for equality of variance for Educational Credentials showed the significance value greater than 0.05 ($F = 0.470, p > 0.05$) on Table 10. The data did not violate the equality of variance assumption in this study. Meanwhile, Independent t-Test results indicated significant difference on satisfaction level between Bachelor's degree of education (M = 4.49; SD = 0.481) and Diploma of education (M = 4.32; SD = 0.511) with t(418) = 5.720, $p < 0.05$ on Transformational leadership of Senior Assistants of Co-curriculum.

Table 9 Independent T-test for satisfaction level of sports advisors on transformational leadership of senior assistants of co-curriculum based on location of schools

Location of schools	N	Mean	SD	Levene's test for equality of variance		T-test for equality of means		
				F	*Sig*	*t*	*df*	*Sig*
Urban	334	4.40	0.489	8.729	0.103	−5.523	418	0.000
Rural	86	4.72	0.409					

Table 10 Independent T-test for satisfaction level of sports advisors on transformational leadership of senior assistants of co-curriculum based on educational credentials

Educational credentials	N	Mean	SD	Levene's test for equality of variance		T-test for equality of means		
				F	*Sig*	*t*	*df*	*Sig*
Bachelor's degree of education	352	4.49	0.481	0.470	0.493	2.720	418	0.007
Diploma of education	68	4.32	0.511					

4 Discussion

Results of the descriptive statistic data analysis has showed very high mean scores for perspectives of Sports Advisors on Transformational leadership practiced by Senior Assistants of Co-curriculum in Sports Management in schools. These results are in line with Abdullah [1], Saad [41], Amin [2], Jaafar [21], Ghazali [13] who also showed high mean scores on Transformational leadership practised by leaders in schools. Therefore, most of the Senior Assistants of Co-curriculum in primary schools have often practised Transformational leadership and they are role models in Transformational leadership with high characteristics of Transformational leadership based on the perspectives of Sports Advisors. Moreover, creative ideas, clear goals and persistences shown by Senior Assistants of Co-curriculum through Transformational leadership practices have brought changes and also increased the effectiveness of sports management in schools.

Independent t-Test analysis for gender shows non significant difference for perspectives on transformational leadership of Senior Assistants of Co-curriculum between male and female Sports Advisors. In addtion, mean value for perceptions of male Sports Advisors is just slightly higher than female Sports Advisors on Transformational leadership of Senior Assistants of Co-curriculum. This result is in line with a study by Ramli and Hamdi [39] which also showed that there is no significant difference for teachers' perceptions based on their gender on Transformational leadership. This study is contrary with the result of a research conducted by Mandell and Pherwani [32] as there was a significant difference among gender for emotional competence on transformational leadership style whereby female respondents showed higher mean score than male repsondents for emotional competence.

Besides, the descriptive statistic result shows satisfaction level of female Sports Advisors are slightly higher than male Sports Advisors on Transformational leadership practised by Senior Assistants of Co-curriculum. However, Independent t-test shows that there is no significant difference in gender for satisfaction level of Sports Advisors on Transformational leadership. This result is also supported by Hassan and Wahab [16], Wiyono [45], Gupta and Gehlawat [14] who showed that there is no significant difference on level of satisfaction based on gender in education.

Therefore, this study have proven that gender is not a factor that can influence the perspectives and satisfaction level of Sports Advisors on Transformational leadership of Senior Assistants of Co-curriculum. This is because all the teachers have gone through the process and strengthen their teaching professionalism in line with the introduction of Malaysian Education Development Plan (PPPM) 2013–2025. Moreover, all the Sports Advisors are treated fairly and they followed the standard operating procedures of sports management in schools [32].

In Independent t-Test analysis for school location, rural schools shows significant difference with city schools for perspectives of Sports Advisor on Transformational leadership practiced by Senior Assistants of Co-curriculum. This result indicates that perspectives of Sports Advisors on Transformational leadership in rural primary schools are better than city primary schools. The result of this study is supported by studies by Tahir [43] which also proved that principals' leadership styles are different based on location and type of schools. So, school's culture in different location can influence the leadership style of a principal [1]. This situation is similar as Senior Assistants of Co-curriculum in city primary schools emphasised less on sports development compared to rural schools. This is due to city primary schools emphasised more on academic achievements. Thus, location of school has great influence to the perspectives of Sports Advisors on transformational leadership of Senior Assistants of Co-curriculum. However, the result of this study is different from a study by Muhammad [36] who indicated that there is no significant difference among rural schools and urban schools in transformational leadership practices.

Moreover, this study also find out that there is significant difference among rural schools and urban schools on satisfaction level of Sports Advisors on Transformational leadership practised by Senior Assistants of Co-curriculum. This study indicates that satisfaction level of Sports Advisors in urban primary schools are highly motivated by Transformational leadership practiced by Senior Assistants of Co-curriculum than.

Sports Advisors in rural primary schools. The result of this study is in line with studies conducted by Sharif et al. [42], Lim [30] who have found there is significant difference on levesl of satisfaction among rural schools' teachers and urban schools' teacher on leadership practices in school. This study shows significant difference on levels of satisfaction based location of school because there are more sports competitions organized in city areas than rural areas. In additon, advantage on location benefits school sports teams in city to have more opporturnity to participate actively in sports competition compared to rural schools and also in terms of cost, time and transportation. So, environmental factor (Lim 1997) such as oppurtunity to participate in sports competition causes the Sports Advisors in urban primary schools to be highly motivated to conduct the sports activities in schools compared to Sports Advisor in rural primary schools. Besides, the sports fund allocation for urban primary schools are also higher than rural schools because they have more pupils in urban schools. Thus, higher sports fund allocation also encourages Sports Advisor in urban primary schools to be more motivated and excited to implement more sports activities in schools. However, this study is different with study by Kadir

[23] who showed that there is no significant difference in teachers' job satisfaction based on location of schools.

In addition, Sports Advisors with bachelor's degree of education have been found to achieve higher mean scores than Sports Advisors with diploma of education in perspectives and satisfication level on Transformational leadership practised by Senior Assistants of Co-curriculum. This study also find out that the Sports Advisors with bachelor's degree of education show significant differences in perspectives and levels of satisfication on Transformational leadership using Independent t-Test when compared to the Sports Advisors with diploma of education. This study is in line with the study of Ramli and Hamid [39], Deal and Peterson [10], Kadir [22] who showed that there is signficant difference in teachers' perceptions on leadership style practised by principals based on educational credentials. However, this study is different from Wiyona who reported that there is no signficant difference in the perspectives on principal's leadership style based on educational credentials. Besides, this study is also contrary with Catania and Randall [5] who showed that there is no significant relationship between educational credentials and level of job satisfaction for teachers.

On the other hand, this study is supported by Emad [12],Gupta and Gehlawat [14] who indicated that teachers with bachelor's degree of education also showed significant difference in levels of satisfaction when performing tasks in schools compared to teachers with diploma of education in which, teachers with bachelor's degree of education are more satistified than teachers with diploma of education. The results of this study shows that Sports Advisors have significant differences in the perspectives and levels of satisfication on Transformational leadership practised by Senior Assistants of Co-curriculum based on their educational credentials because those who have bachelor's degree of education in primary are more knowledgeable and skillful to assist Senior Assistants of Co-curriculum in sports management. This can be shown when postgraduate teachers from Institute of Teacher Education Malaysia are required to participate in sports activities during five years of training in college and this gives them the advantages to manage sports in primary schools. On the other hand, teachers who graduated with diploma of education in higher education institutions or Institute of Teacher Education Malaysia have limited exposure and experiences in participation and management of sports activities with only one or one and a half year training in college. Therefore, postgraduate Sports Advisors are more motivated than Sports Advisors with diploma of education due to their advantages in knowledges, skills and experiences during training.

5 Conclusion

Transformational leadership is a leadership style that is suitable and often practised by Senior Assistants of Co-curriculum in primary schools in Selangor. Moreover, this leadership style also provides high level of job satisfaction among Sports Advisors in conducting sports activities to achieve schools' goals. Therefore, it is effective for Senior Assistants of Co-curriculum to use Transformational leadership to increase Sports Advisors' motivation in conducting sports activities regardless of gender. However, Senior Assistants of Co-curriculum need to consider the differences in school locations and educational credentials before practising Transformational leadership in primary schools. This is because the location of schools and educational credentials have different impact on perspectives and satisfication level of Sports Advisors on Transformational leadership applied by Senior Assistants of Co-curriculum in sports management in primary schools. Thus, Transformational leadership is suitable to be applied by Senior Assistants of Co-curriculum in school sports management to fulfill Malaysian Education Development Plan (PPPM) and to support the new learning structure of curriculum in primary schools through the introduction of Primary school Standard-Based Curriculum (KSSR). Therefore, Transformational leadership is suggested as alternative leadership style to guide and motivate Sports Advisors in conducting sports activities for pupils in primary school.

References

1. Abdullah MZ (2009) Kepimpinan transformasi pengetua dan hubungannya dengan kepuasan kerja guru. Universiti Malaya, Tesis Sarjana
2. Amin TM (2010) Amalan kepemimpinan transformasional pengetua dan keupayaan kepemimpinan guru sekolah menengah di Pahang. Universiti Malaya, Tesis Sarjana
3. Awal A (2009) Amalan Komunikasi Guru Besar dan Kepuasan Kerja Guru Sekolah Rendah Daerah Samarahan. Universiti Kebangsaan Malaysia, Sarawak. Tesis Sarjana Pendidikan
4. Babbie ER (2001) The practice of social research, 9th edn. Thompson Learning, USA
5. Catania G, Randall R (2013) The relationship between age and intrinsic and extrinsic motivation in workers a maltese culture context. International Journal of Art and Science 6(2):31–45
6. Chan, Y. F. (2006). Penilaian keberkesanan program latihan pengurusan dan kepimpinan sekolah/ The evaluation of the effectiveness of school leadership and management training programme. Shah Alam: UPENA, UiTM.
7. Cheng YC (2005) New paradigm for re-engineering education: Globalization, localization, and individualization. Springer, Asia-Pacific Educational Research Association. Netherlands, pp 213–316
8. Chua Yan Piaw (2006) Kaedah dan statistik penyelidikan: kaedah pendidikan. McGraw Hill, Kuala Lumpur
9. Creswell, J. W. (2012). Educational research: Planning, conducting and evaluating quantitative and qualitative research. Pearson
10. Deal, T., & Peterson, K. (2000). Eight roles of symbolic leaders. Fullan, M (pytg) Educational leadership: The Jossey-Bass reader series, 202 – 217. San Francisco: Jossey-Bass.
11. Del Favero M (2006) An examination of the relationship between academic discipline and cognitive complexity in academic deans' administrative behavior. Res. Higher Educ. 47(3):281–315
12. Emad. (2014) Teacher motivation: A study of work motivation of the primary stage teachers in Jordan. American Journal of Applied Psycology 3:57–61
13. Ghazali N (2007) Tahap amalan kepimpinan transformasi dalam kalangan pengetua sekolah menengah Bandar Tangkak. Universiti Teknologi Malaysia, Tesis Sarjana

14. Gupta M, Gehlawat M (2013) Job satisfaction and work motivation ofsecondary school teachers in relation to some demographic variables: A comparative study. Journal of Educationia Confab 2(1):10–19
15. Mohd H (2006) Pengupayaan kepemimpinan pengetua danpenglibatan guru dalam aktiviti kokurikulum. Fakulti Pendidikan, Universiti Malaya, Kertas Projek Sarjana
16. Hassan, N. & Wahab, J. A. (2017). Kepuasan kerja dalam kalangan guru di sekolah menengah zon Bangsar. Jurnal Seminar Serantau, 401–408.
17. Herzberg F, Mausner B, Synderian B (1959) The Motivation to Work. John Wiley & Sons Inc., New York
18. Hoy, W. K. & Miskel, C. G. (2013). Educational Administration: M c Graw- Hill International
19. Idris, N. (2010). Penyelidikan Dalam Pendidikan. Kuala Lumpur: McGraw-Hill Education.
20. Ismail SH (2000) Satu kajian mengenai hubungan pengurusan dengan pencapaian dalam bidang kokurikulum. Universiti Utara Malaysia, Tesis Sarjana
21. Jaafar S (2008) Kepemimpinan transformasional pengetua di sebuah sekolah menengah berprestasi tinggi di Hulu Selangor. Universiti Malaya, Tesis Sarjana
22. Kadir SA (2015) Kepimpinan transformasi guru besar di daerah Bulukumba Sulawesi Selatan Indonesia. Universiti Teknologi Malaysia, Tesis Doktoral
23. Kadir SR (2013) Hubungan antara kepuasan kerja dengan motivasi guru sekolah menengah di daerah Sandakan. Universiti Malaysia Sabah, Tesis Sarjana
24. Kaur G, Sidana JJ (2011) Job Satisfaction of College Teachers of Punjab with Respect to Area. Gender and Type of Institution. Edutracks 10(11):27–35
25. Kementerian Pendidikan Malaysia. (2010). Pelaksanaan KSSR Tahap 1 Mulai 2011. Surat Pekeliling Ikhtisas. Bil 9/2016. Putrajaya: Pusat Perkembangan Kurikulum.
26. Kementerian Pendidikan Malaysia. (2011). Buku Panduan Pelaksanaan Dasar Satu Murid Satu Sukan (1M1S). Putrajaya: Bahagian Sukan.
27. Kementerian Pendidikan Malaysia. (2012). Laporan Tahunan 2012 Jemaah Nazir dan Jaminan Kualiti. Putrajaya: Jemaah Nazir dan Jaminan Kualiti.
28. Kementerian Pendidikan Malaysia. (2013). Pelan Pembangunan Pendidikan Malaysia 2013–2025 (Pendidikan Prasekolah hingga Lepas Menengah). Putrajaya: Kementerian Pendidikan Malaysia.
29. Leithwood, K & Jantzi, D. (2006). Linking leadership to student learning: The contribution of leader efficacy. Educational Administration Quarterly.
30. Lim TS (1995) Kepuasan kerja di kalangan guru-guru Sekolah Rendah Jenis Kebangsaan Cina di negeri Kelantan. Universiti Utara Malaysia, Tesis Master
31. Malek AHA (2005) Faktor-faktor penghalang pengurusan kokurikulum di sebuah sekolah menengah Melaka. Universiti Malaya, Tesis Sarjana Pendidikan
32. Mandell B, Pherwani S (2003) Relationship between emotional intelligence and TL style: A gender comparison. J Bus Psychol 17(3):387–404
33. Mansor R, Esa A (2009) Hubungan gaya kepimpinan guru besar dengan kepuasan kerja guru di tiga buah sekolah rendah di Seremban. Universiti Tun Hussein Onn, Seminar Kebangsaan Pengurusan Pendidikan PKPGB
34. Maslow AH (1970) Motivation And Personality. Harper & Row Publishers, New York
35. Mcgregor D (1960) The Human Side Of Enterprise. McGraw Hill Inc., New York
36. Muhammad L (2006) Pelaksanaan Kepemimpinan Pengajaran di kalangan Pengetua Sekolah Menengah di Negeri Selangor. Universiti Kebangsaan Malaysia, Tesis Dr. Falsafah
37. Newman, W. (2003). Social research methods: qualitative and quantitative approaches (5th ed.). U.S.A: Pearson Education, Inc.
38. Rahim AAA (2015) Pengurusan kokurikulum dan badan beruniform. Penerbit Fajar Bakti, Kuala Lumpur
39. Ramli N, Hamid JA (2006) Gaya kepimpinan pengetua wanita dan hubungannya dengan tahap motivasi guru sekolah menengah. Jurnal Pendidikan Jilid 31:53–69
40. Robbins SP, Judge T (2007) Organizational behavior. Pearson/Prentice Hall, Upper Saddle River

41. Saad MA (2008) Gaya kepemimpinan transformasional pengetua dan hubungannya dengan kepuasan kerja guru di sebuah sekolah dalam daerah Maran. Universiti Malaya, Pahang. Tesis Sarjana Kepengetuaan
42. Sharif S, Omar AT, Mondus S (2013) Organisasi pembelajaran dan hubungannya dengan kepuasan kerja. Jurnal Pusat Penataran Ilmu & Bahasa 19:183–212
43. Tahir LM (2008) Analisis Kepimpinan Pengetua Sekolah Menengah Di Johor. Universiti Teknologi Malaysia, Tesis Doktoral
44. Tang KN (2011) The effect of transformational leadership on school culture in Male's Primary Schools Maldives. Online Journal of Social and Behavioral Sciences 30:2575–2580
45. Wiyono BB (2016) Comparison of teachers' work motivation based on gender, age, educational level, work duration, rank, and school level. Journal of Education and Social Sciences 3:61–66

Development of Motivation Model Towards Anthropometric, Soccer Skills, Maturity and Physical Fitness Using Machine Learning

Ahmad Nadzmi, Ahmad Bisyri Husin Musawi Maliki, Mohamad Razali Abdullah, Rabiu Muazu Musa, Izwan Syahril, Mohd Syaiful Nizam Abu Hassan, Shahrulfadly Rustam, Jorrye Jakiwa, and Syed Kamaruzaman Syed Ali

Abstract Research in soccer has shown that players' technical, tactical, physical, and psychological abilities are required to meet the requirements of the competition. This study uses machine learning to develop a motivation model based on anthropometric, fitness, and soccer skills. Data were collected from 223 young Malaysian athletes consisting of Malaysia's Sport School soccer athletes who play in various positions (defender, midfielder and forward) aged 13 to 17 years old who participated in this study. Athletes are required to complete the study's instrument, which consists of the anthropometric component test, Task and Ego Orientation in Sport Questionnaire (TEOSQ), technical skill component and physical fitness test. Data analysis was carried out using hierarchical agglomerative cluster analysis (HACA) and discriminant analysis (DA). Hierarchical agglomerative cluster analysis is used to divide groups according to their homogenous psychological attributes of the athletes and

A. Nadzmi · A. B. H. M. Maliki (✉) · S. Rustam · J. Jakiwa
Defence Fitness Academy, National Defence University of Malaysia, Kem Sungai Besi, 57000 Kuala Lumpur, Malaysia
e-mail: bisyri@upnm.edu.my

M. R. Abdullah
East Coast Environment Research Institute, University of Sultan Zainal Abidin, Gong Badak Campus, 21300 Terengganu, Malaysia

R. M. Musa
Department of Basic Knowledge and Liberal Studies, Universiti Malaysia Terengganu, 21300 Kuala Nerus, Terengganu, Malaysia

I. Syahril
Faculty of Sport Science, Sultan Idris Education University, 35900 Perak Darul Ridzuan, Tanjong Malim, Malaysia

M. S. N. Abu Hassan
Faculty of Applied Social Sciences, Universiti Sultan Zainal Abidin, Gong Badak Campus, 21300 Kuala Nerus, Terengganu, Malaysia

S. K. Syed Ali
Faculty of Education, University of Malaya, Jalan Profesor Diraja Ungku Aziz, 50603 Kuala Lumpur, Malaysia

© The Author(s), under exclusive license to Springer Nature Singapore Pte Ltd. 2023
S. F. Syed Omar et al. (eds.), *Innovation and Technology in Sports*, Lecture Notes in Bioengineering, https://doi.org/10.1007/978-981-99-0297-2_10

discriminant analysis used for determining the differences in player performance. Three groups formed and successfully discriminated three groups on 13 independent variables with 79.82% (forward stepwise) total variance resulting with Machine Learning method (Artificial Neural Network) 67 athletes predicted with potential. A group tends to have the taller player because of the highest significance in height variables than others. From the result, all groups show their characteristics with unique attributes and need to intervene to characterize their training program based on the group's performance.

Keywords Cluster analysis · Discriminant analysis · Motivation orientation · Position

1 Introduction

Soccer is a highly complex sport influenced by physical, psychological, tactical and technical factors [1]. Research in soccer has demonstrated that players' technical, tactical, physical and psychological players to meet their needs during competition [2]. Therefore, careful observation of the athletic necessities to match play and components encouraging efficacious performance can guarantee that accuracy and rational assessments are taken for establishing the physiological, physical, tactical, technical and psychological components of preparing systematic talent recognition, identification and development training programs towards the development of the players. Normally, coaches and instructors feel difficult to choose an appropriate element of a soccer performance model that will fit a given set of multilateral factors [3].

The athletes should be motivated by two main sources, according to research. First, they may be motivated intrinsically (do sports activities for pleasure, fun or other self-determined reasons). Second, they may have been motivated by extrinsic factors (obtaining benefits, tangible and material such as money or trophies, or social rewards (prestige, public knowledge), or to avoid punishment). (Goal orientation is a determinant factor of an athlete's success in sport [4]. Psychological factors such as goal orientation, concentration and anxiety must be well controlled to produce the best performance in achieving this goal. When athletes' motor skills gaps are getting smaller today, greater mental resilience is required as competition becomes more intense. In sports, especially golf, success and failure are often associated with motivation, attention and arousal [5]. Thus, an athlete's mental preparation is essential before and after a competition [6].

Relevant parties do not seriously address these mental and psychological aspects because of inadequate service from sports psychology in many sports associations in Malaysia. There is not enough psychological training for all athletes to improve their mental strength. There is no scientific research analysis or a detailed report on that particular part, except only a general report in a local newspaper when they are defeated in a competition. This shows that research or documentation in sport psychology is still lacking in Malaysia [7]. This study investigates the dispositional of goal orientation on players' performance in different soccer positions.

2 Methodology

2.1 Players

223 soccer players were selected at the age of (17.40 ± 19.9 years old). The player came from different positions, such as defender, midfielder and forward. The player came from different states around Malaysia under Malaysia Sports School, where the generation of the new sport develops skills and discovers new talent. All the players and coaches who participated in the study obtained written consent.

2.2 Experimental Approach to the Problem

To answer the entire objective of this study, the researcher needs to do a fitness assessment of the performance variables following the standard protocols for the fitness evaluation.

2.3 Anthropometric Component

The Player had been tested for age, weight, sitting height, bicep, triceps subscapular, spiliac upper body circumference (MUAC), calf circumference (CC) and maturity.

2.4 Motivation Component

Using the Task and Ego Orientation Sport Questionnaire (TEOSQ), ego and task were collected. The questionnaire has been translated to Bahasa Malaysia using back translation to make sure the entire player is easy to understand. The questionnaire has been read and explained to the player; there is no fee to participate in the study.

2.5 Technical Skill Component

Player needs to perform various technical skill tests such as running with the ball, juggling (foot), juggling body, speed dribbling, long pass, short pass, shooting Top Right, shooting Top Left, shooting from the pass, and heading mid and side. All the players were given 3 trials, and the highest score was used for collecting data.

2.6 Fitness Component

Fitness components include sit reach, sergeant jump, V sit up, speed 5-m, speed 10-m, speed 20 m and V02max. The player giving one minute to perform the test, and the highest score is used for data.

2.7 Data Analysis

By using the XLSTAT add-ons system, researchers used three methods of analysis. First is principal component analysis (PCA), then analysis continued with Discriminant Analysis (DA) and finally finished with artificial neural network (ANN).

A study stated that the principal component analysis is a commonly used analysis to reduce data from many variables to a smaller set of underlying factors that summarize the essential information contained in the variables [8]. The purpose is to obtain important information from the data schedule and state this information as a new set of orthogonal variables called principal components [9]. In this study, PCA was conducted on the 31 variables and summarised into 21 variables only.

Afterwards, hierarchical agglomerative cluster analysis (HACA) was assigned to separate any homogeneity that was the same as others by using task and ego orientation as variables. HACA is a robust method to identify and categorize components or subjects (observations/population) into clusters with more excellent homogeneity within the class and more significant heterogeneity among classes with regard to a predetermined selection criterion [10]. Moreover, Ward's technique utilizing Euclidean distances as a resemblance in HACA has shown to be very effective [11]. The findings were also shown by dendrogram divide by cluster and their homogeneity.

The discriminant analysis (DA) controls the variables that separate among two or more joined groups/clusters [11]. A descriptive discriminant analysis was conducted to identify which variables best discriminate the previously obtained clusters. Discriminant analysis is robust for these derived rate variables [12]. Three groups for relative performance patterns (three sampling groups represent low performance, medium performance, and high performance) were obtained and selected from HACA. Validation of discriminant models was conducted using the leave-one-out method of cross-validation [13]. The DA was put into the raw data using misclassified mode, predefined mode, corrected group (ANNs), backward stepwise and forward stepwise. Analysis using raw data of the performance of players with goal orientation results by HACA and next using predefined that computerized by excel XLSTAT to analyse predefined data. Using only significant parameters from PCA for performance and cluster from predefined group to run analysis data for the corrected group show significant higher for machine learning, analysis data for backward stepwise and forward stepwise.

Artificial Neural Networks (ANN) are usually considered as tools which can help to analyse cause-effect relationships in complex systems within a big-data framework. Neural network is a powerful computational data model that is able to catch and represent complex input/output relationships. The motivation for the development of neural network technology comes from the desire to develop an artificial system that could perform "intelligent" tasks similar to those performed by the human brain. With applying 100 hidden neurons in hidden layers, Rectified Linear Unit (ReLU) activation, and combined with Adam Optimization Algorithm Solver, this model was chosen to achieve this objective research.

3 Results

Table 1 exhibits the summary statistic of athletes. It shows the total number of 223 athletes in soccer. The athletes' minimum, maximum, mean and standard deviation scores are projected.

From the PCA result, out of the thirty principal components (PCs) generated, only eight PCs with eigenvalues >1 were selected for the feed-forward ANN input selection parameter representing 71.68% of the total variance. Nevertheless, Table 2 highlighted the factor loading after the varimax rotation method in the PCA. Furthermore, the standardized VFs with absolute values equal to or greater than 0.70, as the selection edge is considered solid and stable, specify moderate to strong loadings on the extracted factors in the current study. However, it can be seen from Table 2 out of thirty parameters, only 21 parameters were identified as the most significant across all variables. Due to the transformations of a new data set, the output from this analysis was used as input for further analysis in HACA and DA.

Based on the technical skill-related components extracted by PCA, selection of the categorical dependent component was used in HACA, which led to the identification of HTHE (High Task, High Ego), LTME (Low Task, Moderate Ego) and MTLE (Moderate Task, Low Ego) groups. Figures 1 and 2 shows the result of the HACA.

Table 3 shows the discriminant analysis conducted for further analysis. DA is applied using cluster analysis defined by HACA in order tool for goal orientation of the player. HACA was performed using task and ego orientation to define different clusters for goal orientation. Misclassified mode, predefined mode, Corrected mode, and backward stepwise and forward stepwise mode methods were selected to perform the DA. The precision results showed that misclassified mode, with a total of 40.36%.

Table 1 Summary statistics of athletes

Variable	Observations	Minimum	Maximum	Mean	Std. deviation
Weight	223	26.90	90.50	56.51	9.47
Height	223	128.60	190.60	166.06	8.05
Sitting height	223	38.60	98.40	86.92	5.57
Biceps	223	2.80	12.30	4.10	1.14
Triceps	223	4.50	24.80	7.77	2.26
Sbs. capul	223	4.50	28.80	8.30	2.33
Sp. iliac	223	4.50	45.40	8.11	3.63
MUAC	223	2.30	34.60	24.73	2.89
CC	223	3.70	44.00	34.95	3.57
Maturity	223	1.00	5.00	3.85	0.79
S&R	223	0.00	27.00	13.25	5.18
SJ	223	2.69	198.00	64.30	14.85
V. Sit Up	223	2.00	7.00	5.90	0.96
505A	223	1.64	2.98	2.37	0.25
Speed 5 m	223	0.36	1.55	0.79	0.16
Speed 10 m	223	1.00	2.21	1.53	0.19
Speed 20 m	223	1.50	4.08	2.85	0.29
FI	223	-1.55	6.43	0.48	0.84
VO2mx	223	29.93	63.73	47.38	7.68
Run w/ball	223	1.93	7.58	4.55	1.12
Juggling (foot)	223	5.00	52.00	38.83	11.23
Juggling body	223	3.00	9.00	7.39	1.97
Speed dribbling	223	16.91	29.30	20.95	2.33
Long passing	223	0.00	12.00	3.85	2.78
Short passing	223	1.00	15.00	10.00	3.53
Shooting top right (Dead Ball)	223	0.00	18.00	3.70	2.95
Shooting top left (Dead Ball)	223	0.00	13.00	3.57	2.91
Shooting from a pass (Foot)	223	0.00	26.00	8.18	5.57
Heading (mid_post)	223	0.00	18.00	9.02	4.52
Heading (side_post)	223	0.00	18.00	7.25	4.24

While the predefined mode with a total of 82.96%, the Corrected group reported a total of 93.72% for backward stepwise (11 independent variables) and forward stepwise (11 independent variables), both with a total of 92.83% (Table 4).

Using Orange software to analyze data for machine learning shows that Artificial Neutral Network have higher results from analysis data compared to other machine learning.

Table 2 Factor loading of PCA analysis result

	D1	D2	D3	D4	D5	D6	D7	D8
Weight	0.75							
Height	0.80							
Sitting height	0.68							
Biceps		0.89						
Triceps		0.91						
Sbs. capul		0.87						
Sp. iliac		0.91						
MUAC								
CC								
Maturity	0.68							
S&R								
SJ	0.72							
V. Sit Up								0.89
505A						−0.80		
Speed 5 m				0.88				
Speed 10 m				0.87				
Speed 20 m				0.84				
FI								
VO2mx					0.71			
Run w/ball								
Juggling (foot)					0.68			
Juggling Body						0.67		
Speed Dribbling					−0.66			
Long Passing								
Short Passing					0.67			
Shooting TR (Dead Ball)			0.73					
Shooting TL (Dead Ball)			0.75					
Shooting From a Pass (Foot)			0.63					
Heading (md_post)								
Heading (sd_post)							0.72	
Eigenvalue	6.12	4.57	2.42	1.95	1.38	1.30	1.14	1.02
Variability (%)	13.75	14.04	6.97	9.64	8.97	4.61	4.38	3.96
Cumulative %	13.75	27.79	34.76	44.40	53.38	57.99	62.36	66.32

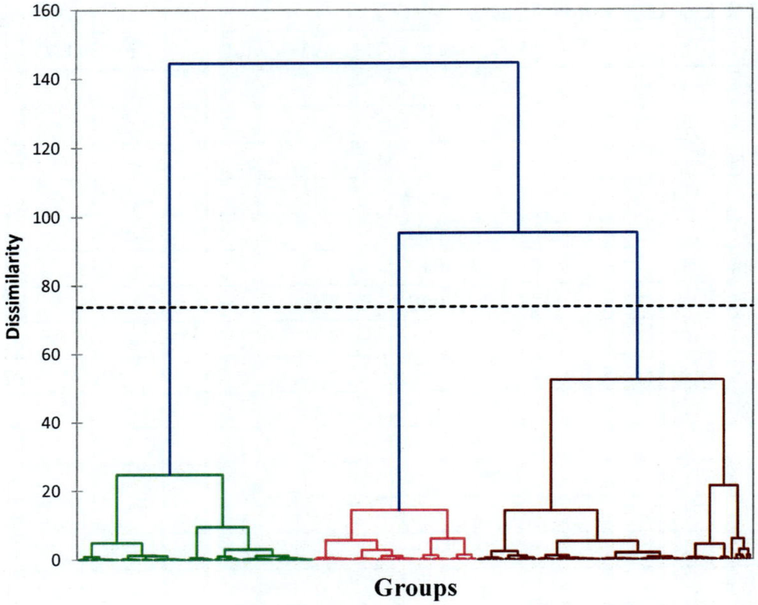

Fig. 1 Result of HACA analysis

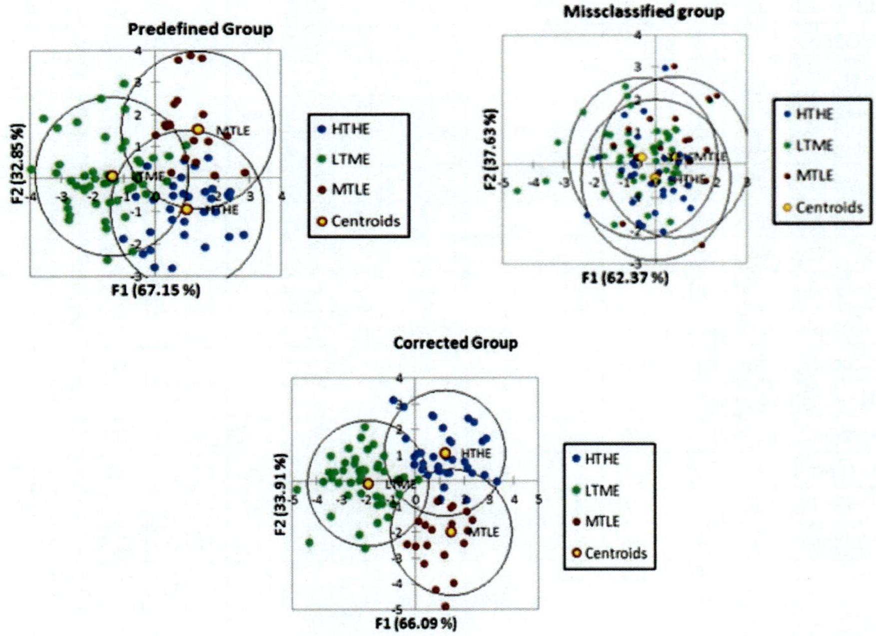

Fig. 2 Observations chart by HACA analysis results

Table 3 Confusion matrix for the cross-validation results of DA

Sampling groups	% Correct	Group assigned by DA		
		HTHE	LTME	MTLE
Misclassified mode				
HTHE	38.46	30	33	15
LTME	49.45	33	45	13
MTLE	27.78	17	22	15
Total	40.36	80	100	43
Predefined mode				
HTHE	87.50	70	5	5
LTME	82.00	11	82	7
MTLE	76.74	8	2	33
Total	82.96	89	89	45
Corrected group				
HTHE	95.51	85	1	3
LTME	93.26	4	83	2
MTLE	91.11	3	1	41
Total	93.72	92	85	46
Backward stepwise (11 independent variables)				
HTHE	94.38	84	1	4
LTME	92.13	5	82	2
MTLE	91.11	3	1	41
Total	92.83	92	84	47
Forward stepwise (11 independent variables)				
HTHE	94.38	84	1	4
LTME	92.13	5	82	2
MTLE	91.11	3	1	41
Total	92.83	92	84	47

Table 4 Evaluation result of Machine Learning

Method	AUC	CA	F1	Precision	Recall
Neural network	0.974	0.876	0.876	0.876	0.876
Naive Bayes	0.8	0.639	0.642	0.656	0.639
Logistic regression	0.933	0.812	0.811	0.811	0.812

Table 5 Artificial neutral network with P-value under 0.05

Variables	P-value
Height	0.001
Sitting height	0.000
Maturity	0.002
V sit up	0.000
505	0.000
Juggling body	0.000
Speed dribbling	0.000
Shooting TR	0.000
Shooting TL	0.000
Shooting from passing (foot)	0.000
Heading	0.000
Task	0.004
Ego	0.000

Artificial neural network show the significant parameters shown in Table 5. Showing that a p-value of < 0.05 significant are from height, sitting height, maturity, V-sit up, 505, juggling body, speed dribbling, shooting TR and TL, shooting from passing, heading, task and ego orientation.

4 Discussion

This present study identifies the factor contributing to player performance in different clusters using Artificial Neutral Network (ANN) technique. Based on the technical skill-related components extracted by PCA, selection of the categorical dependent component were used in HACA, which led to the identification of HTHE (High Task, Low Ego), LTME (Low Task, Moderate Ego) and MTLE (Moderate Task, Low Ego) groups.

Psychological factors such as goal orientation, concentration and anxiety must be well controlled to produce the best performance. When the competition becomes more intense, greater mental resilience is required because motor skills gaps among athletes today are getting smaller. Success and failure in sports, mainly golf is often associated with motivation, attention and arousal [14]. Thus, mental preparation for athletes before and during the competition is essential [5]. Williams et al. (1995) [15] indicated that the performance-related feedback of an individual's sports ability could have implications for an athlete's motivational orientation regarding the task and ego involvement.

Findings showed significant effects for goal orientation that involved task and ego among different goal profiles in the two fundamentals area in goal orientation: task and ego. The task is to achieve mastery over the task through applying relevant skills and ego, which is the motivation to pursue and realize that the ego-oriented goal is fuelled by competition. This study considered that cluster 1 (low task and medium ego-LT/ME) had more excellent goal orientation in both task and ego and athletes in cluster 2 (high task and high ego-HT/HE) and compared to cluster 3 (moderate task and low ego-MT/LE). Results also suggest that it is possible that athletes in cluster 2 would probably benefit during adversities in competition due to reasonable control over themselves, leading to more excellent goal performance. Furthermore, athletes in cluster 2 (high task and high ego) tend to invest more in task and ego orientations; the players focus more on task orientation whereby portrayed as an intrinsically motivated state that leads to persistence when failing, perseverance when faced with difficulties and a fulfilling sense of achievement if the task was mastered, improvement was experienced, and ability was successfully expressed thus the player gaining more on motivation which is aimed at being the best rather than doing one's best (ego). The driving force is to demonstrate a superior and higher ability compared to others rather than one's ability, irrespective of how it compares to others [16, 17]. Positive competition can be generated because of the benefits of having better control over the negative energy, which is unproductive. These findings support the suggestions by [18–22] that moderate to high task and ego orientation patterns can complement a competitive situation.

5 Conclusion

This study research investigates whether goal orientation impacts player performance and has been clarified using machine learning of artificial neural network as a result of confirmation. 32 parameters have been tested for 223 athletes using PCA and have been minimised to only 21 parameters from performance parameters. Using HACA, 3 clusters, HTHE, LTME, and MTLE, have been identified that different the athletes by task and ego parameters. Next, using DA to combine cluster and performance parameters that contribute to athlete's results achievements by using DA research get to narrow down the performance that contributes the biggest to the study and combine the goal orientation from athletes. Lastly, using machine learning to identify the parameter that contributes the most to athletes shows that from 32 to only 13 parameters, including task and ego orientation. Show that only 13 parameters contribute to athletes' performance with goal orientation of athletes.

Summarize it, showing that results from HTHE and 13 parameters need to sharpen so that athletes can perform much better and even more incredible, resulting in an athlete with better achievement not just in psychology but also in their performance. It can help the player to have a better future in order to become a professional player in future.

References

1. Reilly T, Williams AM, Nevill A, Franks A (2000) A multidisciplinary approach to talent identification in soccer. J Sports Sci 18(9):695–702
2. Abdi H, Williams LJ (2010) Principal component analysis. Wiley Interdiscipl Rev Comp Stat 2(4):433–459
3. Gastin PB, Meyer D, Huntsman E, Cook J (2015) Increase in injury risk with low body mass and aerobic-running fitness in elite Australian football. Int J Sports Physiol Perform 10(4):458–463
4. Barbanti V (2013) The multicomponent anthropometric model for assessing body composition in a male pediatric population: a simultaneous prediction of fat mass, bone mineral content, and lean soft tissue. J Obes
5. Majzub R, Muhammad TA (2011) Goal orientation and achievement of junior in golfers Malaysia. Procedia Soc Behav Sci 15:1644–1649
6. Zifonun G, Hoffmann L, Strecker B, Ballweg J (1997) Grammatik der deutschen Sprache. Walter de Gruyter
7. Majzub RM, Bataineh MZ, Ishak NM, Rahman S (2011) The relationship between locus of control and academic achievement and gender in a selected higher education institution in Jordan. In: Proceedings of the 8th WSEAS international conference on education and educational technology (Corfu Island: WSEAS Press). pp 215–220
8. Thorpe MG, Milte CM, Crawford D, McNaughton SA, Lawrence M, Ball K, Worsley A, Nanayakkara J, Margerison C, Hart MJ, Torres S (2017) Predictors of change in dietary patterns determined by principal component analysis in Australians >55 years. J Nutrit Intermedi Metab 8:73
9. Abdi H, Williams LJ (2010) Principal component analysis. Wiley Interdiscipl Rev Comput Stat 2(4):433–459
10. van de Wetering M, Francies HE, Francis JM, Bounova G, Iorio F, Pronk A, van Houdt W, van Gorp J, Taylor-Weiner A, Kester L, McLaren-Douglas A (2015) Prospective derivation of a living organoid biobank of colorectal cancer patients. Cell 161(4):933–945
11. Abdullah MR, Maliki M, Husin AB, Musa RM, Kosni NA, Juahir H, Haque M (2016) Multi-hierarchical pattern recognition of athlete's relative performance as a criterion for predicting potential athletes. J Young Pharm 8(4)
12. Parker TS, Chua L (2012) Practical numerical algorithms for chaotic systems. Springer Science & Business Media
13. Chua TS, Tang J, Hong R, Li H, Luo Z, Zheng Y (2009) NUS-WIDE: a real-world web image database from National University of Singapore. In: Proceedings of the ACM international conference on image and video. ACM, p 48
14. Wann DL (1997) Sport psychology. Pearson College Division
15. Williams L, Gill DL (1995) The role of perceived competence in the motivation of physical activity. J Sport Exerc Psychol 17(4):363–378
16. Breiger J, Cumming SP, Smith RE, Smoll F (2015) Winning, motivational climate, and young athletes' competitive experiences: some notable sex differences. Int J Sports Sci Coach 10(2–3):395–411
17. Steinberg G, Grieve F, Glass B (2001) Achievement goals across the lifespan. J Sport Behav 24(3):298–306
18. Hodge K, Petlichkoff L (2000) Goal profiles in sport motivation: a cluster analysis. J Sport Exerc Psychol 22(3):256–272
19. Kassim M, Rustam S (2021) Measuring physical fitness for military cadets' officers. https://doi.org/10.1007/978-981-15-5856-6_55
20. Kassim M, Othman N, Ujang E (2020) The level of physical fitness among first year female students in national defence university of malaysia. J Phys Confe Ser 1529(2). https://doi.org/10.1088/1742-6596/1529/2/022072

21. Kassim M, Ujang E, Othman N (2020) The association between physical activity and basic fitness tests in female junior reserve officer training unit (ROTU), national defence university malaysia (NDUM). J Phys Conf Ser 1529(2). https://doi.org/10.1088/1742-6596/1529/2/022073
22. Atan SA, Kassim M (2020) Development of a soccer-specific running protocol for young soccer players. https://doi.org/10.1007/978-981-15-3270-2_11

Development of Motor Performance Index: A Preliminary Study Among 7 Years Old Malaysian Kids

Ahmad Nadzmi, Ahmad Bisyri Husin Musawi Maliki, Rabiu Muazu Musa, Mohamad Razali Abdullah, Mohamad Amirur Rafiqi Zainoddin, Intan Meily Puspitasari, Nur Faizatul Amira Jibril, Nur Amirah Nawi, Izwan Syahril, Mohd Syaiful Nizam Abu Hassan, Shahrulfadly Rustam, Jorrye Jakiwa, and Syed Kamaruzaman Syed Ali

Abstract The development of technology has increasingly reduced the practice of physical and motor activity use and it can lead to some health-related quality life problems such as social dysfunction and disease problems. The purpose of this research is to identify the development of motor performance index among 7 years old Malaysian kids. Datasets were collected from 1998 participants in research aged 7 years old in primary schools all around Malaysia. The participants completed multiple physical fitness tests (anthropometrics, standing broad jump, twenty-meter speed, sit and reach and hand wall toss). Data interpretation was carried out using Principal Component Analysis (PCA), Discriminant Analysis and Machine Learning. It was found that there is a small number of male kids that only have high physical fitness performance but female kids have a huge number that have a high physical fitness performance, also male kids have a dominant in some physical motor component and anthropometric and also female dominate some of the physical motor component.

A. Nadzmi · A. B. H. M. Maliki (✉) · S. Rustam · J. Jakiwa
Defence Fitness Academy, National Defence University of Malaysia, Kem Sungai Besi, 57000 Kuala Lumpur, Malaysia
e-mail: bisyri@upnm.edu.my

M. R. Abdullah
East Coast Environment Research Institute, University of Sultan Zainal Abidin, Gong Badak Campus, 21300 Terengganu, Malaysia

R. M. Musa
Department of Basic Knowledge and Liberal Studies, Universiti Malaysia Terengganu, 21300 Kuala Nerus, Terengganu, Malaysia

M. A. R. Zainoddin · I. M. Puspitasari · N. F. A. Jibril · N. A. Nawi · M. S. N. Abu Hassan
Faculty of Applied Social Sciences, Universiti Sultan Zainal Abidin, Gong Badak Campus, 21300 Kuala Nerus, Terengganu, Malaysia

I. Syahril
Faculty of Sport Science, Sultan Idris Education University, 35900 Perak Darul Ridzuan, Tanjong Malim, Malaysia

S. K. Syed Ali
Faculty of Education, University of Malaya, Jalan Profesor Diraja Ungku Aziz, 50603 Kuala Lumpur, Malaysia

© The Author(s), under exclusive license to Springer Nature Singapore Pte Ltd. 2023
S. F. Syed Omar et al. (eds.), *Innovation and Technology in Sports*, Lecture Notes in Bioengineering, https://doi.org/10.1007/978-981-99-0297-2_11

As a result, these models have the potential to reduce the number of kids with poor motor development. Furthermore, time and efforts can be saved because it is much easier to have concentrated parameters or those that have been extensively proven.

Keywords Motor performance · Primary school · Talent identification · Physical fitness · Health sustainability

1 Introduction

Motor performance is comprised of components linked to fitness as well as basic motor abilities. Children and adolescents' motor performance also crucial to their psychological and social development, and it has been acknowledged as a significant predictor of both their present and future health [1, 2]. An important health measure that is favorably correlated with physical activity and may possibly be connected to injuries is motor performance [3, 4]. From this statement, motor performance is a crucial part that needed to get a high attention from authorities that contributes to children's health development. At age 7, the development of motor skills is at its peak, but it is dependent on a number of things, including the quantity of training and learning the person has received [5].

Nowadays, children live through multiple encounters with children of the previous generation. Important inequalities can be seen in terms of increased experimentation opportunities, more time spent on television and video games, and lack of humanistic self-growth due to the rapid and widespread development and access to new media technology due to the internal setting [6]. Unrestricted overuse of technology among kids can lead to negative things such as social media app addiction and online games [2].. Throughout the disabled, children are becoming less active because of a sedentary lifestyle and lead to low efficiency of exercise [7]. Low fitness efficiency can lead to a variety of diseases, and obesity is one of today's key disease issues among children.

Worldwide, nearly 170 million kids are obese [8]. 12.7% of children in Malaysia are obese and Malaysia is the second-highest in Asia after Brunei. [9]. Obesity has short-term and long-term impacts mostly on health and social problems [10]. Since this risk of childhood obesity is less likely, childhood obesity problems can lead to a number of health concerns as children grow up. A big concern, leading to chronic disease, disability and post-hearing death, is the obesity epidemic.

In Malaysia, obesity among school students has shown a large gap in obesity among students living in urban and rural areas, where the average obesity among rural students is lower. Another study found that male primary school students were even more overweight than female students. In addition, students who still practice and compete in school sports showed lower rates of overweight and obesity than students who do not practice and do not participate in school sports [11].

Obesity also giving the consequences effect on the health-related quality life. Obesity in childhood and adolescence can decrease the quality of health-related quality life and lead to an inimical effect on their life including their social function [8, 9]. Beauchamp & Anderson [10] stating social function is described as the way a person functions through social skills and interactions with others in a social setting. Interference on the social function because of obesity will make the person become psychological distress, social isolation and reduced self-esteem and with the interference to the social function will possibly lessen the individual quality of social life [12].

It is important to understand how motor performance developments works in daily life especially children. A few studies stating that showing the relation between motor performance and physical activity levels as well as healthy weight status in children, improving motor performance and this relation may be beneficial to health [13–16]. Motor performance is not only developed for only better physical health but also in mental health as stated from research stating children who engaged in more motor performance activities in childhood had better cardiovascular and skeletal muscle health, less adiposity, and improved levels of anxiety, depression, and academic achievement [17] as the children development is strongly emphasized on two aspects of health, physical and mental health.

By developing motor performance index, it will help the authorities that responsible to take care of the children's health especially teachers at the school and the child's parent on monitoring the children's health by creating a statistical baseline that can help those authorities to monitor the children's weight status and motor performances. The objective of this study is to develop a motor performance index among 7 years old Malaysian kids based on their anthropometry and the result of a physical fitness test.

2 Methodology

2.1 Area of Study

This study examined Malaysia's Physical Fitness Among kids results. This included 217 primary schools in peninsular Malaysia, with 370 primary schools participating in Malaysia Borneo. Such schools include National Schools (SK), Religious Schools (SA), Chinese National Type Schools (SJK(C)) and Tamil National Type Schools (SKJ(T)). The age of these kids is 7 years old. There are 786 males in peninsular Malaysia, 213 in Malaysia Borneo, 806 females in peninsular Malaysia, 193 in Malaysia Borneo.

2.2 Participating and Testing Procedure

Out of 587 primary schools in Malaysia, total of 1998 kids (999 males and 999 females) were participated in this research. Several anthropometric components are tested (weight, height, and BMI) and four motor subscales are tested, such as power, flexibility, coordination, and speed. Procedures for kid's anthropometric measurements and motor fitness tests were conducted as follows.

Parents, guardians, school administrators, and participants are granted this type of consent to clarify certain issues like study methods, research goals, and others. Participants agreeing to participate must voluntarily assist in data collection research. Some of the main components to be fulfilled by the parents under the consent form include participants personal data and contact details; emergency contact information; health information and parent consent assurance. It is only for research purposes that this knowledge is available. Not all sensitive information is disclosed by the investigator.

2.3 Ethic Clearance

Informed written consent from participants was obtained by the writers. Participants were mandatory due to the requirement of physical education subject in Malaysia's school teaching and learning session. The authors value human subjects' privacy rights as an ethical research team. Therefore, the data submitted does not classify participants and has been entirely anonymous and contains no data to identify participants.

2.4 Anthropometrics

The instruments used are like a stadiometer and a weighting scale for the simple weight and height measurement [18]. The height feature was calibrated to the nearest 0.1 cm and kg was calibrated for the weight feature. When standing on the flat wall with the rear position and the arms crossed with the palms facing the researcher, the height is determined by the foot to the head. The stadiometer is the instrument needed.

2.5 Standing Broad Jump (SBJ)

The participants will be standing on the ground behind a line of slightly separated legs. With the arms swinging back and the knees bent rhythmically to about 90 degrees forward, a two-footer departure and landing was used [2]. Without falling backwards, participants must attempt to jump to their feet [19]. If the participants make a double jumping error before jumping, this test (SJB) will be cancelled. Three trials were accepted and the most far-reaching variables were taken into account.

2.6 Sit and Reach (SAR)

The participants are seated on the floor with straight legs, with heels flat against the sitting area and touching the rack and the investigator keeps both sides of the knees flat against the floor, if need be. In the case of hands-on tops and palms facing each other, it is then possible to approach the participants easily with fingertips that move the measuring glass as far as possible across the measuring line [20]. The investigator must guarantee that there are no twitching movements and that the fingertips stay parallel to the thighs. It is appropriate to take readings of 0.5 cm in multiples.

2.7 Hand Wall Toss (HWT)

The gap is approximately 1 m from the ground between the markers. Behind the marker line, the participants will stand straight, facing the wall. The ball is thrown with one hand by pressing the underarm against the wall and attempting to catch the ball with the opposite hand. Then the ball is thrown back to the wall and first-hand caught. This test proceeds with 10 attempts. It will register the number of shots captured.

2.8 Twenty-Meter Speed Test (20MR)

A reported chance for participants to run a single maximum sprint over a given length. The gap was 20 m between the start and the finish. Time started to count when either foot had reached the starting point, and either foot had hit the finishing point. Before the experiment, participants must ensure that the starting position is consistent without any modification, starting from a stationary pose with a foot back to the floor [10]. The time required for each split distance (20 m) to run was measured using a stopwatch [6].

2.9 Principal Component Analysis (PCA)

In this study, researchers used PCA to reveal essential motor performance components for kids. The components selected by researchers include weight, height, BMI, power, flexibility, coordination and speed for males and females. The researcher used PCA to eliminate the less important variables with a lower load factor. The attribute is considered non-significant if it has a separate eigenvalue of the data set. Factor scores results after the varimax rotation will be used for evaluating motor performance index for kids. The research is planned to classify the most important among 7 years old Malaysian kids. This analysis will assist researchers in identifying key sources of physical fitness by the scope of the study. The research would be carried out separately, based on gender differences.

2.10 Index

An index is a method of structured monitoring of the results of any group of properties. Indexes are often produced for the calculation of other financial or economic data such as interest rates, inflation or output from development. Variables that formed high factor loading will be used in index analysis. Range, score and clustering will appear as an index result. The index can also be used to monitor 7 years old kid's motor performance. From the result of PCA after varimax rotation, factor scores for every observation in the dominant factors and the variability percentage of variance will be used to determine the motor performance for kids. After that, the negative score will transform to positive scores without changing the patterns with maximum value 1. The scores split up into three groups (high, low and moderate) by using univariate clustering.

Determine Motor Performance Index Score.

(a1n X b1) + (a2n X b2)

a = Factor scores after varimax rotation

b = Variability percentage of variance after varimax rotation

n = Sample of observation.

2.11 Discriminant Analysis

Based on index analysis, researchers need discriminant analysis to validate index results that were formed. Index result as the dependent variable while the independent variable was including age, weight, height, BMI, power, flexibility, coordination and speed. Based on the independent variable above, DA will discriminate them into three subgroups which are low, moderate and high. Variables are counted step by step in the forward step mode, starting with the greatest significant variable until no significant adjustments have been made. In the step-by-step reverse mode, variables are removed step by step starting with the least important variable until no significant changes have been achieved. The correct percentage of classification will be determined based on results from the confusion matrix table. From this table, the researcher will find the most dominant factor for this research.

2.12 Machine Learning

Machine learning is an evolving branch of computational algorithms that are designed to emulate human intelligence by learning from the surrounding environment [21]. The machine learning method is way easier to use, low costs and can analyze more specific data samples. From the result of Discriminant Analysis, Machine Learning will establish group prediction and differentiate the F-value from the group. Machine learning also will notify the researcher which kids have the potential to move from the actual group to the predicted group.

Several machine learning models were chosen such as kNN, Tree, Random Forest, SVM, Logistic Regression, Naive Bayes, and Neural Network as to identify the kids motor performance potential based on their class. Any model that can show highest precision will be chosen to be reported in this research findings.

3 Results

The outcome of this research will be to predict a gender-based motor performance index for Malaysian children aged 7 years. The PCA was evaluated to test the necessary components with a value greater than 1 (eigenvalue >1.0) prior to the main analysis. Figures 1 and 2 indicate that, due to higher own values (>1), PCA identified two components as the most important. For varimax rotation, further research was performed by adding two additional latent variables to the necessary understanding.

The PCA pattern for males is seen in Table 1 after the varimax rotation, while the PCA pattern for females is shown in Table 2. With a combined variance of 53.49 percent, the result of the variance can be seen at 32.60 percent for D1 and D2 (20.89 percent). Males and females each have three main components that surpass the physical characteristics of the loading threshold factor (loading threshold factor

Fig. 1 Scree plot of descriptive eigenvalue for male

Fig. 2 Scree plot of descriptive eigenvalue for female

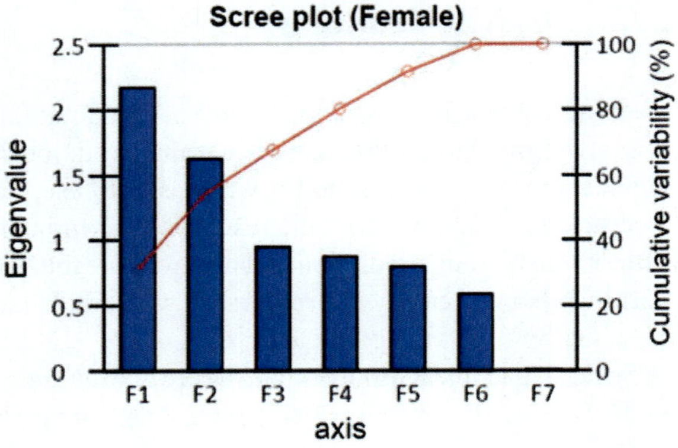

>0.65), which is weight, height, and BMI. The three-factor represents a powerful achievement in the exercise. The health output of children was important to evaluate because it was correlated with weight, height and BMI values. The epidemic of obesity can lead to a number of health issues as children grow up, according to previous studies. While the various components are described as the domain's motor fitness attribute, the power and speed of the second factors are clarified. This outcome demonstrates the success factors in gender, weight, height and BMI for fitness. Based on the cumulative variance percent in the factor loading pattern after varimax rotation in the table below, showing that the value is 54.2% and it can still relevant because in the humanities science field, the explained variance is generally set as low as 50–60%, so based on the range set, the cumulative variance percent is still can be accepted [22] (Table 2).

Furthermore, further study was calculated by the acquisition of the fitness index by incorporating the performance of the PCA. Three separate categorical fitness sets are shown in Table 3, including a low, moderate and high fitness index for

Table 1 Factor loading pattern after Varimax rotation (factor loading set at >0.65) for male

	• D1	• D2
• WEIGHT (kg)	• 0.9899	• –0.0487
• HEIGHT (cm)	• 0.6750	• 0.0895
• BMI (kg/m2)	• 0.8611	• –0.0988
• POWER (cm)	• –0.2271	• 0.7142
• FLEXIBILITY (cm)	• 0.0120	• 0.4267
• COORDINATION (no.)	• 0.1253	• 0.5764
• SPEED (sec)	• 0.0183	• –0.6744
• Eigenvalue	• 2.2818	• 1.4624
• Variability (%)	• 32.5965	• 20.8916
• Cumulative %	• 32.5965	• 53.4880

Table 2 Factor loading pattern after Varimax rotation (factor loading set at > 0.65) for female

–	– D1	– D2
– WEIGHT (kg)	– 0.9933	– 0.0014
– HEIGHT (cm)	– 0.6227	– 0.0904
– BMI (kg/m2)	– 0.8623	– 0.0492
– POWER (cm)	– 0.0967	– 0.7913
– FLEXIBILITY (cm)	– 0.1585	– 0.4613
– COORDINATION (no.)	– 0.0722	– 0.5538
– SPEED (sec)	– 0.1079	– 0.6889
– Eigenvalue	– 2.1704	– 1.6297
– Variability (%)	– 31.0051	– 23.2821
– Cumulative %	– 31.0051	– 54.2872

males based on the most dominant components. 187 male children are in an index-dependent high health index group, while 490 male children are in a moderate group and 322 male children are in a population with a low fitness index. The disparity in exercise intensity of the participants was further assessed based on the results of the health index. Based on the most dominant components, Table 4 reveals three distinct categorical fitness sets, including a low, moderate and high female fitness index. A high health index group based on an index is 158 female children, while a moderate group is 458 female children and a low fitness index group is 383 female children.

The precision of the above studies is presented in this section and the classification of training samples is given (N = 1998). To reiterate, the training dataset was used to match the model as the reference. Training dataset classification findings were obtained by discriminant analysis. Malaysian children in the high, moderate, low range of 7-year-old classes were treated as dependent variables, whereas relative success variables were treated as independent variables.

Table 3 Index status of fitness index for male

Status	Freq	Cum.Freq	%	Cum.%	Group Range	Group
0.00	322	322	32.23	32.23	0.00 ≤ Low ≥ 0.43	Low
0.44	490	812	49.05	81.28	0.44 ≤ Moderate ≥ 0.60	Moderate
0.60	187	999	18.72	100	0.60 ≤ High ≥ 1.00	High

Table 4 Index status of fitness index for female

Status	Freq	Cum Freq	%	Cum. %	Group range	Group
0.00	383	383	38.34	38.34	0.00 ≤ Low < 0.39	Low
0.39	458	841	45.85	84.19	0.39 ≤ Moderate < 0.58	Moderate
0.58	158	999	15.82	100.00	0.58 ≤ High ≤ 1.00	High

The specificity of the training dataset classification among the male group analyzed (N = 999) was shown in Table 4 below. The regular DA mode allocated 187, 322 and 490 males in the high, low and intermediate grades, with 91.79% of classification correctness (seven discriminant variables) respectively. Forward stepwise DA mode allocated 143, 297 and 559 males in the strong, low and moderate groups, respectively, with a classification correctness of 95.30% (seven discriminant variables). Finally, the DA backward mode allocated 143, 297 and 559 males in the high, low and moderate groups, respectively, with the correctness of the classification being 95.30% (seven discriminant variables). Results had confirmed the ability of newly developed motor performance index to classify male according to their parameters.

The specificity of the training dataset classification among the female group analyzed (N = 999) was seen in Table 5 below. The regular DA mode allocated 458, 383 and 158 females in the strong, low and moderate groups, respectively, with a classification correctness of 93.39% (seven discriminant variables). DA's forward stepwise mode allocated 508, 365, and 126 females in the strong, medium, and moderate grades, with 95.60% of classification correctness (seven discriminant variables) respectively. Stepwise DA mode allocated 508, 365 and 126 females in the strong, low and moderate grades, respectively, with a classification correctness of 95.60% (seven discriminant variables). The newly created motor performance index's ability to distinguish females according to their parameters was verified by the findings (Table 6).

Table 5454 shows the evaluation result of machine learning based on the selected model for both gender. It shows that for male evaluation, logistic regression model shows the highest precision with value 0.982, and for female also showing logistic regression model as having the highest precision with value 0.986 (Tables 7 and 8).

Table 5 Confusion matrix results for male

DA mode	Category	High	Low	Moderate	Total	% Correct
Standard	High	141	0	46	187	75.40
	Low	0	292	30	322	90.68
	Moderate	2	4	484	490	98.78
	Total	143	296	560	999	91.79
	High	121	0	22	143	84.62
Forward stepwise	Low	0	281	16	297	94.61
	Moderate	5	4	550	559	98.39
	Total	126	285	588	999	95.30
	High	121	0	22	143	84.62
Backward stepwise	Low	0	281	16	297	94.61
	Moderate	5	4	550	559	98.39
	Total	126	285	588	999	95.30

Table 6 Confusion matrix results for female

DA mode	Category	High	Low	Moderate	Total	% Correct
Standard	High	450	8	0	**458**	98.25
	Low	26	357	0	**383**	93.21
	Moderate	232	0	126	**158**	79.75
	Total	**508**	**365**	**126**	**999**	**93.39**
Forward stepwise	High	499	9	0	**508**	98.23
	Low	15	350	0	**365**	95.89
	Moderate	20	0	106	**126**	84.13
	Total	**534**	**359**	**106**	**999**	**95.60**
Backward stepwise	High	499	9	0	**508**	98.23
	Low	15	350	0	**365**	95.89
	Moderate	20	0	106	**126**	84.13
	Total	**534**	**359**	**106**	**999**	**95.60**

Table 7 Evaluation result of machine learning model for male

Model	AUC	CA	F1	Precision	Recall
Logistic regression	0.999	0.982	0.982	0.982	0.982
Neural network	0.999	0.980	0.980	0.980	0.980
SVM	0.995	0.943	0.943	0.943	0.943
Random forest	0.959	0.856	0.856	0.857	0.856
kNN	0.931	0.824	0.823	0.827	0.824
Tree	0.854	0.808	0.808	0.808	0.808
Naive Bayes	0.908	0.768	0.770	0.775	0.768

Table 8 Evaluation result of machine learning model for female

Model	AUC	CA	F1	Precision	Recall
Logistic Regression	0.999	0.986	0.986	0.986	0.986
Neural network	0.999	0.982	0.982	0.982	0.982
SVM	0.996	0.948	0.948	0.948	0.948
Random forest	0.961	0.869	0.869	0.873	0.869
kNN	0.949	0.856	0.855	0.858	0.856
Tree	0.866	0.815	0.815	0.815	0.815
Naive Bayes	0.929	0.800	0.802	0.806	0.800

Based on the Machine Learning Confusion Matrix Table on the male table, the overall prediction accuracy is 99% and for female, the overall prediction accuracy is 97%. For male prediction, there is only one kid from moderate class motor performance predicted that are potentially to achieve a high motor performance capability. For female prediction, there is one kid predicted that are potentially dropped her motor performance from high performance to low performance but also there is one kid also predicted that potentially increase her motor performance level from low performance to high performance (Tables 9 and 10).

Table 8 shows the analysis of differences and significance between the seven variables (weight, height, BMI, power, flexibility, coordination and speed. All variables for males and females were significant respectively with the group high, low and moderate. Based on the ANOVA result, there is a significant difference between males and females. Male significantly dominate females in the physical attribute for weight with f-value (242.638), height (617.784), BMI (329.497), coordination (88.058) and speed (59.167) while female significantly dominate male in physical attributes for power with f-value (89.405) and flexibility (69.529) (Table 11).

As shown in Table 9, the researchers managed to prove the prediction on the development of motor performance index among 7 years old Malaysian kids by using PCA, Index, DA and Machine Learning to monitor 7 years old kids. This would also help to improve the development of motor performance among 7 years old. Therefore, these models can decline the population of kids who have low development of motor performance. Moreover, the time and energy consumed can be reduced as it is much easier to have focused parameters or the ones which have been significantly proven [23, 24].

Table 9 Result of male prediction in machine learning

			Predicted		
		High	Low	Moderate	Σ
	High	186	0	1	187
Actual	Low	0	322	0	322
	Moderate	0	0	490	490
	Σ	186	322	491	999

Table 10 Result of female prediction in machine learning

			Predicted		
		High	Low	Moderate	Σ
	High	457	1	0	458
Actual	Low	1	382	0	383
	Moderate	0	0	158	158
	Σ	458	383	158	999

Table 11 Summary statistics of motor performance index for male and female

	Male (f-value)	Female (f-value)
Weight (kg)	242.638*	181.722*
Height (cm)	617.784*	567.779*
BMI (kg/m2)	329.497*	293.355*
Power (cm)	36.925*	89.405*
Flexibility (cm)	27.646*	69.529*
Coordination (no.)	88.058*	71.238*
Speed (sec)	59.167*	49.961*
SIGNIFICANT	(p = 0.000, N = 999)	(p = 0.000, N = 999)

4 Conclusion

Based on this study, the expansion of the motor performance index of 7-year-old Malaysians shows that the ability of the newly developed motor performance index to classify both sexes depend on its parameters. This study will help parents or school authorities to improve motor skills at the age of 7 years. Therefore, these models have the potential to reduce the population of children with low motor performance development.

Acknowledgements In collaboration with the National Sports Institute (ISN), Sultan Zainal Abidin University (UniSZA) has obtained data on physical fitness among children. A grant for this research was given by the National Sports Institute (ISN). The full funding for this research is given by this institute. For the success of this study, the researchers would like to thank parents, guardians, school authorities and participants for their good cooperation.

References

1. Larsen LR, Kristensen PL, Junge T, Rexen CT, Wedderkopp N (2015) Motor performance as predictor of physical activity in children: the CHAMPS studyDK. Med Sci Sports Exerc 47:1849–1856
2. Larsen LR, Kristensen PL, Junge T, MØller SF, Juul-Kristensen B, Wedderkopp N (2016) Motor performance as risk factor for lower extremity injuries in children. Med Sci Sports Exerc 48:1136–43
3. Lubans DR, Morgan PJ, Cliff DP, Barnett LM, Okely AD (2010) Fundamental movement skills in children and adolescents: review of associated health benefits. Sports Med 40:1019–1035
4. Smith JJ, Eather N, Morgan PJ, Plotnikoff RC, Faigenbaum AD, Lubans DR (2014) The health benefits of muscular fitness for children and adolescents: a systematic review and meta-analysis. Sports Med 44:1209–1223
5. Jakiwa J, Suppiah PK (2020) The differences between gross motor performance amongst children according to ethnic and age chronology. Malaysian J Move Health Exerc 9(1):159
6. Nadzmi A, et al (2021) Comparison between gross motor skills performance and 7-years-old children with controlled effect of residential area. In: First international conference on science, technology, engineering and industrial revolution (ICSTEIR 2020). Atlantis Press

7. Azahari H, Abdullah MR, Husin AB, Eswaramoorthi V, Juahir H, Musa RM, Zawi K (2019) Physical fitness and anthropometric characteristics in school children: a comparison of urban and rural areas in east coast of Peninsular Malaysia Indian. J Public Heal Res Dev 10:619–623
8. Juahir H, Ghazali A, Bisyri A, Maliki HM, Sultan U, Abidin Z, Zolkipli HM, Juahir H, Adiana G, Zainuddin N, Ismail A, Maliki ABHM, Hussain NI, Kamarudin MKA, Toriman ME, Mokhtar M (2018) Spatial assessment and the most significant parameters for drinking water quality using chemometric technique: a case study at malaysia water treatment plants Mohd khairul amri Kamarudin spatial assessment and the most significant parameters for drinking water quality using chemometric technique: a case study at malaysia water treatment plants artic. Int J Eng Technol 7:115–122
9. Kassim M, Ujang E, Othman N (2020) The association between physical activity and basic fitness tests in female junior Reserve Officer Training Unit (ROTU), National Defence University Malaysia (NDUM). J Phys Conf Ser 1529:022073
10. ABHM, Abdullah MR, Juahir H, Muhamad WSAW, Nasir NAM, Musa RM, Mat-Rasid SM, Adnan A, Kosni NA, Abdullah F, and Abdullah NAS (2018) The role of anthropometric, growth and maturity index (AGaMI) influencing youth soccer relative performance. In: IOP conference series: materials science and engineering, vol 342. Institute of Physics Publishing
11. Abu Hassan MSN, Ismail NH, Abd Halim NS, Abd Aziz M, Mohd Sidik NH, Kamarudin MKA, Abu Bakar N, Maliki ABHM (2019) Research on male and female youth perceptions on the use of pornographics displays in social media 285. J Recent Technol Eng
12. Maliki ABHM, Abdullah MR, Juahir H, Abdullah F, Abdullah NAS, Musa RM, Mat-Rasid SM, Adnan A, Kosni NA, Muhamad WSAW, Nasir NAM (2018) A multilateral modelling of Youth Soccer Performance Index (YSPI). In: IOP conference series: materials science and engineering, vol 342, p 012057. Institute of Physics Publishing
13. Lubans DR, Morgan PJ, Cliff DP, Barnett LM, Okely AD (2010) Fundamental movement skills in children and adolescents review of associated health benefits. Sports Med 40:1019–1035. https://doi.org/10.2165/11536850-000000000-00000
14. D'Hondt E et al (2014) A longitudinal study of gross motor coordination and weight status in children. Obesity 22:1505–1511. https://doi.org/10.1002/oby.20723
15. Ortega FB, Ruiz JR, Castillo MJ, Sjöström M (2008) Physical fitness in childhood and adolescence: a powerful marker of health. Int J Obes 32(1):1–11
16. Jibril NFAB, et al (2021) Comparison of body mass index class against motor performance towards 7-year-old boys in Malaysia. In: First international conference on science, technology, engineering and industrial revolution (ICSTEIR 2020). Atlantis Press
17. Buttitta M, Iliescu C, Rousseau A, Guerrien A (2014) Quality of life in overweight and obese children and adolescents: a literature review. Qual Life Res 23:1117–1139
18. Zeller MH, Inge TH, Modi AC, Jenkins TM, Michalsky MP, Helmrath M, Courcoulas A, Harmon CM, Rofey D, Baughcum A, Austin H, Price K, Xanthakos SA, Brandt ML, Horlick M, Buncher R (2015) Severe obesity and comorbid condition impact on the weight-related quality of life of the adolescent patient. J Pediatr 166:651-659.e4
19. Beauchamp MH, Anderson V (2010) SOCIAL: An Integrative Framework for the Development of Social Skills. Psychol Bull 136:39–64
20. Rosema S, Crowe L, Anderson V (2012) Social function in children and adolescents after traumatic brain injury: a systematic review 1989–2011. J Neurotrauma 29:1277–1291
21. El Naqa I, Murphy MJ (2015) What is machine learning? Springer, Cham
22. Taherdoost HAMED, Sahibuddin S, Jalaliyoon NEDA (2022) Exploratory factor analysis; concepts and theory. Adv Appl Pure Math 27:375–382
23. Kassim M, Othman N, Ujang E (2020) The level of physical fitness among first year female students in national defence University of Malaysia. J Phys Conf Ser 1529:022072
24. Kassim M, Rustam S (2020) Measuring physical fitness for military cadets officers. Adv Intell Syst Comput 1183:554–561

An Analysis of Ranking for Football Teams in Malaysia Super League Based on Football Rating System

Nazim Razali and Aida Mustapha

Abstract The analysis of football has always offered great interest and attract many people among experts, researchers, pundits, and fans. Football data have been observed and studied from various perspective aiming for several objectives whether for predicting matches results or goals as well as analyzed team or player performance. This paper presents an analysis and discussion for team ranking in Malaysia Super League (MSL) 2021 based on football rating system. The football dataset is limited to seven seasons of MSL football data between 2015 and 2021, however, the study mainly focuses on final ranking in MSL league table for season 2021 based on football rating system consist of Elo rating, pi-rating and Poisson model. Each of the football rating system presented have their own unique calculation to introduce the football team rating whether as whole team for every round of league's matches (Elo rating), rating while home or away (Pi-rating) and rating for attack and defense (Poisson model). The football rating system mainly rely on the number of goals scored, goals conceded and match results whether at home and away to be evaluated for rating the football team strength in term of attack, defense, or team as whole. Thus, the findings show that Johor Darul Ta'zim Football Club successfully become prominent football club that dominate MSL 2021. Moreover, it is suggested to include other tier of Malaysian Football Leagues (MFL) as well as adjustment of football rating system for optimization to suit the MFL environment.

Keywords Malaysia Super League · Elo Rating · Poisson model · Pi-rating · Football rating system

N. Razali
Faculty of Computer Science and Information Technology, Universiti Tun Hussein Onn Malaysia, Parit Raja, Malaysia

A. Mustapha (✉)
Faculty of Applied Sciences and Technology, Universiti Tun Hussein Onn Malaysia, Parit Raja, Malaysia
e-mail: aidam@uthm.edu.my

© The Author(s), under exclusive license to Springer Nature Singapore Pte Ltd. 2023

S. F. Syed Omar et al. (eds.), *Innovation and Technology in Sports*, Lecture Notes in Bioengineering, https://doi.org/10.1007/978-981-99-0297-2_12

1 Introduction

The sports analytics and sports big data already move along as the time flown with rapid growth of information technology [1]. According to [2], sports analytics can be defined data management, analytic models and information systems that assist the entity such as managers, coaches, trainers to make decision in gaining the competitive advantage. Its cover both performance analysis and predictive analysis. The development of sports analytic in association football has always offered great interest and attract many people among experts, researchers, pundits, and fans all around the world. As most popular sports in the world, football data have been observed and studied from various perspective aiming for several objectives whether for predicting matches results [3] or goals [4] as well as analyzed team [5] or player performance [6].

This paper attempt to analyze and discuss team ranking in Malaysia Super League (MSL) 2021 based on football rating system. The football rating system can be related with performance of a football team as whole regardless the individual player performance. Thus, it will give a picture about current or previous football team performance in football leagues competition. Hopefully, exploring football rating system in MSL through this paper may assist and give insight to local researchers so that they manage develop new or improve existing sports analytics in Malaysian football whether for performance analysis and predictive analysis. The remaining paper are organized as follows: Sect. 2 presents background study on Malaysian football and sports analytics; Sect. 3 describes the methodology or workflow of studies; Sect. 4 describe the football data of MSL; Sect. 5 an overview and introduction to football rating system such as Poisson model, Elo Rating and Pi-rating; Sect. 6 analyze and discuss the findings of team ranking in Malaysia Super League (MSL) 2021 based on football rating system; Finally, the conclusion in Sect. 7.

2 Background

The association football or also known as football or soccer also have high popularity in Malaysia. The history of football in Malaysia started at the end of nineteenth century during British colonization in Malaya (name before independence of Malaysia) until formally organized in 1936 as Football Association of Selangor which limit to states until its several extensions to central football association such as Malayan Football Association (MFA), Football Association of Malaya (FAM), and Football Association of Malaysia (FAM) in 1960 with the first Prime Minister of Malaysia, Tunku Abdul Rahman become its president and Malaysia football leagues has grown since then [7].

There are 3-tier well known league in Malaysia or also known as M-League which are the Malaysia Super League or Liga Super Malaysia (MSL), Malaysia Premier League or Liga Perdana Malaysia (MPL) and Malaysia FAM League or Liga FAM Malaysia which later known as Malaysia M3 League or Liga M3 Malaysia. The Malaysia Super League (MSL) is the top tier in M-League consist of 12 teams that made up of state teams and clubs which contest for champion title as well as to secure AFC Champion League slot, AFC Cup slot and avoiding relegation to MPL [7–9].

There are several academic literatures related to Malaysia Football have been produced past few years. Interestingly, there are increase in the number of works related football in Malaysia after year 2013 which may relate to revolution and privatization of football in Johor carried out by Tunku Ismail Ibni Sultan Ibrahim (the State Crown Prince of Johor). According to [2], only six works on sports analytics for MSL were published between 2010 and 2018 through SCOPUS database by keyword on query of "Malaysia Super League", "Malaysia" AND "Sports Analytics" and "Malaysia" AND "Football" and five of them are produced after 2013. The review by [2] focusing on diverse fields of study such as goal pattern analysis, stadium attendance prediction, decision support system, and football team ranking prediction in MSL table.

There are also literatures found via Google Scholar [10] where also focuses various fields of study such as analysis on ball possession [11], analysis on player satisfaction and performance [12], player physiology [13], coaching management [14, 15] and talent identification [16]. As a result, the focus of this paper is to study and discuss the team rankings in the Malaysia Super League (MSL) 2021 based on a football rating system.

3 Methodology

The methodology section present the workflow of research, dataset, football rating system, data visualization, analysis and finally report of findings of research. The MSL football data first will be collected and cleaned through data preprocessing phase before be calculated to specific football rating system. Then, the dataset was visualized for analysis and reporting of findings. Figure 1 shows the research workflow of ranking for football teams in MSL based on football rating system. Note that the football data are extracted via online from [17] which later will be explained in brief in Sect. 4.

The football rating system calculation consist of three parts which are the implementation of three football rating system for MSL football team namely Elo rating, Pi-rating and Poisson model for MSL season 2015 to season MSL 2021. However, the final findings for analysis and discussion only focuses on MSL 2021.

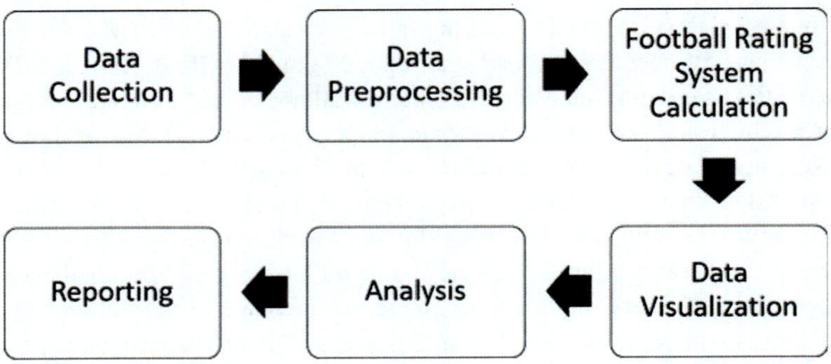

Fig. 1 A research workflow of ranking for football teams in MSL based on football rating system

4 Football Data

The football dataset only limited to 7 seasons of Malaysia Super League (MSL) from season 2015 to 2021. The dataset is collected and examined via online through [17] which consists of total 858 instances that divided into 6 basic data features such as date (Date), name of home team (HomeTeam), name of away team (AwayTeam), home team scored (FTHG), away team scored (FTAG) and the matches results in term of home win, draw or away win (FTR). Table 1 shows the description of MSL dataset used for the studies.

The football rating system calculation consist of three parts which are the implementation of three football rating system for MSL football team namely Elo rating, Pi-rating and Poisson model for MSL season 2015 to season MSL 2021. However, the final findings for analysis and discussion only focuses on MSL 2021. Noted that there are some changes occurs in MSL season and MSL 2020 only played as a single round robin (11 matches) due to interruption of Covid-19 pandemic. Figure 2 shows there are several changes occur in MSL season 2015 to season 2021. The name of football team that changes several times are updated to most recent name and needed

Table 1 The description of MSL dataset used for the studies

Features	Abbreviation	Description	Datatype	Example
Date	Date	The date of the match	Date	31/1/2015
Home team	HomeTeam	The team playing at home	Nominal	JDT
Away team	AwayTeam	The team playing at away	Nominal	Pahang FA
Home scored	FTHG	The number of goals scored by home team	Numeric	2
Away scored	FTAG	The number of goals scored by away team	Numeric	0
Matches results	FTR	The match results in term of home win (H), draw (D) and away win (A)	Nominal	H

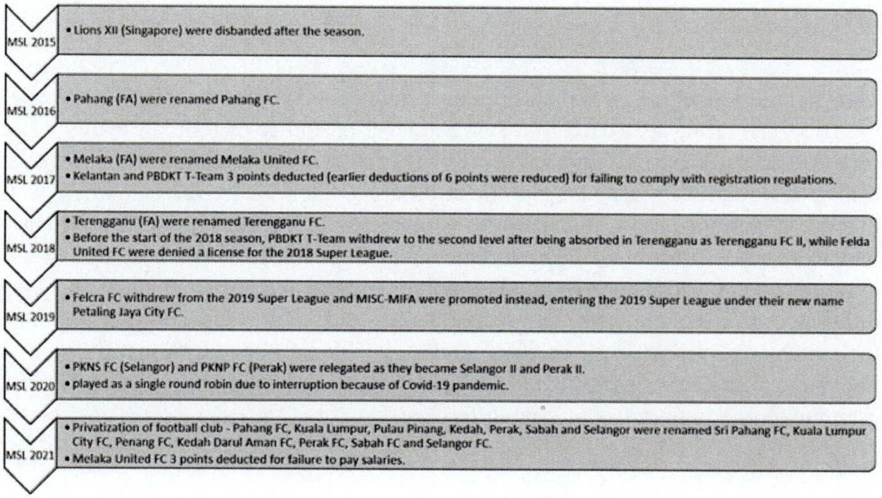

Fig. 2 The changes occur in MSL season 2015 to season 2021

to be consistent since the name of team will store the value of rating during football rating system calculation. However, the final findings of football rating system only concentrate on MSL season 2021.

5 Football Rating System

In this section, the football rating system such as Elo rating, Pi-rating and Poisson model will briefly be described.

5.1 Elo Rating

Basically, the Elo rating was introduced Arpad Elo [18] for rating chess player. In 1997, Bob Runyan [19] takes an initiative to extend and improvise the Elo rating to suit with association football. The Elo rating has become one of prominent football rating system thenceforth to rate football club strength. The Elo rating used in this paper will be based on work of [20] which also known as result-based Elo rating. The Elo rating consider the number of goals, match result and home advantage to rate the abilities of football team.

5.2 *Pi-Rating*

The [21] introduced a new football rating system called pi-rating through their work in 2013. The rating was used to assess the level of ability of football teams based on the relative disparities in goal scores between two competing football team in a match. In summary, pi-rating proposition are influenced by three football facts that are commonly used by experts to capture football team current performance which are home advantage, the importance of updated results versus outdated results during current performance estimation and the importance of win result versus goal difference. Consequently, the pi-rating operates differently depending on whether the match is played at home or away, and it considered newer information such as home and away performance to be more relevant than previous information, as well as reducing the importance of goal difference.

5.3 *Poisson Model*

The Poisson model was introduced by Maher [22] or also known as Maher model. He developed the model using Poisson distribution based on the number of goals scored and conceded while playing at home or away to rate the attacking strength and defending strength of the football club in English Premier League (EPL). Later, this Maher model has become foundation to another improvised Poisson Model such as [23–26].

6 Analysis and Discussion

In this work, the Malaysia Super League (MSL) dataset of 858 instances from season 2015 to season 2021 was employed. The dataset comprises six features, including the date, the name of the home team, the name of the away team, the home team's score, the away team's score, and the match results (home win, draw, away win), and it was calculated to rate the team's overall performance strength using three proposed football rating systems such as Elo rating, Pi-rating, and Poisson model. All the football rating system are calculated from MSL 2015; however, the final findings of football rating system only concentrate on MSL season 2021. All the football rating system have considered home advantage and there are several changes occurs in MSL dataset as described in Sect. 4. Table 2 shows the final ranking of the football teams in MSL 2021 based on football rating system.

Figure 3 revealed that SPFC, PRK, UiTM rating decrease significantly as the progress in MSL round and vice versa for PNG and KLC. Meanwhile, SEL seem inconsistence in performance when there is fluctuation in rating. JDT and KDA show

Table 2 The final ranking of the football teams in MSL 2021 based on football rating system

Football teams	Abbreviation	Actual	Elo rating	Pi-rating (Home)	Pi-rating (Away)	Poisson attack rating	Poisson defense rating
Johor Darul Ta'zim	JDT	57 (1)	1832.27 (1)	1.6411 (1)	1.4509 (1)	0.46 (1)	0.79 (1)
Kedah Darul Aman FC	KDA	43 (2)	1634.62 (2)	0.9436 (4)	0.7852 (3)	0.27 (2)	0.10 (4)
Penang FC	PNG	41 (3)	1519.69 (5)	0.1537 (8)	0.2330 (6)	−0.01 (8)	−0.70 (11)
Terengganu FC	TFC	38 (4)	1586.29 (4)	0.9724 (3)	0.6768 (4)	0.13 (5)	0.03 (6)
Selangor FC	SEL	36 (5)	1597.38 (3)	1.0140 (2)	0.8603 (2)	0.26 (3)	0.06 (5)
Kuala Lumpur City FC	KLC	33 (6)	1505.21 (8)	0.5752 (5)	0.0938 (7)	0.05 (7)	−0.16 (9)
Petaling Jaya City FC	PJC	24 (7)	1499.41 (9)	−0.1163 (10)	−0.5836 (12)	−0.28 (11)	0.17 (3)
Melaka United FC	MUFC	21 (8)	1518.77 (6)	0.3049 (7)	0.2765 (5)	0.06 (6)	−0.03 (7)
Sabah FC	SBH	19 (9)	1451.18 (12)	−0.0589 (9)	−0.3739 (10)	−0.26 (9)	−0.20 (10)
Sri Pahang FC	SPFC	18 (10)	1510.54 (7)	0.4028 (6)	−0.1063 (8)	0.15 (4)	0.14 (2)
Perak FC	PRK	16 (11)	1487.39 (10)	−0.2039 (11)	−0.3771 (11)	0.06 (6)	0.03 (6)
UiTM FC	UiTM	13 (12)	1460.12 (11)	−0.2845 (12)	−0.2611 (9)	−0.27 (10)	−0.10 (8)

Noted that, the MUFC points in MSL 2021 was deducted by 2 points due to failure for paying salaries. Table 2 revealed that the JDT successfully achieved 1st placed for all football rating system and match with the actual position in final ranking of MSL 2021. Meanwhile, the new promotion football team, PNG and KLC successful to avoid relegation. Although the rating value on PNG show that PNG worth only to only achieved mid table finish in MSL 2021, PNG amazingly shock others by successfully achieved 3rd placed in MSL 2021 and almost securing ticket to Asian Football Confederation (AFC) Cup by differ 2 points from 2nd place, KDA. The PRK and UiTM was unfortunate to have been relegated to MPL and both team records lowest value in Elo rating and pi-rating. The rating SEL and SPFC seem promising, however, the final position for SEL and SPFC in MSL 2021 quite disappointing and SPFC almost relegated to MPL which differ by 2 points from PRK

consistence in rating performance by securing 1st and 2nd place for MSL 2021 based on Elo rating which are same with actual final ranking in MSL 2021.

Pi-rating is differ from Elo rating. Elo rating measure football team strength as a whole but pi-rating measure football team at home and at away which means that a football team will have 2 rating according to venue of match whether played at

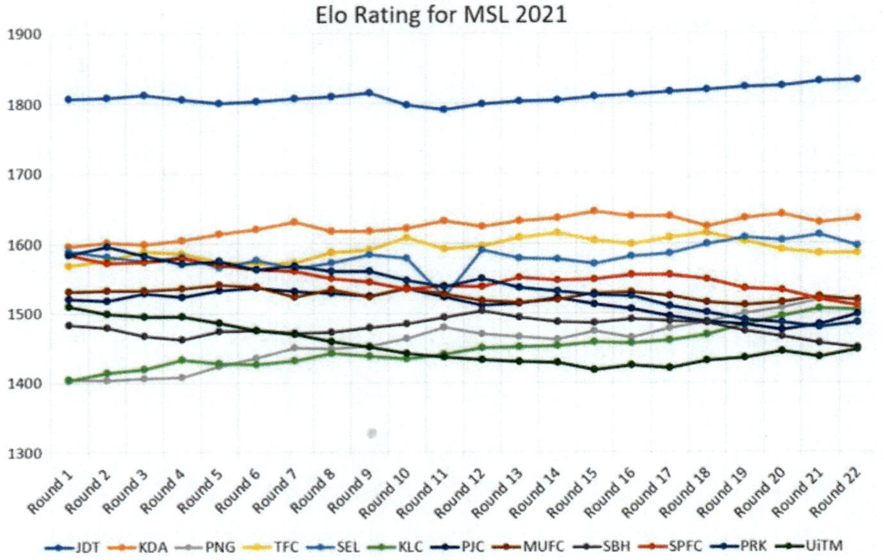

Fig. 3 The Elo rating for MSL season 2021

home or away. Figure 4 shows pi-rating at home for MSL season 2021. The Fig. 4 revealed that KDA, TFC and SEL competing to have high rating at home while JDT the highest remain unchallenged. PRK rating at home very disappointing as the chart show the rating decrease significantly while PNG and KLC show the opposite in the rating chart. The pi-rating at away show in Fig. 5 also revealed almost the same in Fig. 4, however, it seems all football team have disadvantage during played at away cause by the effect of home advantage by opponent team.

Like pi-rating and Elo rating, Poisson model also differ from them. Poisson model generate football team attacking strength and defending strength rating compared to pi-rating calculated and divided into venue and Elo rating calculated as whole rating. Figure 6 show majority of football team have consistent in attacking strength except 4 teams which are PNG, PJC, UiTM and SBH. The rating of PNG increase by round while PRK rating slowly decrease by round and the rating UiTM and SBH extremely inconsistent. For Poisson model on defense in Fig. 7 show that almost all team have consistent defense rating except UiTM and SBH. The UiTM gained low in rating for both attack and defense which make them obtained last spot in actual ranking of MSL 2021. Meanwhile, PRK have consistent and top 6 value in attack and defense rating. PRK still not manage to survive MSL 2021.

Overall, JDT obtained the highest value of rating in all football rating system while UiTM obtained lowest value of rating in all football rating system. It seems the constraint in financial can influence football team performance as both relegated PRK and UiTM facing financial problems [27, 28]. MUFC also facing 2 points deducted after failure to pay salaries in MSL 2021. However, there are some disappointments in SEL because SEL rating in all football rating system position them in top contender

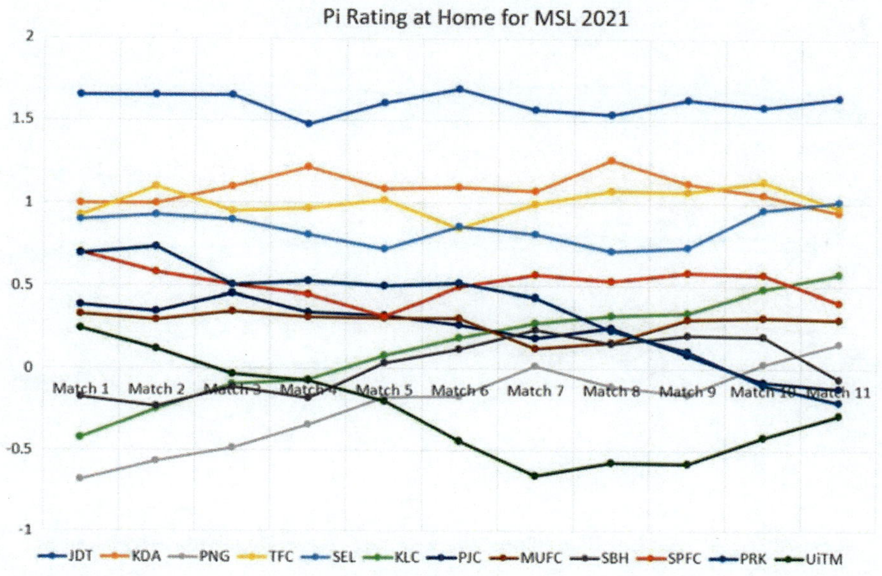

Fig. 4 The Pi-rating at home for MSL season 2021

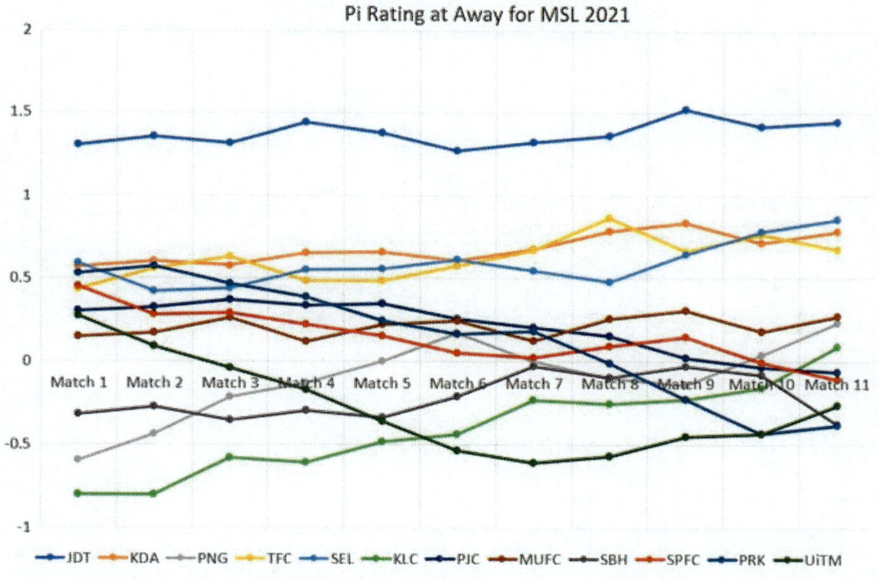

Fig. 5 The Pi-rating at away for MSL season 2021

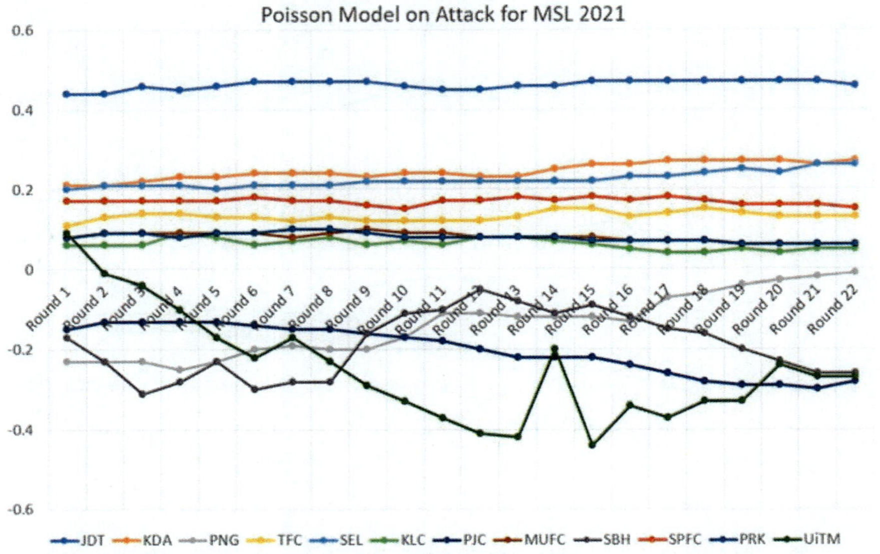

Fig. 6 The Poisson model on attacking strength for MSL season 2021

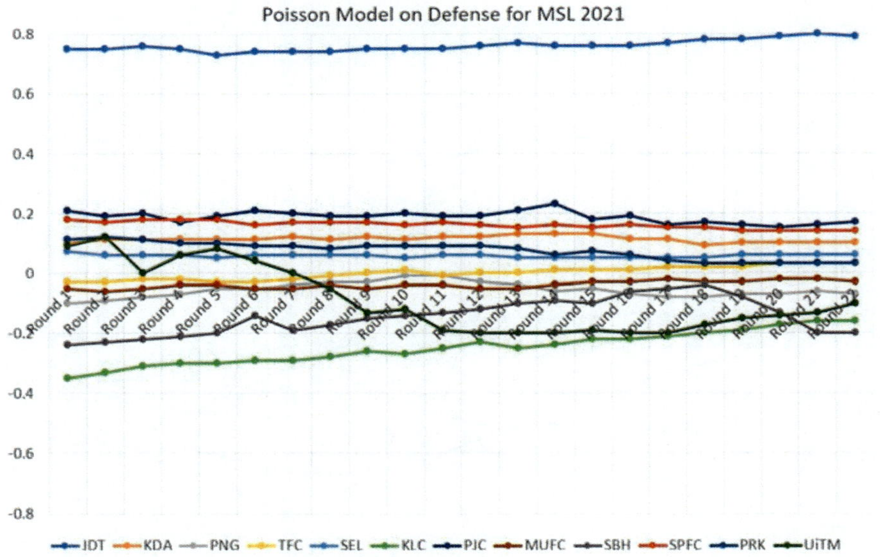

Fig. 7 The Poisson model on defending strength for MSL season 2021

to challenge top 3 in MSL 2021. Though anything can happen in football. There are many variations of factors can influence football team performance despite of good rating such as sudden management or financial problems and too many injuries among key players in middle of football season.

7 Conclusion

The paper presents the an analysis and discussion for team ranking in Malaysia Super League (MSL) 2021 based on football rating system. The football rating system mainly rely on the number of goals scored, goals conceded, goals difference and match results whether at home and away to be evaluated for rating the football team strength in term of attack, defense, or team as whole. Thus, the findings shows that Johor Darul Ta'zim Football Club (FC) successfully become prominent football club that dominate MSL 2021. The newcomer of Penang FC after its promotion surprising took 3rd places in MSL 2021 since its relegation to Malaysia Premier League (MPL) season 2017 surely impressive despite the fact Penang FC have low value in football rating system. Amazingly, the team only 2 points differ with Kedah Darul Aman Football Club in order to secure ticket to group stage for Asian Football Confederation (AFC) Cup. Although there are several football teams have high value of rating (strong team), the teams still failed to secure top position in MSL 2021 and vice versa. Moreover, it is suggested to include other tier of Malaysian Football Leagues (MFL) as well as adjustment of football rating system for optimization to suit the MFL environment.

References

1. Bai Z, Bai X (2021) Sports big data: management, analysis, applications, and challenges. Complexity
2. Razali N, Mustapha A, Soeed K, Othman M (2019) A review of sports analytics on malaysia super league (MSL). J Adv Res Dyn Control Syst 11(5 Special Issue):820–826
3. Dubitzky W, Lopes P, Davis J, Berrar D (2019) The open international soccer database for machine learning. Mach Learn 108(1):9–28
4. Wheatcroft E, Sienkiewicz E (2021) A probabilistic model for predicting shot success in football. arXiv
5. Li Y, Ma R, Goncalves B, Gong B, Cui Y, Shen Y (2020) Data-driven team ranking and match performance analysis in Chinese Football Super League. Chaos Solit Fract 141
6. Pappalardo L, Cintia P, Ferragina P, Massucco E, Pedreschi D, Gionnotti F (2019) PlayeRank: data-driven performance evaluation and player ranking in soccer via a machine learning approach. ACM Trans Intell Syst Technol 10(5):1–27
7. FAM History Page https://www.fam.org.my/history Accessed 07 Jan 2022
8. AFL About Page https://the-afl.my/?page_id=2083 Accessed 07 Jan 2022
9. MFL Homepage https://www.malaysianfootballleague.com/ Accessed 07 Jan 2022
10. Google Scholar https://scholar.google.com/ Accessed 07 Jan 2022

11. Miswan MS, Aznan EAM, Ismail Z, Jamaludin M, Mohd Kassim AF (2018) Ball possession of a successful team (Johor Darul Ta'zim FC) in Malaysia super league 2015: a case study. Malaysian J Movement Health Exerc 7(2):117–126
12. Zid A, Amiruddin R, Auzair SM, Janor H (2021) Comprehensive performance measurement system (CPMS) and satisfaction of Malaysian football players. Curr Psychol
13. Miswan MS, Sulaiman N (2017) Selected physiological profile among Malaysian 3rd tier fam league football players. Int J Asian Soc Sci 7(5):372–380
14. Karim ZA, Nadzalan AM (2017) Malaysia football coaches: development characteristics. Int J Acad Res Bus Soc Sci 7(9):305–312
15. Ramalu RR, Abd Karim Z, Elumalai G (2021) Malaysian state sports schools football coaching process key themes development: constant comparison method in data analysis. Pertanika J Soc Sci Human 29(3):1535–1549
16. Hanapiah KS, Hashim A, Karim ZA (2019) A Multidimensional assessment approach of talent identification in male youth malaysia football players. Int J Acad Res Bus Soc Sci 9(11):33–49
17. Stokkermans K (2015) Malaysia. http://www.rsssf.com/tablesm/malay2015.html#super Accessed 29 Mar 2022
18. Elo AE (1978) The rating of chess players. Past and Present. Arco Pubishing, New York
19. Runyan B (2022) World football elo ratings. https://eloratings.net/ Accessed 29 Mar 2022
20. Hvattum LM, Arntzen H (2010) Using ELO ratings for match result prediction in association football. Int J Forecast 26(3):460–470
21. Constantinou AC, Fenton NE (2013) Determining the level of ability of football teams by dynamic ratings based on the relative discrepancies in scores between adversaries. J Quant Anal Sports 9(1):37–50
22. Maher M (1982) Modelling association football scores. Stat Neerl 36(3):109–118
23. Dixon MJ, Coles SG (1997) Modelling association football scores and inefficiencies in the football betting market. J R Stat Soc Ser C Appl Stat 46(2):265–280
24. Rue H, Salvesen O (2000) Prediction and retrospective analysis of soccer matches in a league. J R Stat Soc Ser D (The Statistician) 49(3):399–418
25. Angelini G, De Angelis L (2017) PARX model for football match predictions. J Forecast 36(7):795–807
26. Ley C, de Wiele TV, Eetvelde HV (2019) Ranking soccer teams on the basis of their current strength: a comparison of maximum likelihood approaches. Stat Model 19(1):55–73
27. Azharie F Bos Gaurus in financial crisis, https://www.nst.com.my/sports/football/2021/01/659539/bos-gaurus-financial-crisis Accessed 07 Jan 2022
28. Msar S UiTM FC kembali 'bernyawa'. https://www.hmetro.com.my/arena/bola-sepak/2022/02/812957/uitm-fc-kembali-bernyawa Accessed 07 Jan 2022

Machine Learning Approach for Malaysia Super League Football Match Outcomes Prediction Based on Elo Rating System

Nazim Razali, Aida Mustapha, Amira Qistina Aiman A. Aziz, and Salama A. Mostafa

Abstract Predicting football match results or goals is unexceptionally buzzworthy. Football prediction can be classified into two clusters which are statistical and machine learning. Despite successfully introducing numerous statistical and machine learning techniques to predict football match outcomes, flaws still exist. This paper attempt to present football matches outcomes prediction models based on an Elo rating system and machine learning algorithms using limited data of football matches result for Malaysia Super League. The dataset used for the prediction is the MSL football data which consists of 7 seasons played between 2015 to 2021 that contain several basic features such as date, name of home team, name of away team, home team scored, away team scored and the matches results (Win, Draw, Loss). The football data were calculated to Elo rating that rate the strength of MSL football team before are divided into training and testing set. Machine learning (ML) algorithms such as Naïve Bayes (NB), Logistic Regression (LR), Support Vector Machine (SVM), and Random Forest (RF) have been selected in this paper to predict football matches outcomes for MSL 2021. The accuracy and average of Rank Probability Score (RPS) are used as performance matric to evaluate the prediction models. Based on the comparative analysis conducted, all the models were able to predict the outcomes for more than 50% accuracy of the matches except RF which only obtained 49.24% accuracy. The NB is the best ML algorithm compared to SVM, LR and RF for predicting MSL football matches outcomes by achieved highest accuracy of 54.55% and lowest value of average RPS by 0.2025.

Keywords Malaysia Super League · Prediction · Elo Rating · Machine Learning

N. Razali · S. A. Mostafa
Faculty of Computer Science and Information Technology, Universiti Tun Hussein Onn Malaysia, Parit Raja, Malaysia

A. Mustapha (✉) · A. Q. A. A. Aziz
Faculty of Applied Sciences and Technology, Universiti Tun Hussein Onn Malaysia, Parit Raja, Malaysia
e-mail: aidam@uthm.edu.my

© The Author(s), under exclusive license to Springer Nature Singapore Pte Ltd. 2023

S. F. Syed Omar et al. (eds.), *Innovation and Technology in Sports*, Lecture Notes in Bioengineering, https://doi.org/10.1007/978-981-99-0297-2_13

1 Introduction

Football is the world's most renown sports with millions of fans across the globe. Sports analytics in association football always become hot topic be discussed whether in term of performance or prediction. Predicting football match results or goals is unexceptionally buzzworthy. Football prediction can be classified into two clusters which are statistical and machine learning. Despite successfully introducing numerous statistical and machine learning techniques to predict football match outcomes, flaws still exist. For example, limited accessible data sample obtained from unpopular football league has become a constraint for researchers.

In Asia, Malaysia Super League (MSL) has rising as one of prominent football leagues that gained recognition in Asian Football Federation (AFC) competition where Malaysia is ranked 20th in the AFC Club Competitions Ranking [1]. The winner of the MSL will qualify for following season's AFC Champions League group stages and the winner of the Malaysia FA Cup will qualify for the following season's AFC Champions League play-off slots. If a club lost during the play-off slots and were unable to reach the group stages, the club will play in the AFC Cup play-off slots. Despite its successful reputation as rising football league in recent year, there are still lack of research paper related to MSL whether in term of performance analysis or predictive analysis [2]. Moreover, there are constraint in availability of online football data except basic historical match results that consists of round, date, time, home and away team name, and number of goals score by both team which are not very informative for predictive analysis.

As the result, this paper attempt to present football matches outcomes prediction models based on an Elo rating system and machine learning algorithms using limited data of football matches result for Malaysia Super League. The remaining paper are organized as follows: Sect. 2 presents related work on sports analytics in MSL; Sect. 3 describes the experiments in brief including the MSL dataset, Elo rating, Machine Learning (ML) algorithms and performance metric used; Sect. 4 is about results and discussion; Finally, the conclusion in Sect. 5.

2 Related Work

Football is most popular sport in Malaysia. Despite of inconsistencies in its performance, Malaysia was formerly the king of Southeast Asian football and widely respected in the Asian continent [3]. As the Malaysian football in Asian and world arena seem deteriorate as the time flows, various initiative programs related to the development of football talent in the nation has been carried out my government started the establishment of Program Tunas Cemerlang in 1989, Program Sukan Teras (2006), Program Sukan Prestasi Tinggi Sekolah (2008) and finally a roadmap for 2014–2020 called Phase 1 National Football Development based on the philosophy or DNA of Malaysian football was introduced in 2013 before another roadmap for 2019–2030 called "Malaysian Way" to provide a new approach to the holistic development of talent to cater the demand of new era of modern football were introduced in 2018 because its seem Malaysia football team still failed to dominate Asia and international level [4]. Despite of popularity of football in Malaysia, there are still lack of research regarding Malaysia football league.

A review by [2], show only 6 works on sports analytics for MSL from 2010 to 2018 published that focuses on various field of research such as pattern of goals analysis, stadium attendance prediction, decision support system and football team ranking prediction in MSL table. The work [5, 6] and [7] is quite similar to this paper which focusing on football prediction. [6] implemented Bayesian expectation maximization for generalized Bradley Terry model that divided into two which include ties score and without ties score to estimate the football team's rankings of 12 MSL football team season 2015. Meanwhile, [5] proposed decision support system (DSS) called E-compare of Soccer Tournament Structure in order to provide statistical results tables for reporting all the results of all match and probability positions for assisting the process of decision making. Then, the DSS was improved in their other work in 2018 [7] by extending the effect of changing the weightage values point for win, draw and lose in MSL.

Later, research related to MSL showed several works, however, it does not focus on football prediction. [3] present a development pathway for Malaysia football coaches whether "A", "B" and "C" licensed coaches meanwhile [4] proposed a multidimensional assessment approach of talent identification in Male Youth Malaysia Football Players. Thus, this paper focuses on the development of football matches outcomes prediction models for MSL.

3 Experiment

The experiment section present the workflow of experiment, dataset, Elo rating system, Machine Learning (ML) algorithms and finally performance metric that will be used in this experiment. The MSL football data first will be extracted and cleaned before be calculated to Elo rating. Then, the dataset was learned and tested using ML algorithms for MSL football matches outcomes prediction. Figure 1 shows the machine learning workflow for football matches outcomes prediction based on Elo rating system for MSL.

The modeling process of the proposed football matches outcomes prediction model consist of two parts which are the implementation of Elo rating system for MSL football team rating system and the application of ML algorithms such as Naïve Bayes (NB), Logistic Regression (LR), Support Vector Machine (SVM), and Random Forest (RF) on the Elo rating system for prediction.

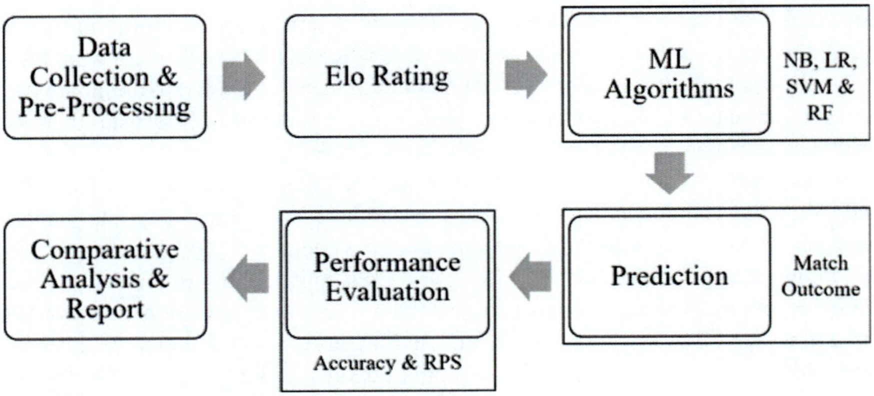

Fig. 1 A machine learning workflow for football matches outcomes prediction based on Elo rating system for MSL

3.1 Dataset

The dataset of Malaysia Super League (MSL) was extracted via online through worldfootball.net [8] and flashscore.com [9] which contain 6 features such as date, name of home team, name of away team, home team scored, away team scored and the matches results (Win, Draw, Loss) of 7 seasons of MSL between 2015 to 2021. Overall, the experiment is set to predict MSL 2021 matches results using ML algorithms based on Elo rating system. The learning set consists of 726 instances (MSL 2015-MSL 2020) and the testing set consists of 132 instances (MSL 2021). Note that the dataset was revised based on [10] since there are some changes occurring in every MSL season such as disbanded, reconstructed and withdrawal of teams, failed to comply regulations (registration and financial) and privatization of teams. In addition, MSL for season 2020 only played for eleven rounds of matches due to COVID-19 pandemic [11]. Figure 2 shows the excerpt of dataset used for the experiment.

3.2 Elo Rating

The Elo rating applied in this experiment is Elo rating based on result adopted from [12] where the rating of team is representing the current strength of team based on match result and rating of opponent team in a match. The rating may fluctuate depending on match results of team for every match played.

Date	HomeTeam	AwayTeam	FTHG	FTAG	FTR
31/1/2015	JDT	Pahang FA	2	0	H
7/2/2015	Young Lions	PDRM FA	5	3	H
7/2/2015	Perak FA	Sime Darby FC	2	0	H
7/2/2015	Terengganu FA	Selangor FA	2	0	H
7/2/2015	Felda United FC	Sarawak FA	3	3	D
7/2/2015	ATM FA	Kelantan FA	0	2	A
14/2/2015	Felda United Fc	Sime Darby FC	3	0	H
14/2/2015	Pahang FA	Selangor FA	1	1	D
14/2/2015	PDRM FA	JDT	1	0	H
14/2/2015	Sarawak FA	Terengganu F/	3	1	H
14/2/2015	Perak FA	ATM FA	2	0	H
15/2/2015	Kelantan FA	Young Lions	2	0	H
21/2/2015	Young Lions	Perak FA	2	2	D

Fig. 2 Excerpt of MSL dataset used in the experiment

3.3 Algorithms

There are four types of Machine Learning (ML) algorithms that have been used in this experiment which are Naïve Bayes (NB), Logistic Regression (LR), Support Vector Machine (SVM), and Random Forest (RF) on the Elo rating system for prediction. All the ML algorithms are carried out using Waikato Environment for Knowledge Analysis (WEKA) [13] by default option.

- **Naïve Bayes (NB)** is a simplified Bayes' theorem that assumes all the features are not related and independent.
- **Logistic Regression (LR)** by default is a predictive analysis for binary class of dependent variable. Multinomial logistic regression is an improved logistic regression which can handle multiclass classification problems.
- **Support Vector Machine (SVM)** is a supervised machine learning algorithm that can handle both predictive analysis whether for classification and regression problems.
- **Random Forest (RF)** is a supervised machine learning that is made up of decision tree that can handle both predictive analysis whether for classification and regression problems.

3.4 Performance Metric

There are two performance metrics that have been applied for comparative analysis of ML algorithms for MSL prediction based on Elo rating [14]:

- **Accuracy** can be defined as in Eq. (1), where the total number of correctly predicted results is divided by the total number of actual observed results. The higher the value of score indicates better accuracy.

$$Accuracy = \frac{\text{The total prediction results}}{\text{The total observed results}} \tag{1}$$

- **Rank Probability Score (RPS)** can be defined as in Eq. (2), where r is the number of potential outcomes, p_j is the forecasted probability of outcome j and pe_j is the actual probability of outcome j. The smaller value of score indicates better forecast accuracy.

$$RPS = \frac{1}{r-1}\sum_{i=1}^{r-1}\left(p_j - e_j\right)^2 \tag{2}$$

4 Results and Discussion

The Malaysia Super League (MSL) of 7 season from 2015 to 2021 consist of 858 instances was used as dataset in this paper. The dataset contains 6 features such as date, name of home team, name of away team, home team scored, away team scored and the matches results (Win, Draw, Loss) and was calculated to rate the team overall performance strength in term of rating system called Elo rating. Then, the Elo rating data were divided into two set of data for training and testing purposes. The training set data was model using Machine Learning (ML) algorithms such as Naïve Bayes (NB), Logistic Regression (LR), Support Vector Machine (SVM), and Random Forest (RF) before using the testing set for prediction. The performance metrics were accessed via accuracy and average RPS on all prediction models. Table 1 shows the comparative analysis results of the ML prediction models using Elo rating for MSL in terms of accuracy and rank probability score.

Based on the comparative analysis conducted, all the models were able to predict the outcomes for more than 50% accuracy of the matches except RF which only obtained 49.24% accuracy and the best model is NB which have the highest accuracy of 54.55%, follow closely by LR by 53.03% and SVM by 52.27%. For RPS, NB successfully gained the lowest value by 0.2025. Although, LR successfully overcome

Table 1 The comparative analysis results of the ML algorithms based on Elo rating in term accuracy and RPS

ML Algorithms	Accuracy (%)	RPS (AVG)
NB	54.55	0.2025
LR	53.03	0.2167
SVM	52.27	0.2186
RF	49.24	0.2360

SVM in accuracy performance, LR only gained RPS of 0.2167 which is slightly better than SVM that obtained 0.2186 in RPS value. Lastly, RF only succeed to achieved 0.2360 in RPS. Several factors that might influence the accuracy and RPS obtained has been observed. Thus, NB is the best ML algorithm compared to SVM, LR and RF for predicting MSL football matches outcomes by achieved highest accuracy of 54.55% and lowest value of RPS by 0.2025. Overall, the predictive performance was quite low because the prediction solely relied on Elo rating as data feature. There are many factors that can influence and impact either negative or positive to the football matches outcomes prediction. The high rating team does not always win over low rating team. For example, the low rating team can overcome high rating team because desperate to avoid relegation from the leagues or high rating team not using their key players because of rotational system or injury or international called up or to rest them for other domestic or international competition.

5 Conclusion

The paper presents the ML approach for MSL match outcome prediction based on Elo rating. In conclusion, NB is the best ML algorithm compared to LR, SVM and RF for predicting MSL football matches outcomes by achieved highest accuracy of 54.55% and lowest value of RPS 0.2025 meanwhile RF is the least accurate 49.24% and achieved high value in RPS of 0.2360. The findings of this paper hopefully may give insight and promote more research on Malaysia Football League whether in term of performance analysis or predictive analysis. Several factors have been observed for future works. The football data is limited to only certain limited features such as the football team involves in MSL, and the number of goals scored in a match that later are calculated into Elo rating. Therefore, it is recommended to study the impacts of other features to the outcomes of the matches in MSL such as considering low tier Malaysia football league matches in Elo rating such as the Malaysia Premier League as well as adjustment of Elo rating to follow the MSL environment in term of home advantage and K-factor. Moreover, alternative of other football rating system besides Elo rating and advanced machine learning algorithms as well as incorporate more football data as features to build richer prediction model may improve the predictive performance.

References

1. AFC Club Competition Ranking (2022) https://footyrankings.com/crank2022/ Accessed 16 Mar 2022
2. Razali N, Mustapha A, Soeed K, Othman M (2019) A review of sports analytics on malaysia super league (MSL). J Adv Res Dyn Control Syst 11(5 Special Issue):820–826
3. Karim ZA, Nadzalan AM (2017) Malaysia football coaches: development characteristics. Int J Acad Res Bus Soc Sci 7(9):305–312

4. Hanapiah KS, Hashim A, Karim ZA (2019) A multidimensional assessment approach of talent identification in male youth malaysia football players. Int J Acad Res Bus Soc Sci 9(11):33–49
5. Yusof MM, Khalid R, Sulaiman T, Hamid MSA, Mansor R, Rosli NC (2016) E-compare of soccer tournament structure. Int Rev Manag Mark 6(8):294–298
6. Nor SBM, Mahmud Z (2016) Teams ranking of Malaysia super league using Bayesian expectation maximization for Generalized Bradley Terry Model. In: 4th international conference on quantitative sciences and its applications, ICOQSIA
7. Yusof MM, Khalid R, Hamid MSA, Mansor R, Sulaiman T (2017) Modelling the effect of changing point systems to teams' competition standing in a Malaysian soccer super league. In: 1st International Conference on Computing, Technology, Science and Management in Sports, ICoTSM
8. Malaysia Super League https://www.flashscore.com/football/malaysia/super-league/archive/ lA2ccessed9 Mar 2022
9. Malaysia Super League https://www.worldfootball.net/history/mas-super-league/ Accessed 29 Mar 2022
10. Stokkermans K (2015) Malaysia 2015, http://www.rsssf.com/tablesm/malay2015.html#super Acessed 29 Mar 2022
11. Fai OK (2020) Malaysia Super League (Liga Super Malaysia) 2020 restart: fixtures, format change, results, table, top scorers and everything you need to know, https://www.goal.com/en-my/news/malaysia-super-league-msl-2020-return-fixtures-format-change/1uag8k97dv0z71mbcvwfkyts9i#Top Accessed 29 Mar 2020
12. Hvattum LM, Arntzen H (2010) Using ELO ratings for match result prediction in association football. Int J Forecast 26(3):460–470
13. Frank E, Hall MA, Witten IA (2016) The WEKA workbench. online appendix for data mining: practical machine learning tools and techniques. Morgan Kaufmann, Fourth Edition
14. Dubitzky W, Lopes P, Davis J, Berrar D (2019) The open international soccer database for machine learning. Mach Learn 108(1):9–28

The Reliability and Validity of React-Run Agility Test Assessment System

Muhammad Najib Abdullah, Che Fai Yeong, Asha Hasnimy Mohd Hashim, Eileen Lee-Ming Su, Kang Xiang Khor, C. Yang, Haohui Huang, and Hisyam Abdul Rahman

Abstract Badminton is known as one of the fastest racket sports in the world. Therefore, it is important for the player to have excellent reactive agility to stay competitive in a badminton match. The purpose of this study was to investigate the validity and reliability of the React-Run, a new reactive agility assessment system for badminton players. Six subjects were divided into two groups, three male amateur players with experience in representing badminton clubs for competitions and three recreational players, who used to play badminton, were recruited for the experiment. The measurements were performed with the use of the React-Run system, where the experiment was conducted on an actual badminton court. The result of eight averages (in milliseconds) indicates reactive agility from different sensor locations (node) (i.e., Front, Front-Right, Right, Back-Right, Back, Back-Left, Left, and Front-Left) were stored and analyzed. The mean, standard deviation, and range were calculated for each outcome variable. Pearson correlation method and an independent t-test analysis were used to evaluate the construct validity of the React-Run system meanwhile the

M. N. Abdullah (✉)
Faculty of Engineering, Universiti Teknologi Malaysia, Johor Bahru, Malaysia
e-mail: najibbinabdullah5609@gmail.com

C. F. Yeong · E. L.-M. Su
Faculty of Engineering, School of Electrical Engineering, Universiti Teknologi Malaysia, Johor Bahru, Malaysia

A. H. Mohd Hashim
School of Education Faculty of Social Sciences and Humanities, Universiti Teknologi Malaysia, Johor Bahru, Malaysia

K. X. Khor
Techcare Innovation Sdn. Bhd, Johor Bahru, Malaysia

C. Yang
Bristol Robotics Laboratory, University of the West, Bristol, England

H. Huang
Department of Automation, Shanghai Jiao Tong University, Shanghai, China

H. Abdul Rahman
Faculty of Electrical and Electronic Engineering, Universiti Tun Hussein Onn Malaysia, Batu Pahat, Johor, Malaysia

© The Author(s), under exclusive license to Springer Nature Singapore Pte Ltd. 2023

S. F. Syed Omar et al. (eds.), *Innovation and Technology in Sports*, Lecture Notes in Bioengineering, https://doi.org/10.1007/978-981-99-0297-2_14

reliability (within-subject variation) was established using Cronbach's alpha value. Results indicated that React-Run system showed good construct validity and reliability as it was statistically able to distinguish the performance of amateur and recreational badminton players (p-value < 0.05) and had internal consistency between the node sensors' positions (Cronbach's alpha value > 0.9).

Keywords Reactive agility · Badminton · Performance · Court sport · Racket sport · Reaction time

1 Introduction

Badminton is one of the most popular sports in the world, with an estimated 200 million people playing badminton [1]. The shuttle speed in badminton can reach up to 493 km/h as reported in Guinness world record [2]. A previous study has revealed that a badminton player has 0.1 s to react to a shuttle returned by their opponent [3], therefore, it is important for the player to have excellent reactive agility to stay competitive in a match [4]. Agility is defined as "a rapid whole-body movement with change of velocity or direction in response to a stimulus" which basically consisted of two important factors, i.e. reaction time and change body direction speed [5].

Advanced technologies, such as movement sensors and high definition camera, have been extensively used in measuring the reaction time of athletes when performing sports activities [5, 6]. However, due to the complexity and high implementation cost, professional trainers prefer to use conventional changes of direction test, such as the shuttle run agility test and Illinois agility test [4, 7, 8]. However due to pre-planned movements in the test procedures, this test does not accurately reflect the real situations in badminton competition [5, 9].

A new system, named React-Run was developed to assess the reactive agility performance for badminton players. The system had achieved medium rates of construct validity ($R^2 = 0.69$) [10]. The confirmation of construct validity was performed by comparing measured reaction-time in milliseconds against the distance travelled of badminton players within a badminton court area. Therefore, the primary aim of this study is to extend the construct validity study and reliability of this React-run system tested with different skill levels of badminton players. This study is supported by ethical approval from FRGS committee with research grant number [R.J130000.7851.5F187].

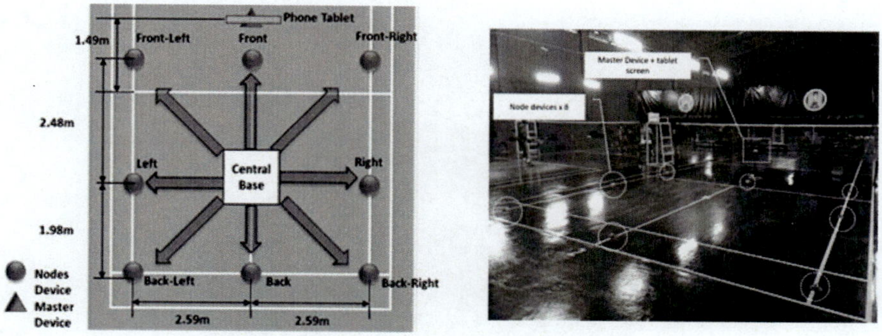

Fig. 1 React-run agility test system setup in Badminton academy

2 Materials and Methods

2.1 Experimental Design

The React-run system has eight custom sensors called nodes and is used with an Android mobile application. These eight nodes were positioned around corners and edge points on a badminton court area (5.18 m × 4.46 m). These points were denoted as "Front-Left", "Front", "Front-right", "Right", "Back-right", "Back", "Back-left" and "Left" (see Fig. 1). Each node would be activated randomly during the test activity. An 8-inch display tablet was placed 3.47 m in front of the central base position, on top of an adjustable tripod. This study was conducted in a badminton academy during the weekly training session for the amateur players.

Each node was equipped with an LED and buzzer to provide notification to the participant during the test conducted (see Fig. 2). The whole set of activities (i.e., start-test, randomized instructions, end-test) were controlled by a custom-coded Android mobile application.

The results will be eight averages of durations (i.e., Front-Left, Front, Front-right, Right, Back-right, Back, Back-left and Left) in milliseconds presented at the end of a session. In total, 24 sets of data were available to be exported from the application for analysis (see Fig. 3).

2.2 Participants

Six male badminton players were recruited as the subjects for this study. The subjects were categorized into two groups, amateur or recreational players. Three subjects, who received training at least twice a week, competed at state-level badminton competitions or beyond and had at least 2 years' experience in competitive badminton tournaments, were categorised as amateur players. The other three subjects, who did

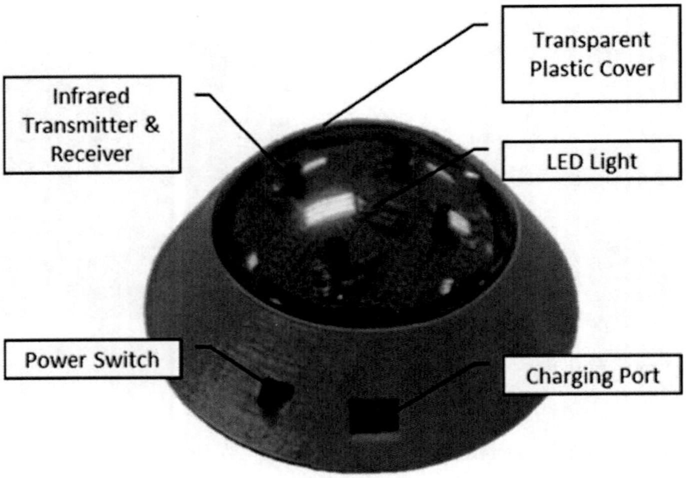

Fig. 2 Node sensor

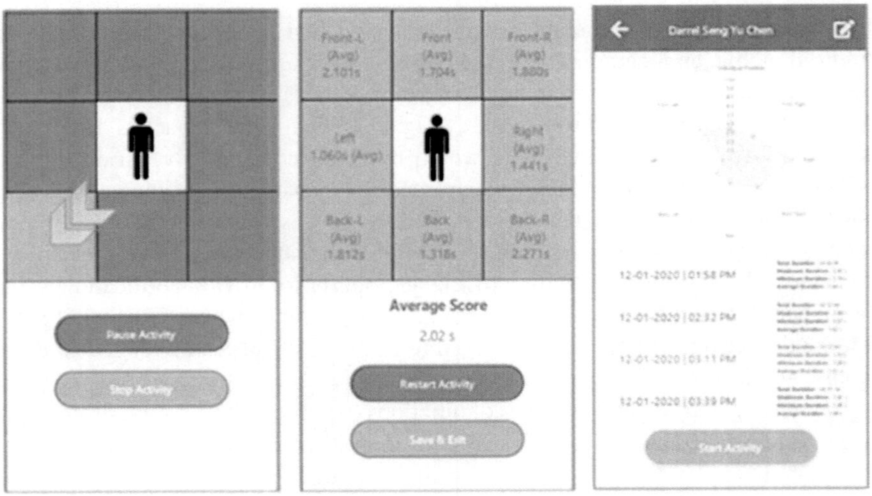

(a) (b) (c)

Fig. 3 **a** Run direction instruction as provided by the mobile application. **b** Scores of the subject were presented at the end of each session. **c** The progress of the subject is recorded in the system and presented in infographic format

not undergo badminton training and only played badminton for recreational purposes at least once a week, were categorised as recreational players. The demographics of the players were shown in Table 1. All test subjects recruited were right-handed badminton players and free from musculoskeletal injury within the past 6 months.

Table 1 Descriptive characteristics of the amateur and recreational players

	Amateur	Recreational
Males (n)	3	3
Age (years)	15 ± 2	26 ± 3
Height (cm)	172 ± 7.1	166.2 ± 4
Weight (kg)	56 ± 9.2	61 ± 2
Playing experience (years)	2–3	0
Previous season competitions (range)	2–6	0
Monthly on-court training (range)	4–8	1–2

2.3 Procedure

Prior to testing, all subjects were briefed and given a demonstration of the system and experimental protocols to ensure that they fully understood the test procedures. Then, all subjects were required to provide informed consent to participate in this study and to provide information of their name, age, gender, weight, height, dominant hand and leg during registration.

Before the test started, all subjects were asked to perform 10 min typical warm-up activities to prevent any muscular injury during the test and to ensure they were at optimized performance level. Each subject was instructed to stand and be ready at the central base position. The test would start after 5 s countdown, as displayed on the tablet screen.

During the test, the subject was given instructions through the tablet screen as shown in Fig. 3. The node sensors will be activated (LED lit up and beep sound emitted) randomly during the test and the subject would need to deactivate (turn off) the sensor by running as quickly as possible towards the sensor node position and wave their right-hand at approximately 10 cm height above the sensor.

The subject was then required to return to centre base immediately after deactivating the sensor node. The next target will be activated 1 s after the deactivation of the previous target. The test will end after total of 24 randomized instructions were completed, with each of the 8 targets randomly activated three times during the experimental session.

The reaction-time was measured in milliseconds (i.e. reactive agility performance) for the player to complete each instruction, from node activation to node deactivation, was recorded. The 24 data from eight sensor positions were stored in React run application local database and exported for analysis for each participant. The test procedure is illustrated in Fig. 4.

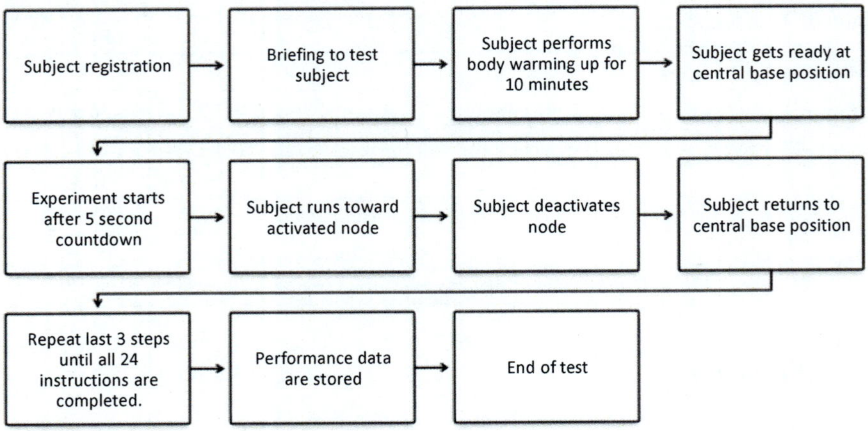

Fig. 4 Test procedure flow chart

2.4 Statistical Analysis

Descriptive statistics (mean, standard deviation, and range) were calculated for each outcome variable. Normality of the data collected was checked using Anderson–Darling test and the probability plot is used to visualize the fit of the normal distribution. The interpretation of p-value > 0.05 indicate that the data is normally distributed; otherwise, the data is not normally distributed.

Construct validity was analysed using an independent t-test analysis to examine the difference of performance between the amateur and recreational players. The reliability (internal consistency) was established using Cronbach's alpha value. Cronbach's alpha is a measure of how reliable something is and runs from 0 to 1. The stronger the internal consistency, the closer Cronbach's alpha coefficient is to 1 [11]. Table 2 presents the rule of thumb to interpret Cronbach's alpha value.

The statistical analysis of the data collected through this experiment was performed using Minitab software version 19.1 on Windows 10.

Table 2 Interpreted Cronbach's alpha value [11]

Alpha value	Interpretation
>0.9	Excellent
>0.8	Good
>0.7	Acceptable
>0.6	Questionable
>0.5	Poor
<0.5	Unacceptable

3 Results and Discussion

Figure 5 shows a subject in action during the experiment. Figure 6 presented the probability plots for the recreational group and the amateur group. The plots showed that the data were normally distributed since the p-value for amateur players and recreational players were 0.417 and 0.471 respectively (p-value > 0.05).

Figure 7 shows the individual plot of mean reaction time for all players in the amateur and recreational groups. The result reveals that the mean reaction time to reach all eight node positions for the amateur badminton player group was (1556 ± 165 ms) and this was 18% faster than the recreational badminton player group with

Fig. 5 A player performing the test activity

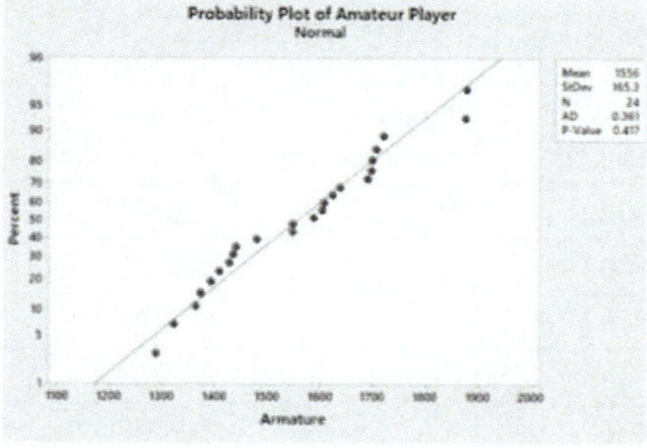

(a)

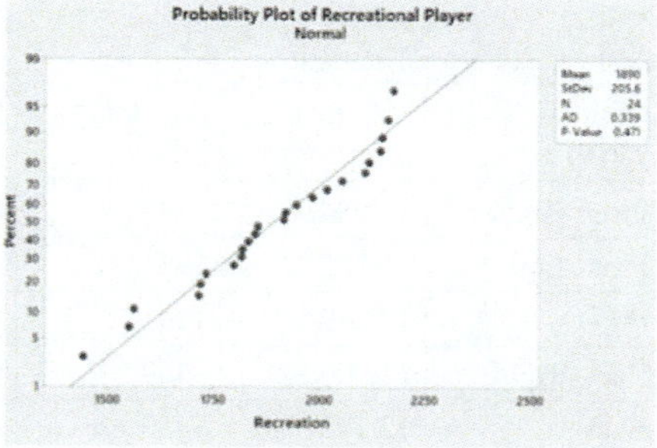

(b)

Fig. 6 **a** Probability plot for amateur badminton players (p-value = 0.417), **b** Probability plot for recreational badminton players (p-value = 0.471)

(1890 ± 206 ms). This difference was statistically significant, with *p-value* <0.001, at 0.05 significant level.

For all node positions, the Cronbach's alpha value was 0.9572 indicating a high internal consistency. Meanwhile, Table 3 shows the result of Cronbach's alpha values if any node position is removed from the analysis.

Agility is a physical ability that involves various mechanisms and is influenced by variety of internal and external factors. Conventionally, agility has been described by the ability to change direction of movement swiftly and precisely [12]. In more recent studies, factors such as initial reaction time, global movement control, dynamic flexibility, and body balance during abrupt change of direction were identified contributing to agility performance [13, 14]. In addition, other study also stated that change in movement direction is influenced by variety of cognitive elements apart from Bio-moto abilities [15, 16].

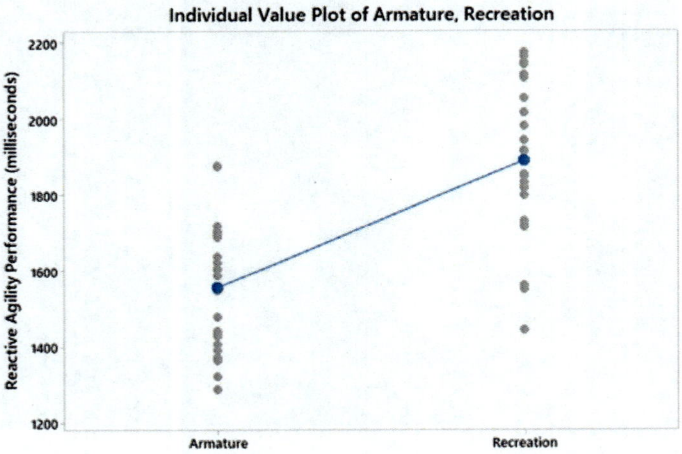

Fig. 7 Individual plot for mean reaction time for amateur and recreational badminton players

Table 3 Item-total statistics

Omitted variable	Adj. total mean	Adj. total StDev	Item-Adj. total corr	Squared multiple corr	Cronbach's alpha
Front-Right	11,984	1341	0.9780	*	0.9427
Front-Left	11,956	1381	0.8063	*	0.9533
Front	12,239	1341	0.8992	*	0.9474
Back-Right	11,906	1369	0.9448	*	0.9459
Back	12,192	1450	0.6869	*	0.9608
Back-Left	11,865	1319	0.9439	*	0.9444
Left	12,245	1410	0.7851	*	0.9551
Right	12,095	1308	0.7892	*	0.9600

Cognitive processes such as analysing, and decision making are important in order to make a wise action during a dynamic, time-constrain, and often under extreme physical stress [17]. In badminton, agility thus can be understood as an open skill which involves a cognitive response to an unplanned stimulus in a dynamic environment [18, 19]. As a result, such movement scenarios that are not pre-planned movement are regarded as reactive agility.

4 Conclusion

Therefore, the React-Run Agility Test Assessment System (React-Run) was designed and developed to test badminton player in reactive conditions. The results from this study revealed that amateur badminton players presented better than recreational player (*p-value* <0.05). This showed that React-run assessment system was able to discriminate performance level in badminton players and demonstrating its construct validity. Additionally, the results from this study demonstrated that the React-run displayed good reliability because the Cronbach's alpha value obtained was 0.9572, <0.9. Thus, it can be concluded that the measuring instrument is "excellent". As can be observed form the alpha values from Table 3, the system achieved its best reliability when all nodes were considered in the analysis.

There were several practical difficulties we encountered during this study. Firstly, the difficulty to recruit the case studies during pandemic outbreak in Malaysia. The small number of participants in this study might not represent the actual population, thus a greater number of test subjects is required for future work. Secondly, the number of repeats for a test should be increased to minimize the experimental noise factor. In overall, this system could be implemented in field as a part of training and assessment program.

Acknowledgements The authors would like to express their gratitude to Ching She Badminton Training Centre, Badminton Academy of Johor and Master Badminton Club (MBC) for the collaboration for this project. This work is supported by the Fundamental Research Grant Scheme (FRGS) and Universiti Teknologi Malaysia research grant [R.J130000.7851.5F187].

References

1. Kwan M, Cheng CL, Tang WT, Rasmussen J (2010) Measurement of badminton racket deflection during a stroke. Sport Eng 12(3):143–153. https://doi.org/10.1007/s12283-010-0040-5
2. Vincent L. Malaysian badminton star breaks smash speed record—Russia (pic). Badminton-Planet.com, Aug 23, 2013. https://www.badmintonplanet.com/badminton-news/2322-malaysian-badminton-star-breaks-smash-speed-record-russia-pic.html. Accessed 7 Feb 2021
3. Bankosz Z, Nawara H, Ociepa M (2013) Assessment of simple reaction time in badminton players. Trends Sport Sci. 20(1):54–61

4. De França Bahia Loureiro L, Costa Dias MO, Cremasco FC, Da Silva MG, De Freitas PB (2017) Assessment of specifcity of the Badcamp agility test for badminton players. J Hum Kinet 57(1):191–198 (2017). https://doi.org/10.1515/hukin-2017-0060
5. Sheppard J, Young W (2006) Agility literature review: classifications, training and testing. J Sports Sci 24(9):919–932. https://doi.org/10.1080/02640410500457109
6. Raya MA et al (2013) Comparison of three agility tests with male service members: Edgren side step test, T-Test, and Illinois agility test. J Rehabil Res Dev 50(7):951–960. https://doi.org/10.1682/JRRD.2012.05.0096
7. Dong M, Lyu J, Hart T, Zhu Q (2019) Should agility training for novice badminton players Be physically or perceptually challenging? J Sports Med Phys Fitness 59(12):2015–2021. https://doi.org/10.23736/S0022-4707.19.09666-X
8. Cinthuja P et al (2015) Physical fitness factors of school badminton players in Kandy district Department of physiotherapy, Faculty of Allied Health Sciences, General Sir John Kotelawala. Eur J Sport Exerc Sci 4(2):14–25
9. De França Bahia Loureiro L, De Freitas PB (2016) Development of an agility test for badminton players and assessment of its validity and test-retest reliability. Int J Sports Physiol Perform **11**(3):305–310. https://doi.org/10.1123/ijspp.2015-0189
10. Najib M, Syafiq M, Fai YC, Khor KX, Su E, Yang C (2021) Development of ReactRun: advance Universal Iot System for agility training in sport and healthcare, no. April. https://doi.org/10.1109/IECBES48179.2021.9398743
11. Gliem JA, Gliem RR (2003) Calculating, interpreting, and reporting Cronbach's alpha reliability coefficient for likert-type scales. Stud Inorg Chem 14(C):349–372. https://doi.org/10.1016/B978-0-444-88933-1.50023-4
12. Sheppard JM, Dawes J, Jeffreys I, Spiteri T, Nimphius S (2014) Broadening the view of agility: a scientific review of the literature. J Aust Strength Cond 22(3):6–26
13. Sporis G, Jukic I, Luka M, Vlatko V (2010) Reliability and factorial validity of agility tests for soccer players, **24**(3):79–686
14. Hulteen RM, Lander NJ, Morgan PJ, Barnett LM, Robertson SJ, Lubans DR (2015) Validity and reliability of field-based measures for assessing movement skill competency in lifelong physical activities: a systematic review. Sport Med 45(10):1443–1454. https://doi.org/10.1007/s40279-015-0357-0
15. Menezes GB et al (2021) Does motor coordination influence perceptual-cognitive and physical factors of agility in young soccer players in a sport-specific agility task? Sport Biomech 1–14. https://doi.org/10.1080/14763141.2021.1995476
16. Delaney JA et al (2015) Contributing factors to change-of-direction ability in professional rugby league players, 29(10)
17. Spasic M, Uljevic O, Coh M, Dzelalija M, Sekulic D (2013) Predictors of agility performance among early pubescent girls. Int J Perform Anal Sport 13(2):480–499. https://doi.org/10.1080/24748668.2013.11868664
18. Horníková H, Jeleň M, Zemková E (2021) Determinants of reactive agility in tests with different demands on sensory and motor components in handball players. Appl Sci 11(14). https://doi.org/10.3390/app11146531
19. Mackala K et al (2020) Evaluation of the pre-planned and non-planed agility performance: comparison between individual and team sports. Int J Environ Res Public Health 17(3). https://doi.org/10.3390/ijerph17030975

Automated Classification of Woodball Swinging Phases from Inertial Measurement Unit Using Least Square Method

Nur Sakinah Mohd Hisam, Ahmad Faizal Salleh, and Mohd Yusoff Mashor

Abstract Woodball is a rising regular sport that garnered a lot of attention from people with different ages and genders due to its easy playstyles and adaptability to play it almost everywhere. Without any teaching and guides from professional, some of novice players may unconsciously practicing wrong swing postures which is the most important fundamental in woodball. Having a good swing posture helps players to dictate how the mallet swings around the body, promotes good balance and helps the body to turn correctly. In this study, we proposed and verified a simple regression method using a portable, miniature Inertial Measurement Unit (IMU) to classify swing phases to assist the novice players to have a correct swing postures as professional players does. The IMU was attached to the player's left hand wrist to collect the angle, acceleration and angular velocity of the swing. The signal data were processed using second-order Butterworth filter with cutoff frequency of 10 Hz, estimated using 2nd order Least Square and the swing phases were classified automatically with labels. The proposed system yields high classification performance with average accuracy of 99% for all swing phases.

Keywords Woodball swing phases · Inertial measurement unit · Least square

1 Introduction

Woodball was created in Taiwan according to International Woodball Federation (IWF) as an outdoor activity in 1990 and was recognized officially as regular sport two years after its establishment. With members from 43 countries, regional competitions as well as woodball championships were held by IWF, an organization that holds the power to authorize the regulations and rules for woodball. The sport which is broad of age and gender and with simple gameplays has spread quickly to various countries such as Japan, Denmark, Korea and Croatia. It is also playable in sand, grass or indoor and this feature makes it even more popular as it allows people to

N. S. Mohd Hisam (✉) · A. F. Salleh · M. Y. Mashor
Faculty of Electronic Engineering Technology, Universiti Malaysia Perlis (UniMAP), 02600 Arau, Perlis, Malaysia
e-mail: sakinahhisam@studentmail.unimap.edu.my

© The Author(s), under exclusive license to Springer Nature Singapore Pte Ltd. 2023

S. F. Syed Omar et al. (eds.), *Innovation and Technology in Sports*, Lecture Notes in Bioengineering, https://doi.org/10.1007/978-981-99-0297-2_15

play it almost anywhere. The skills required to play woodball is almost similar to golf, for example the stance, address position and grip of the mallet [1].

In general, woodball is a sport where a mallet is used to pass a ball through gates. The rules are very much alike to golf; when a player completes a round of woodball from the first to the 12th hole, the game is over. The player who get the lowest number of strokes will be the winner and the final score is determined by the total stroke count during play. This game can be played in single, double or in a team. The equipment required to play woodball are a mallet, wooden ball and a gate (Fig. 1). The mallet which roughly weighing around 800 g is a T-shaped with a handle and a bottle-shaped head. The wooden ball is 9.5 cm ± 0.2 cm and 350 g ± 60 g in diameter and weight, respectively [1]. Two bottles were fixed to the ground as the base pillar for the gate, so that the wooden ball can passes through it for a score.

Woodball can improve the hand–eye coordination and aiming ability of the players as it is highly related to the striking skills required in woodball [2]. In addition, their striking skills can also be applied on different types of target sports such as golf and field hockey due to their similarities on movement patterns of strokes; the tee shot (full swing), the chip shot and putting [3]. Even though this sport is slowly raising to be regular sport in championship, there are still no citations on the swing phases as far as we knew. Hence, according to woodball coaches and players, we can presumed woodball has similar swinging phases with golf.

A full golf swing with clubhead motion was divided into five phases: Before Swing (BF), Backswing (BS), Downswing (DS), Follow-through (FT) and After Swing (AF) as shown in Fig. 2a. Meanwhile, Fig. 2b shows the clubhead speed with four dividing points in the full golf swing: Address (ADD), Backswing Top (BST), Impact (IMP) and Finish (FIN). The location of ADD and BST are when the clubhead speed is at the local minimum before the BS and right after the BS, respectively. In addition, when the clubhead passes the marker on the ground aligned with the golf

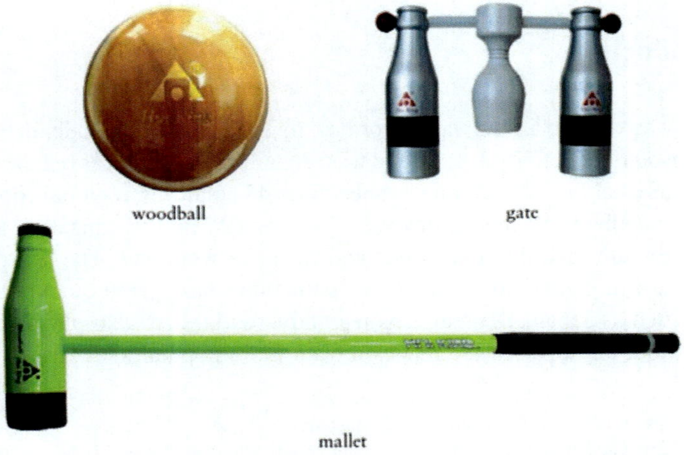

Fig. 1 Equipment used in woodball [1]

ball during the DS, the IMP is marked. FIN is when the clubhead speed reaches the local minimum again after the FT [4].

There are many types for monitoring systems and wearable sensor unit is one of them. It was revealed that it show a very high potential to observe and monitor the ambulatory activities especially in the home environment [5]. They also contribute in the importance of (1) the process of rehabilitation in injured patients such as generating input for health interventions (real-time personalized feedback) to design the treatment plans for the patients and able to follow-up the monitoring [6]; (2) the detection of risk situations especially for elderly patients [7]; and (3) the diagnosis and treatment of patients with neurological diseases [8]. One of popular wearable sensor unit is Inertial Measurement Unit (IMU) sensor. By definition, IMU is an electronic device that combined multiple sensors such as gyroscopes, accelerometers and magnetometers. It also can be equipped with an antenna (wireless technology), or an output pin logged by wire to a base station, or even a secure digital (SD) card [5]. The components mentioned are the commonly used in IMUs to obtain relevant data for gait analysis. The gyroscope can output angular velocity signals meanwhile,

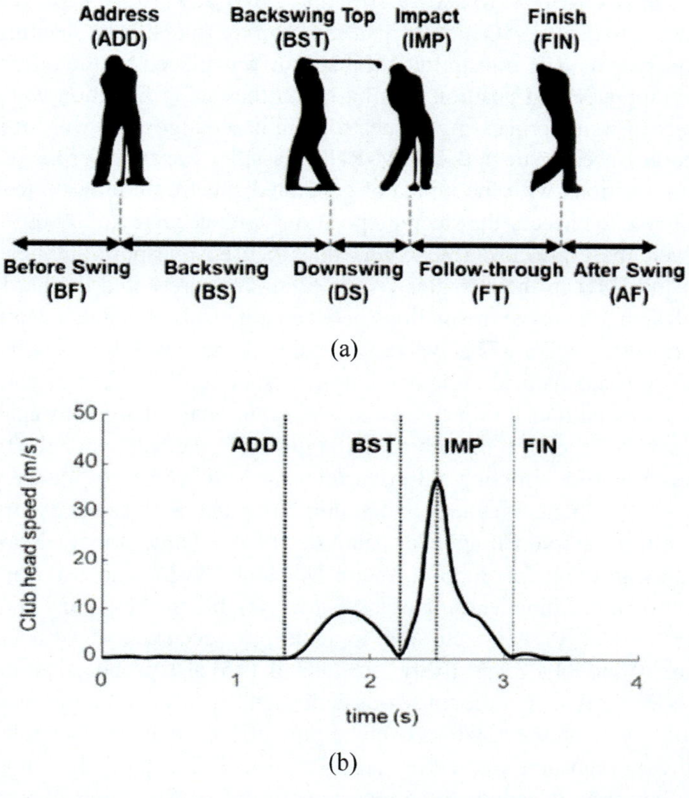

Fig. 2 **a** Schematic diagram of full golf swing representing the phases and dividing points, **b** Clubhead speed with five phases and four dividing points of events [4]

accelerometer can output linear acceleration signals on the three axes in space [9]. In gait analysis, it is important for the portable systems based on body sensors to take note on the parameters such as conformability, precision and usability or transportability [10]. IMU are much lighter, cheaper, smaller, noninvasive, wearable and portable, able to identify human motion without altering natural movement patterns in a wide variety of environments compared to other systems [11].

There are various research journals that focused on the implementation of IMU especially using some deep learning techniques. Añazco et al. [12] proposed the implementation of RNN for the hand gesture recognition system using a single patchable IMU. The IMU consists of IC-based electronic components on an adhesive and stretchable substrate interconnections and was attached to the subject's wrist. Some of other components used in the proposed patchable IMU are a low-cost six-axis inertial sensor, a chip antenna, several electronic components for real-time biometric data acquisition and a Bluetooth low energy (BLE) microprotocol system-on-a-chip (SoC). The classification data were tested with two deep learning models, a RNN model based on GRU units and a RNN model based on Bidirectional LSTM units. The results on both models were validated and achieved average classification accuracy of 96.06% for RNN-BiLSTM and 99.16% for RNN-GRU. Eyobu et al. [13] applied an IMU sensor to capture 3D motion data and proposed a real-time sleeping position recognition system. The transmitter IMU sensor was placed on the left arm of the subject and the sleeping position data based on the facing direction was collected in real-time. In signal processing, a spectrogram based algorithm was implemented and the features were extracted. LSTM-RNN classifier was used for the recognition of sleeping positions with the inputs of generated spectrogram based features and the system able to classify the sleeping positions with accuracy of 99.09%.

Moreover, there are also researchers that focused on sports application even without using IMU in their studies. Wang et al. [14] presented the hardware and software design to analyze the motions between elite, sub-elite and amateur volleyball players during spikes. This wearable sensing device (WSD) was built based on the microelectromechanical systems motion sensors and it is cheaper and required much less computational power compared to conventional videography analysis. The data points associated with the subject failing to spike the ball inside the target area were removed before applying a 3-point filter moving average to reduce the effect of the noise. 12 statistical features and 3 morphological features were extracted and selected using Principle Component Analysis (PCA). Three principal components were used as an input for Support Vector Machine (SVM) and was compared to the conventional machine learning of k-Nearest Neighbour (kNN) and Naïve Bayes (NB). The SVM + PCA system able to achieve an accuracy of 94% while kNN and NB are 90 and 84% respectively. Zhang et al. [15] also proposed almost similar methods as Wang et al., but the study was dedicated to analyze and improve the skills of the golf players on their swings using computerized methods. The golf club was integrated with high-precision strain gage sensor to collect the golf swings signals. PCA was then used to reduce the dimensions of the 1000 features values for each acquired 134 swings. The authors applied two SVM and three kNN classifiers to classify the golf player's swings signals and the best classifier was evaluate on their performance on training and testing accuracy and the time taken for training. Out of five classifiers, linear SVM provided the best accuracy which are 100% accuracy on training and testing and costs an average time of 4 s for training process. Hence, linear SVM was selected.

Furthermore, Jiao et al. [16] investigated a method for classification on the golf swing data using deep CNN. The data were collected from four professional and amateur golf players with each players required to perform several predefined golf swings with shapes. The dataset were then augmented to balance the swing counts in each class by applying some random time translation and data rescaling and were fed into the CNN. tenfold cross-validation was employed and achieved accuracy of 95% which outperforms the conventional method of SVM with 86.8% accuracy. Although there are many researchers doing research on IMU with deep learning, there are hardly any documentations on the woodball especially on its swing phases. The study about swing phases is important to both players and coaches to improve their swing postures, especially to those who learnt woodball through self-educated. This method may unintentionally lead to inaccurate swing postures. The optical motion capture system is commonly used to capture swing motion in sports because of its high accuracy but it is limited to being in a laboratory. IMU presents great advantages for sport analysts to obtain data out of the laboratory due to their features which are miniature in size, non-stationary and has good wireless capability. Thus, this study is intended to focus on the automated classification of woodball swing phases by utilizing a simple algorithm and provide evaluations on the swing performances of novice players. This was performed using Least Square (LS) method which can estimated the IMU signals and automatically classified the signals into its phases.

2 Methodology

2.1 Participants

30 woodball players with different range of playing experiences were volunteered to take part in this study. The players were provided with written informed consent and was confirmed in ahead of time to have no physical limitations or musculoskeletal injuries. The participants which have mean characteristics of; age 37.70 $\pm$ 12.38 years, body mass 72.97 $\pm$ 16.10 kg and height 167.83 $\pm$ 7.79 cm, were briefed on the purposes and the procedure of the experiment. This was to ensure each participants understood the aims of this study and are willing to provide their information for research purpose.

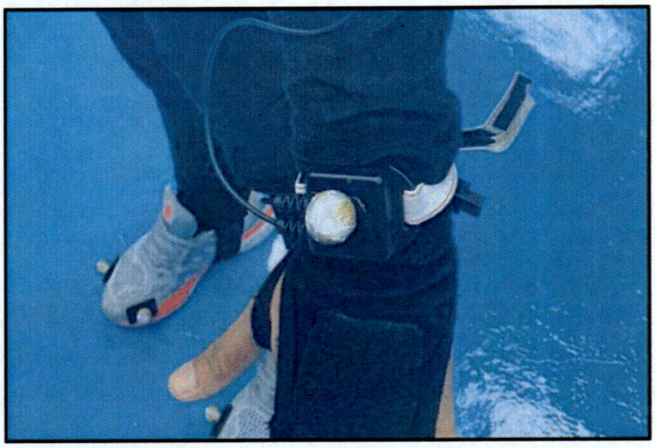

Fig. 3 Placement of IMU at the players' left hand wrist with a retro reflective marker

2.2 Procedures

The sensors used in this studies are a single unit of IMU (MotionNode, USA) and an eight-camera motion analysis system (Qualisys Medical AB, Goteburg, Sweden) which captured the kinematic data at frequency of 500 Hz [17, 18]. The IMU (MotionNode, USA) was fixed at the players' left hand wrist (Fig. 3) while a retro reflective marker for Qualisys Optical Motion Capture System (OMCS) was placed on the top of IMU. This experiment use a medium aluminum mallet with a retro reflective marker positioned at the mallet head [18] to collect the resultant linear speed and acceleration of the swing. Before each data collection session, OMCS was calibrated and only calibrations with average residuals of less than 1.0 mm for each camera were accepted. Players performed warm up and few swinging trials to get used to the experiment. Once the player was ready, both IMU and OMCS started recording the data. The player flexed and extended his/her left hand which resulted in a peak acceleration in the signal and this was intended for the synchronization of data for both IMU and OMCS [18]. Each players performed five trials of full woodball swing motion with one minute rest between each trials.

2.3 Data Analysis

Raw data from IMU and OMCS were collected after each data collection session. The data which consists several parameters were analyzed and the most suitable parameters to be used in the study were taken. Some of the parameters are angle, angular velocity and acceleration. Linear velocity data of the swing obtained from OMCS were used as a reference for the IMU data validation considering their similarities on the shape and envelope of the signals. Both data from IMU and OMCS undergoes signal filtering and the IMU data were estimated using LS before the segmentation and classification process.

Signal Processing. Signal processing performs certain processing technique on the original signal either by enhancing the signal or refining the value of signal outcomes and prepare it for further processing or analysis. In a simple term, these techniques will converts the original raw signal into an easy form for further processing by the user [19]. There are many techniques that can be used for processing such as signal filtering, smoothing, resampling, detrend and others. One of the popular signal processing techniques is signal filtering. A filter is a device that can suppresses unnecessary features or components from a signal and the most commonly used filters are low-pass, high-pass, band-pass and band-stop. There are four classic electronic filters which are Butterworth filter, Chebyshev filter, Bessel filters and Elliptic filter.

In this study, a Butterworth low-pass filter will be used. Low-pass filter is a designable circuit which able to pass only desired signal and modify, reshape or reject all unwanted high frequencies. Butterworth filters are one of the most commonly used digital filters in motion analysis and in audio circuits due to its characteristics. It is a type of signal processing filter designed to have as flat frequency response as possible (no ripples) in the pass-band and zero roll off response in the stop-band [20]. The mathematical equation on the magnitude of squared response of an N-th order for Buttheworth filter, $H_a(s)$ is as in Eq. (1) where Ω_c is the 3 dB cut-off frequency [21].

$$|H_a(j\Omega)|^2 = \frac{1}{1 + (\Omega/\Omega_c)^{2N}} \tag{1}$$

Least Square. LS is a standard regression means to estimate the solutions of over determined systems that is regularly used in data fitting and estimation. This method will identify the best match for the system by reducing the sum of square error [22]. Presumed that there is a certain linear relationship between $p-1$ independent variable $x_1, x_2, \ldots, x_{p-1}$ and random variable y, it satisfies the relationship as in Eq. (2) and can be changed for n group samples $\left(x_{i1}, x_{i2}, \ldots, x_{ip-1}; y_i\right) i = 1, 2, \ldots, n$ to Eq. (3.)

$$\begin{cases} y = \beta_0 + \beta_1 x_1 + \cdots + \beta_{p-1} x_{p-1} + \varepsilon \\ E(\varepsilon) = 0, D(\varepsilon) = \sigma^2 \end{cases} \tag{2}$$

$$y_i = \beta_0 + \beta_1 x_{i1} + \cdots + \beta_{p-1} x_{i,p-1}, n = 1, 2, \ldots, n \tag{3}$$

Equation (4) shows the matrix form of Eq. (3) with equivalent form of as in Eq. (5). Y is the variable observation vector of $n \times 1$ and X is the known design matrix of $n \times p$. β is the unknown parameter vector of $p \times 1$ and ε is the random error vector.

$$\begin{pmatrix} y_1 \\ y_2 \\ \vdots \\ y_n \end{pmatrix} = \begin{pmatrix} 1 & x_{11} & \dots & x_{1,p-1} \\ 1 & x_{21} & \dots & x_{2,p-1} \\ \vdots & \vdots & \vdots & \vdots \\ 1 & x_{i1} & \dots & x_{n,p-1} \end{pmatrix} \begin{pmatrix} \beta_0 \\ \beta_1 \\ \vdots \\ \beta_n \end{pmatrix} + \begin{pmatrix} \varepsilon_1 \\ \varepsilon_2 \\ \vdots \\ \varepsilon_n \end{pmatrix} \quad (4)$$

$$\begin{cases} Y = X\beta + \varepsilon \\ E(\varepsilon) = 0,\ D(\varepsilon) = \sigma^2 I_n \end{cases} \quad (5)$$

The error vector $\varepsilon = Y - X\beta$ was squared and the total error was computed with Eq. (6). The equation was later unfolded to Eq. (7).

$$Q(\beta) = \|Y - X\beta\|^2 = (Y - X\beta)(Y - X\beta) \quad (6)$$

$$Q(\beta) = Y'Y - 2Y'X\beta + \beta'X'X\beta \quad (7)$$

The partial derivative of β is obtained and was set to zero according to the extremum principle of calculus to achieve the system of equation $X'X\beta = XY$. From Eq. (8), the estimated value of β is obtained. Thus, linear regression model can fit the known data better and make more accurate predictions from it [22].

$$\hat{\beta} = (X'X)^{-1}X'Y \quad (8)$$

Classification of Swing Phases. Once the IMU data were filtered and estimated using LS, the swing phases were segmented using peak analysis and classified automatically by the system with reference to the Fig. 2b. The maximum peak of the estimated swing signal was obtained and marked as IMP point. Meanwhile, the second maximum peak was used to find the valley between two maximum peaks and was later pointed as BST point. Both ADD and FIN points were obtained by utilizing the value from BST point to indicate the start and end of the swing signals, respectively. For the purpose of accuracy calculation, the resultant velocity from OMCS and estimated IMU signals were used.

3 Results and Discussion

The signals obtained from both IMU and OMCS were represented in three axes and the raw data of angle, acceleration and angular velocity for IMU were demonstrated in Fig. 4. The X-axis, Y-axis and Z-axis on each parameters were indicated with red solid line, black dashed line and blue-dotted line, respectively.

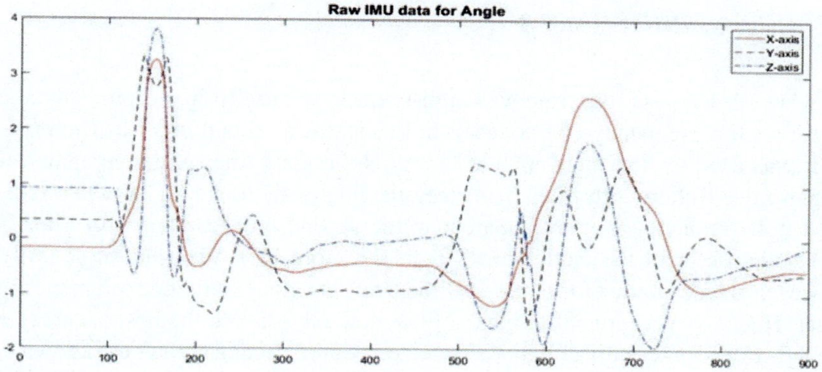

(a)

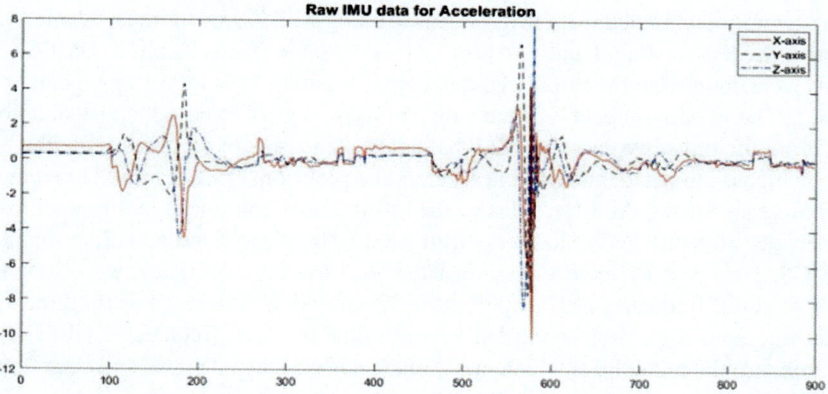

(b)

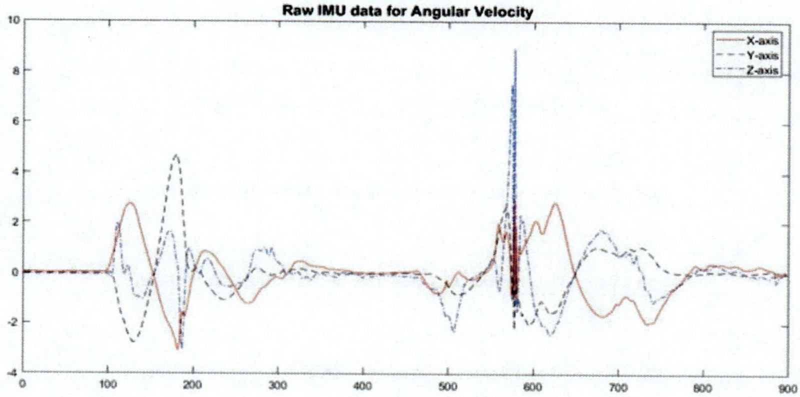

(c)

Fig. 4 Raw IMU data for **a** Acceleration, **b** Angle, and **c** Angular velocity

3.1 Signal Processing

Butterworth lowpass filter removes high-frequency noise from a digital signal and preserves low-frequency components. In this study, a second-order of Butterworth filter was used as it is more inclined towards the ideal filter graph compared to a single-order Butterworth filter. The steep decline of the input signal when crosses the cutoff frequency is more apparent in the second-order Butterworth filter. The higher the order of the Butterworth filter, the more ideal the gain curve will be. However, the accuracy of the filter will be decrease along with the increment of the order. Hence, it was crucial to have a filter that can provide the best accuracy and second-order Butterworth filter was selected. Cutoff frequency is best known as a frequency that keep the boundary between pass and stop band. In low-pass filter, the cutoff frequency is 70.7% of its input voltage. To find the best cutoff frequency, Y-axis for accleration obtained from IMU was used along with Z-axis of linear acceleration from OMCS as we found that Y-axis in IMU is the same axis as Z-axis in OMCS. The tests were conducted for cutoff frequency of 5, 10, and 15 Hz for all signal data as in Fig. 5. The maximum peak of each subplots were aligned to ease the visualization.

From the plots, we can observed that cutoff frequency of 10 Hz provide the best plot which is almost identical to the acceleration plot from OMCS and able to remove unnecessary noises. At 5 Hz, most of the information from the signal was removed especially after the IMP occurred, while, at 15 Hz, it provided a similar signal as OMCS. However, to avoid the inconsistency of the filter accuracy, we decided to choose cutoff frequency at 10 Hz. Figure 6 illustrated the IMU signals for three axes after filtered using second-order Butterworth filter at cutoff frequency of 10 Hz. As mentioned, the red solid line is X-axis, black dashed line is Y-axis and blue-dotted line is Z-axis.

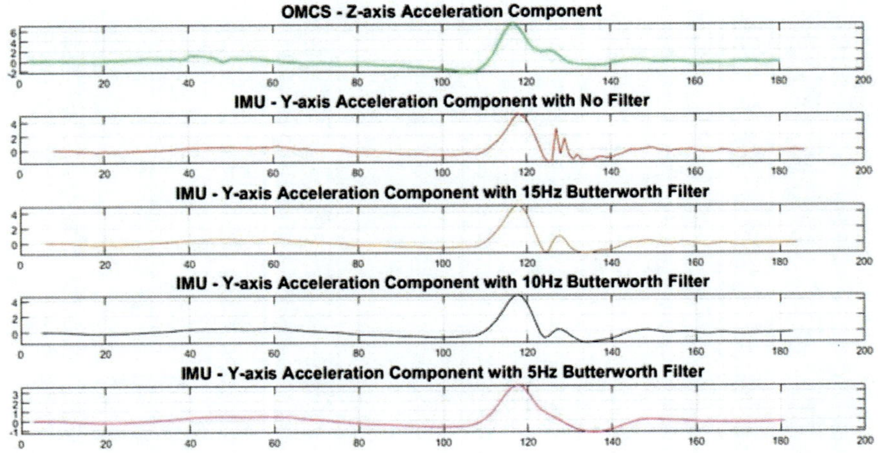

Fig. 5 The comparison of acceleration between OMCS's Z-axis component and IMU's Y-axis component with the filtered signals of IMU at 5, 10 and 15 Hz

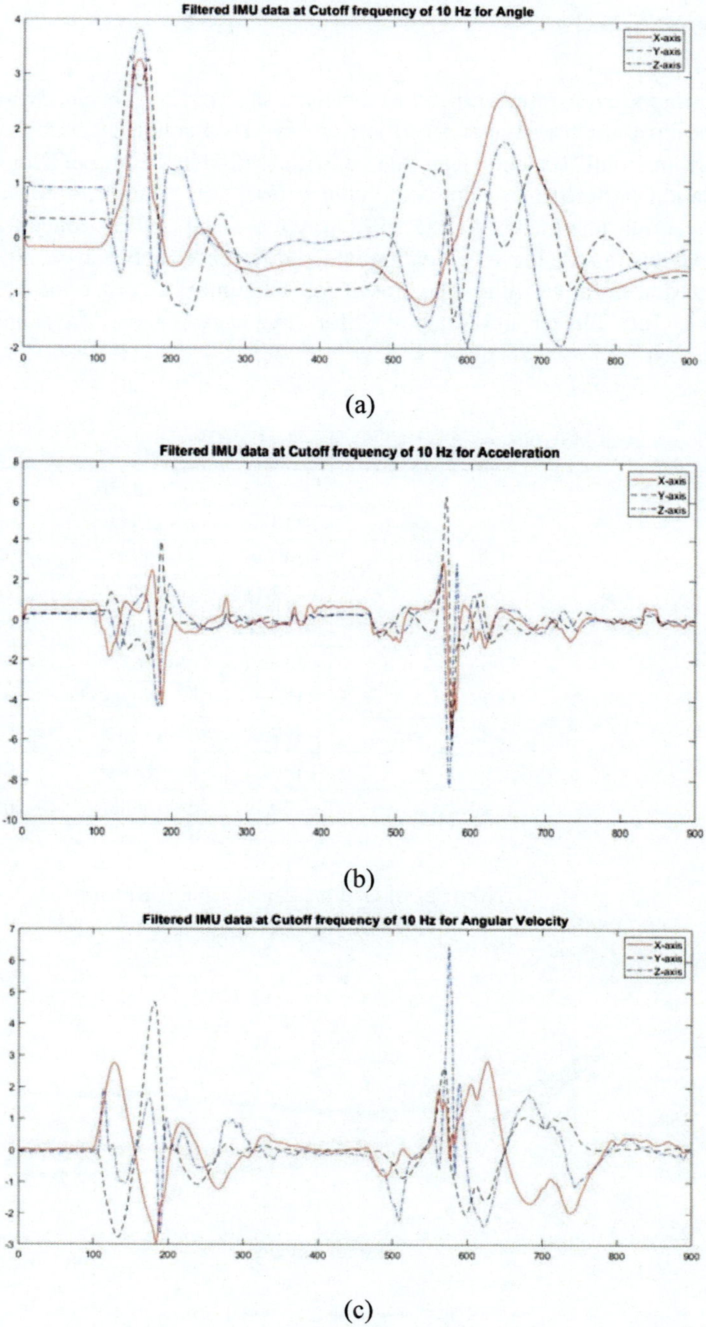

Fig. 6 The filtered IMU signals with second-order Butterworth filter at cutoff frequency of 10 Hz for **a** Angle, **b** Acceleration, and **c** Angular velocity

3.2 Least Square

The most important parameter that must be considered in LS is the number of order. To find the most suitable number of order, the analysis was conducted with the number of order from 1 until 10. The Mean Square Error (MSE) in the unit of decibels (dB) for validation performance using validation dataset was taken into consideration as an evaluation parameter. As the MSE was too small, it was converted in the unit of decibels to ease the visualization on the differences between each value and this resulted in negative values. The result for the number of order and MSE were summarized into Table 1 and was plotted for three axes, X-axis, Y-axis and Z-axis as in Fig. 7.

Table 1 Summary of MSE in dB for LS's order

No. of order	X-axis	Y-axis	Z-axis
1	−78.47	−88.66	−83.41
2	−115.62	−126.09	−135.51
3	−140.15	−159.45	−154.33
4	−142.11	−174.24	−174.99
5	−144.68	−169.35	−189.45
6	−139.67	−176.06	−197.81
7	134.54	−180.26	−197.11
8	−127.17	−177.17	−196.69
9	−129.11	−168.86	−200.14
10	−126.11	−168.86	−200.14

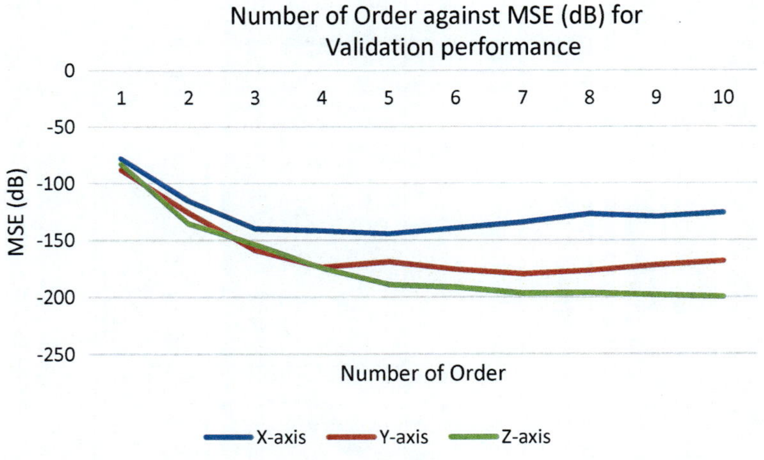

Fig. 7 Number of LS's order against MSE in decibels for validation dataset

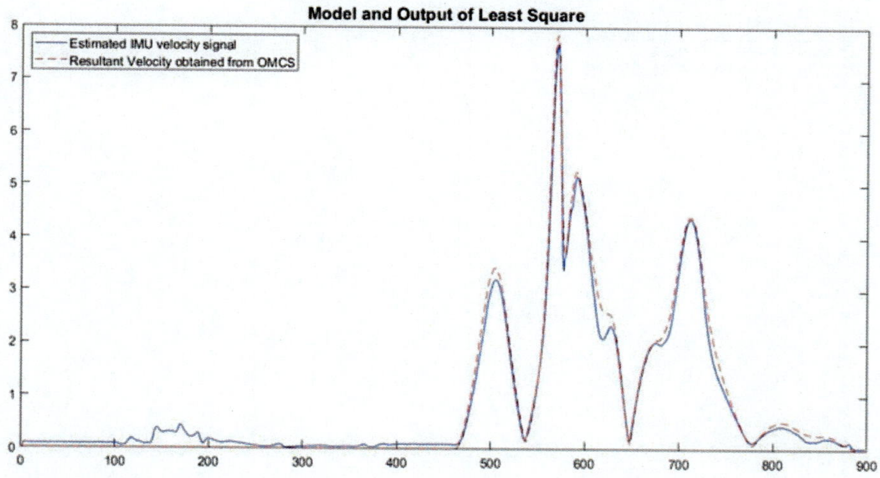

Fig. 8 Estimated IMU signal mapped with resultant velocity from OMCS

From Fig. 7, we can observe that the MSE decrease steadily from 1st order until 10th order. However, the graph started to fluctuate with a slight increase and decrease in their values from 4th order and have a stable trend after that. In theory, the further the values from origin, the better the model will be. To avoid the LS model to be underfitting or overfitting, it is necessary to select the most optimum order. At 3rd order, X-axis and Y-axis plots begins to have a steady trend. We also observed that at 5th order, the values for Z-axis is the lowest, however, the LS model might be overfitting and thus may influence in the performance of the model. The 1st order was omitted as it have the highest MSE values for all three axes and the model might be unable to learn properly. Hence, the 2nd order was selected for the most optimum order of the LS model. Figure 8 shows the estimated IMU after utilizing Butterworth filter at cutoff frequency of 10 Hz and 2nd order LS. The solid blue line indicate the estimated IMU while the dotted red line is the resultant velocity from OMCS. Both of the signals were mapped together to present that LS model is able to estimate the IMU data and produced the signal that has almost similar shape and envelope as reference signal (velocity signal from OMCS).

3.3 *Classification of Swing Phases*

As mentioned in Methodology section, the estimated IMU was later segmented, classified, and labelled automatically by the system. Figure 9 presented the final outcome of the system and Table 2 shows the average accuracy percentage for the classification of each swing phases. The system demonstrated high accuracy on the classification with the percentage more than 99%.

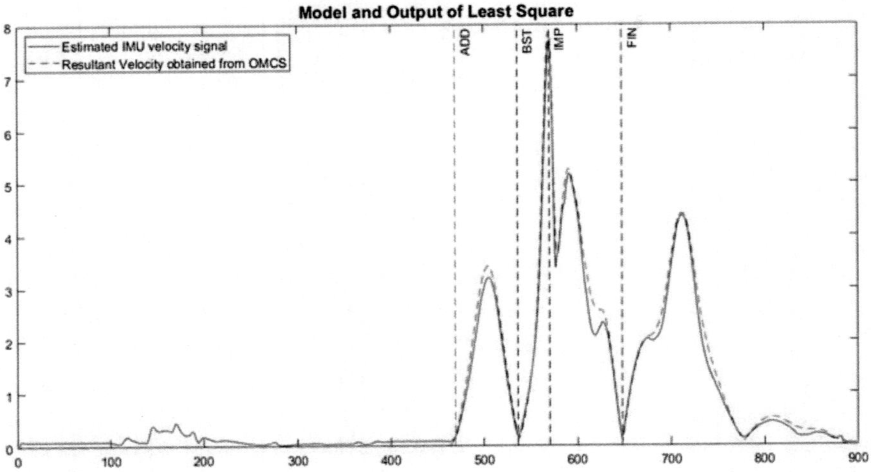

Fig. 9 Classified estimated IMU with labels for each swing phases

Table 2 Average accuracy for classification of woodball swing phases

Swing phases	Average accuracy (%)
Address	99.72
Backswing top	99.43
Impact	99.42
Finish	99.13

4 Conclusion

The purpose of this study was to develop an automated classification system on the woodball swinging phases which supposed to classify the swing phases and provide the analysis on the swinging postures to woodball players and sport analysts. This system was developed through a few development stages, starting from data collection stage until classification stage. 30 woodball players volunteered to take part in this study and both IMU and OMCS data were collected simultaneously. The data undergo signal processing stage by applying second-order Butterworth filter with cutoff frequency of 10 Hz. The filtered signals were then estimated using 2nd order LS model before being classified and labelled according to their respective swing phases. The results proved that the system able to accomplish the aims of this study with high average accuracy percentage of 99.72, 99.43, 99.42 and 99.13% for swing phases of ADD, BST, IMP and FIN, respectively. When the swing from the novice player matched with the professional's swing with high accuracy, we can assumed that the player has mastered the correct swing postures.

Acknowledgements The authors would like to thank Mr. Azan bin Yaakob who is a deputy vice president of Malaysian Woodball Association for kindly cooperating with this study and show a great interest in it. In addition, the authors are very thankful to the 30 volunteered woodball players for actively contributing for the data collection and Centre of Excellence, Sports Engineering Research Centre (SERC), Universiti Malaysia Perlis (UniMAP) for giving permission to conduct this study.

References

1. Chang SH, Lee J (2017) Teaching striking skills in elementary physical education using woodball. J Phys Educ Recreat Danc 88(8):21–27. https://doi.org/10.1080/07303084.2017.1356767
2. Parker M, Steinberg GM (1999) The benefits of assessing eye dominance in motor skill instruction. J. Phys Educ Recreat Danc 70(3):20–21. https://doi.org/10.1080/07303084.1999.10605887
3. Schmidt RA, Wrisberg CA (2008) Motor learning and performance: a situation based learning approach, 4th ed. Human Kinetics
4. Kim M, Park S (2020) Golf swing segmentation from a single IMU using machine learning. Sensors 20(16):4466. https://doi.org/10.3390/s20164466
5. Ribeiro NF, Santos CP (2017) Inertial measurement units: A brief state of the art on gait analysis. In: ENBENG 2017—5th Portuguese Meeting in Bioengineering (ENBENG), January 2017. https://doi.org/10.1109/ENBENG.2017.7889458
6. Beynon S, McGinley JL, Dobson F, Baker R (2010) Correlations of the gait profile score and the movement analysis profile relative to clinical judgments. Gait Posture 32(1):129–132. https://doi.org/10.1016/j.gaitpost.2010.01.010
7. Campo E, Bonhomme S, Chan M, Esteve D (2010) Remote tracking patients in retirement home using wireless multi-sensor system. In: 2th IEEE international conference on e-health networking, application and services, pp 226–230 (2010). https://doi.org/10.1109/HEALTH.2010.5556567
8. Sekine M, Tamura T, Akay M, Fujimoto T, Togawa T, Fukui Y (2002) Discrimination of walking patterns using wavelet-based fractal analysis. IEEE Trans Neural Syst Rehabil Eng 10(3):188–196. https://doi.org/10.1109/TNSRE.2002.802879
9. Liu H (2020) Rail transit collaborative robot systems. In: Robot systems for rail transit application, 1st ed. Elsevier, 2020, Chapter 3, pp 110–112
10. Muro-de-la-Herran A, García-Zapirain B, Méndez-Zorrilla A (2014) Gait analysis methods: an overview of wearable and non-wearable systems, highlighting clinical applications. Sensors (Switzerland) 14(2):3362–3394. https://doi.org/10.3390/s140203362
11. Liu T, Inoue Y, Shibata K (2009) Development of a wearable sensor system for quantitative gait analysis. Meas. J. Int. Meas. Confed. 42(7):978–988. https://doi.org/10.1016/j.measurement.2009.02.002
12. Añazco EV, Han SJ, Kim K, Lopez PR, Kim TS, Lee S (2021) Hand gesture recognition using single patchable six-axis inertial measurement unit via recurrent neural networks. Sensors (Switzerland) 21(4):1–14. https://doi.org/10.3390/s21041404
13. Eyobu OS, Kim YW, Cha D, Han DS (2018) A real-time sleeping position recognition system using IMU sensor motion data. In: IEEE international conference on consumer electronics, pp 1–2 (2018). https://doi.org/10.1109/ICCE.2018.8326209
14. Wang Y, Zhao Y, Chan RHM, Li WJ (2018) Volleyball skill assessment using a single wearable micro inertial measurement unit at wrist. IEEE Access 6:13758–13765. https://doi.org/10.1109/ACCESS.2018.2792220
15. Zhang Z, Zhang Y, Kos A, Umek A (2018) Strain gage sensor based golfer identification using machine learning algorithms. Procedia Comput Sci 129:135–140. https://doi.org/10.1016/j.procs.2018.03.061

16. Jiao L, Wu H, Bie R, Umek A, Kos A (2018) Multi-sensor golf swing classification using deep CNN. Procedia Comput Sci 129:59–65. https://doi.org/10.1016/j.procs.2018.03.046
17. Sinclair J, Currigan G, Fewtrell D, Taylor PJ (2014) Three-dimensional kinematics observed between different clubs during the full golf swing. J Athl Enhanc 3(3) (2014). https://doi.org/10.4172/2324-9080.1000147
18. Cheon M, Khuyagbaatar B, Yeom JH, Kim YH (2020) Analysis of swing tempo, swing rhythm, and functional swing plane slope in golf with a wearable inertial measurement unit sensor. J Mech Sci Technol 34(7):3095–3101. https://doi.org/10.1007/s12206-020-0640-3
19. Polat K, Muthusamy H, Acharya R, Guo Y (2017) Guest editorial: new trends in data preprocessing methods for signal and image classification. Neural Comput Appl 28(10):2839–2841. https://doi.org/10.1007/s00521-017-3202-6
20. Akdemir Akar S, Kara S, Latifoğlu F, Bilgiç V (2013) Spectral analysis of photoplethysmographic signals: the importance of preprocessing. Biomed Signal Process Control 8(1):16–22 (2013). https://doi.org/10.1016/j.bspc.2012.04.002
21. Mitra SK (2006) Digital signal processing. McGraw Hill Companies, New York
22. Kong M, Li D, Zhang D (2019) Research on the application of improved least square method in linear fitting. In: IOP conference series: earth and environmental science, vol 252, no 5 (2019). https://doi.org/10.1088/1755-1315/252/5/052158

Experimental Investigation of Mechanical Properties of Sepak Takraw Ball Based on Different Ball Orientation

Nik Mohd Haikal Mohamed Hassan, Nasrul Hadi Johari, Mohd Hasnun Arif Hassan, Idris Mat Sahat, Mohd Nazeri Omar, and Zulkifli Ahmad

Abstract Sepak Takraw players have intensively practiced heading the ball as the game's primary movement. Repetitive takraw ball heading can result in head injuries such as concussion, internal bleeding, and dizziness. The head injury criterion can be measured practically, yet the mechanical properties of the takraw ball are not well examined. The primary aim of this study is to investigate the mechanical properties of takraw balls at different orientations based on a quasi-static compression experiment of two takraw balls, GE511 and MT908. Each ball has been subjected to compression tests in three distinct orientations of woven layers, namely Orientations 1–3. MT908 has a greater ultimate force and stiffness than GE511. It is also discovered that Orientation 1 has the highest values for both the mean ultimate force that occurred and the mean stiffness, followed by Orientations 2 and 3. These discoveries are particularly relevant to the creation of modern takraw balls and head protectors.

Keywords Takraw ball · Compressive testing · Ball stiffness

1 Introduction

Sepak Takraw is a unique game compared to other sports due to its athletic ball control, particularly using the legs and head. A powerful, high-speed, and accurate serve-ball needs to be made by a player for a decent attacking move [1]. The opponent

N. M. H. Mohamed Hassan (✉) · N. H. Johari · M. H. A. Hassan · I. M. Sahat · M. N. Omar · Z. Ahmad
Faculty of Mechanical and Automotive Engineering Technology, Universiti Malaysia Pahang, 26600 Pekan, Pahang, Malaysia
e-mail: nhaikalhassan5@gmail.com

N. H. Johari
e-mail: nhadi@ump.edu.my

N. H. Johari
Centre for Advanced Industrial Technology, Universiti Malaysia Pahang, 26600 Pekan, Pahang, Malaysia

© The Author(s), under exclusive license to Springer Nature Singapore Pte Ltd. 2023
S. F. Syed Omar et al. (eds.), *Innovation and Technology in Sports*, Lecture Notes in Bioengineering, https://doi.org/10.1007/978-981-99-0297-2_16

will be required to receive a hard serve-ball using his head and players train intensively on ball control and heading. A sufficient ball head from a serve-ball practice session can be as many as 100 balls per session.

Many studies show that purposefully heading a soccer ball is related to mild traumatic brain injury (MTBI) [2]. Insofar as this study is concerned, only soccer and sepak takraw need a purposeful ball head which takraw ball is harder than a soccer ball. The current synthetic ball has been introduced in SEA Games 2007 to replace rattan ball even though has been protested when it was claimed to be more dangerous by Malaysia's team. In addition, a powerful serve comes with a high-speed ball. Thus, repeatedly receiving a takraw serve ball using the head possibly leads to MTBI. A suggestion of the need for protective equipment in Asian Games 2018 on South Korean athletes' sports injury incidence when it shows that sepak takraw is the highest in rate of injury [3].

A Sepak Takraw ball is not bounded the same way that a soccer ball or any other ball is. Unlike normal weave, the takraw ball's weave is made by the warp and weft of 4 non-uniform synthetic polymeric fibers yarn filaments. The filaments are woven with hollow interlacement [4], which makes several big holes on the ball surface, as shown in Fig. 1. This makes the ball unique in that air can go through the ball surface. This airflow has been studied by Mubin et al. [5] for CFD simulation where the flow is through to the front view of three (3) orientations (see Fig. 1).

Several studies were carried out to determine the threshold of head injury, also known as head injury criteria and has been proposed into three categories: based on translational accelerations of the head's center of gravity; translational and rotational accelerations of the head's center of gravity; and stresses and strains in brain tissue [6]. A study [7] based on translational accelerations by its mathematical model shows that a soccer ball's stiffness results in brain translational accelerations. The stiffer the soccer ball, the higher the peak force [8, 9]. As a result, the corresponding quasi-static stiffness of the takraw ball for each orientation is designed and tested in this paper.

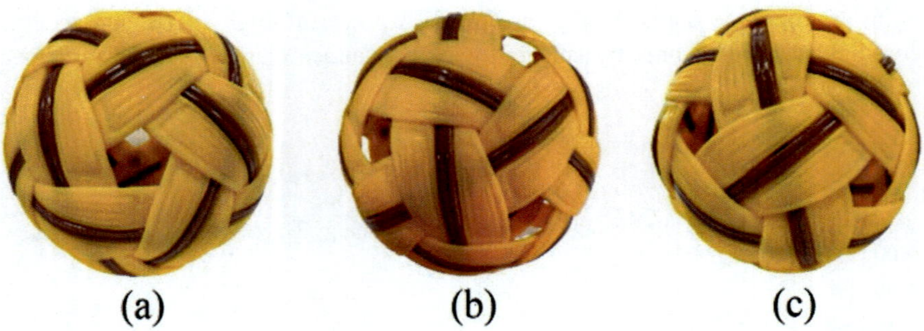

Fig. 1 Three orientation: **a** Orientation 1, **b** Orientation 2, and **c** Orientation 3 [5]

2 Methodology

For the experiments, the International Sepak Takraw Federation (ISTAF)-approved, tournament-specific GE511 and MT908 takraw balls were used. GE511 was constructed from special waterproof synthetic polymeric fibers, whereas MT908 was constructed from rubber-coated synthetic polymeric fibers. As shown in Fig. 2a as well as Fig. 2b, these takraw balls are created by combining four distinct fiber shapes and sizes and shaping them into a circle (Tables 1, 2 and Fig. 3).

Compression tests were performed three times, using three balls for each orientation. The dimensions and weight were calibrated as shown in Table 1 and Table 2. The compression tests were performed using the Instron 3369 universal testing machine (Fig. 3a) according to the ASTM Standard C496-96, which describes the procedure to be followed in splitting the tensile strength of cylindrical concrete specimens [10, 11]. The test speed was 2 mm/min and stopped at 50% of its outer diameter. Figure 3b shows an example of a compression test.

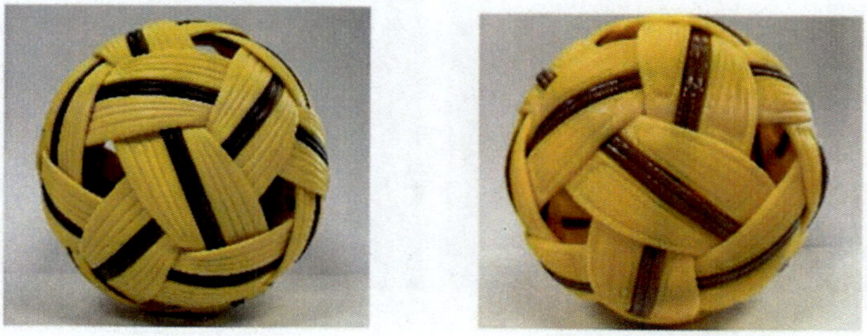

(a) (b)

Fig. 2 Unique Woven Takraw ball **a** GE511 **b** MT908

Table 1 The dimension and weight for GE511

Parameter	Orientation 1			Orientation 2			Orientation 3		
	Ball 1	Ball 2	Ball 3	Ball 1	Ball 2	Ball 3	Ball 1	Ball 2	Ball 3
Weight [g]	172.6	172.7	171.6	173.5	169.8	172.2	172.3	170.1	169.7
Outer diameter [mm]	129.65	129.25	129.7	131.4	131.2	131.5	132.5	132.35	132.65
Mean outer diameter [mm]	129.533			131.367			132.5		
Inner diameter [mm]	122.25	121.85	122.3	124	123.8	124.1	125.1	124.95	125.25
Thickness [mm]	3.7	3.7	3.7	3.7	3.7	3.7	3.7	3.7	3.7

Table 2 The dimension and weight for MT908

Parameter	Orientation 1			Orientation 2			Orientation 3		
	Ball 1	Ball 2	Ball 3	Ball 1	Ball 2	Ball 3	Ball 1	Ball 2	Ball 3
Weight [g]	178.7	178	178.3	179.4	178.8	178.6	178.7	178	177.5
Outer diameter [mm]	129.65	129.7	129.7	130.5	131.2	131.0	132.8	134.6	134.7
Mean outer diameter [mm]	129.683			130.867			134.017		
Inner diameter [mm]	122.05	122.1	122.1	122.9	123.6	123.4	125.2	127.0	127.1
Thickness [mm]	3.8	3.8	3.8	3.8	3.8	3.8	3.8	3.8	3.8

Fig. 3 **a** Universal test machine. **b** Example of compression test until 50% of its outer diameter

3 Result and Analysis

Understanding non-linear compression properties is critical in weaving design. The force–displacement curve explains some properties of the ball. The results delivered from the testing are summarized in Table 3.

Ultimate force and ball compression before failure are discussed in this section, aided by the graphs in Fig. 4a–f, while the ball stiffness to its compression is discussed in Fig. 5a–f. It is clear from Fig. 4a–d that the ultimate force for both GE511 and MT908 is carried when crushing occurs for Orientations 1 and 2. However, for Orientation 3, the GE511 balls did not fail and their maximum force maybe has yet to occur. Interestingly, multiple failures occurred for MT908 at Orientation 3, as seen in epoxy-based sandwich composites with integrally woven fabric [12].

On the other hand, the ultimate force is orientation dependent. Both GE511 and MT908 show that the highest mean ultimate force occurred at Orientation 1, followed by Orientations 2 and 3. The mean ultimate force of MT908 is higher than GE511 for

Table 3 Properties of GE511 & MT908 Takraw ball

Type of ball	Orientation	Ball	Compressive displacement at maximum force [mm]	Mean compressive displacement at maximum force [mm]	% Mean compressive displacement to mean outer diameter	Ultimate force [kN]	Mean ultimate force [kN]	Mean stiffness [N/m]
GE 511	Orientation 1	Ball 1	64.66	64.71	49.956	0.67	0.66	11,606.15
		Ball 2	64.62			0.71		
		Ball 3	64.85			0.61		
	Orientation 2	Ball 1	65.7	65.51	49.866	0.48	0.47	10,188.14
		Ball 2	65.07			0.45		
		Ball 3	65.75			0.48		
	Orientation 3	Ball 1	65.7	64.70	48.833	0.39	0.37	9450.88
		Ball 2	62.08			0.34		
		Ball 3	66.33			0.37		
MT 908	Orientation 1	Ball 1	63.52	64.22	49.523	0.9	0.97	15,058.17
		Ball 2	64.85			1.02		
		Ball 3	64.3			0.98		
	Orientation 2	Ball 1	65.25	65.44	50.0003	0.68	0.64	14,110.90
		Ball 2	65.58			0.62		
		Ball 3	65.48			0.62		
	Orientation 3	Ball 1	51.62	56.34	42.042	0.50	0.52	12,020.09
		Ball 2	63.29			0.55		
		Ball 3	54.12			0.55		

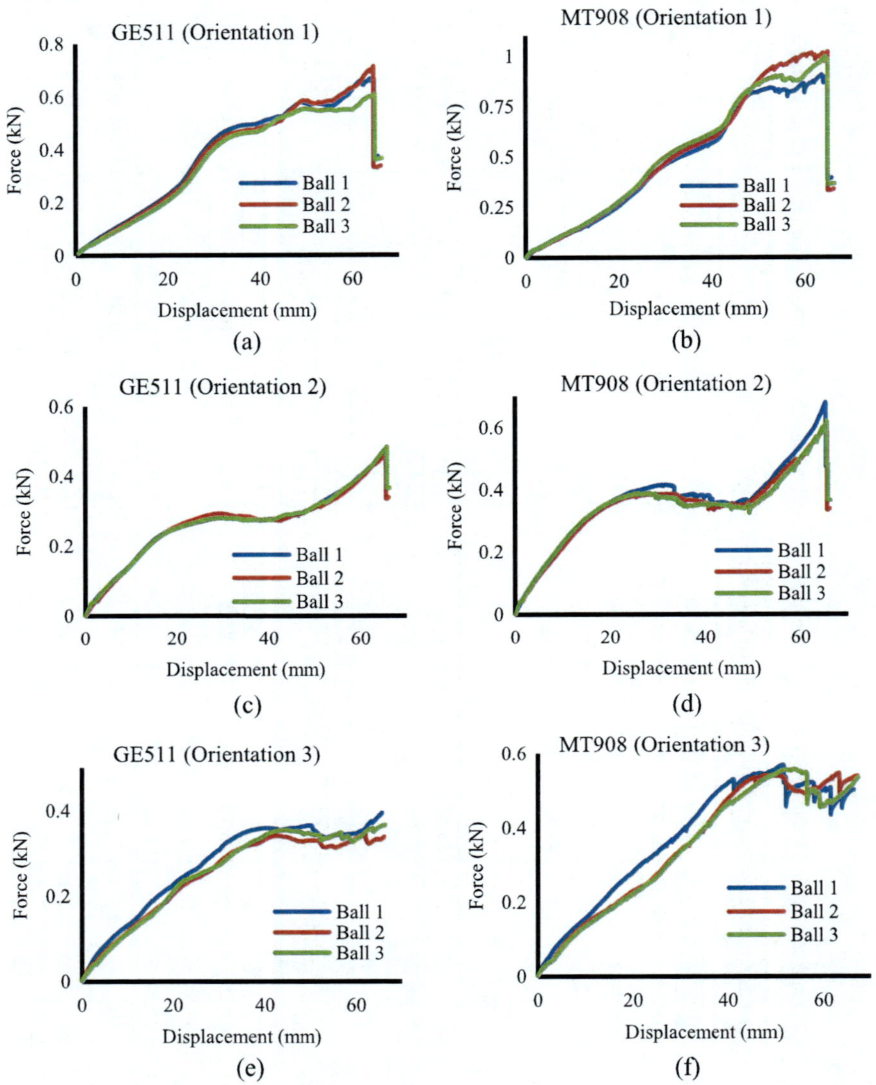

Fig. 4 Force–displacement plot for GE511 and MT908 Orientation 1 (**a**) & (**b**), Orientation 2 (**c**) & (**d**), and Orientation 3 (**e**) & (**f**)

all orientations. This shows MT908 can bear more compressive load prior to failure than GE511 in all orientations. In the case of mean ball stiffness, MT908 ball stiffness is higher than GE511. Interestingly, for both GE511 and MT908, Orientation 1 shows the highest value, followed by Orientations 2 and 3.

Three curves of the force–displacement of Balls 1–3 are plotted in Fig. 4a–f for GE511 and MT908. A multilinear plastic pattern is shown in Orientations 1 and 2 until

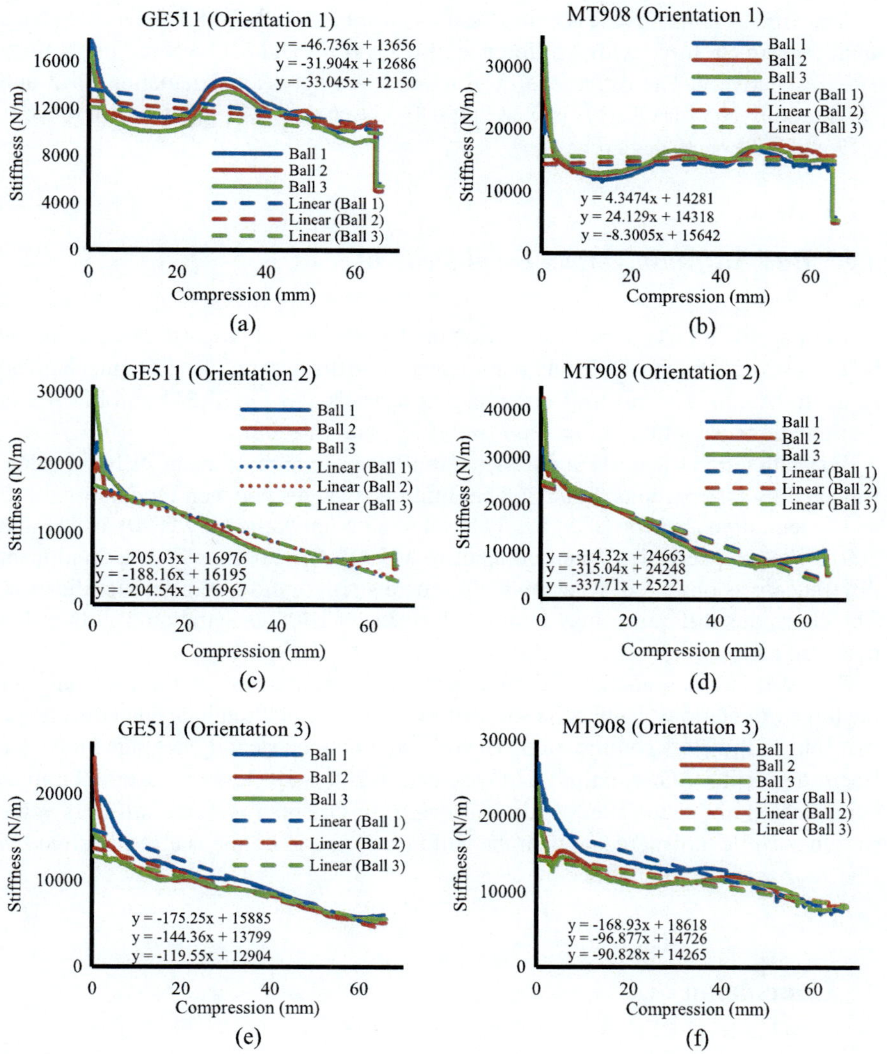

Fig. 5 Stiffness-compression plot for GE511 and MT908 Orientation 1 (**a**) & (**b**), Orientation 2 (**c**) & (**d**), and Orientation 3 (**e**) and (**f**)

fracture. It is noted that the outer diameter is different between orientations; therefore, it is best to calculate the percentage of the Mean Compressive Displacement to its outer diameter. All orientations failed below 50% of their original size. MT908 in Orientation 1 bears the lowest compression with 38.9%, while GE511 in Orientation 3 bears the highest compression with 50.004%. GE511 and MT908 balls show the same behavior between orientations, but the load of failure for MT908 is higher than GE511.

The GE511 balls have less compression capacity than MT908 balls as they have smaller ultimate force with a difference of 0.21, 0.17, and 0.15 kN in Orientations 1–3, respectively. The difference in ultimate force between Orientations 1–2 and Orientations 1–3 for GE511, is 0.33 and 0.44 kN, and for MT908, the difference is 0.19 and 0.3 kN, respectively.

3.1 Ball Stiffness Versus Displacement

To evaluate the effects of ball orientation on the mechanical properties of the takraw ball to estimate brain injury, the parameters of stiffness were used for mechanical characterization. The stiffness-compression of Balls 1–3 for GE511 and MT908 in 3 orientations is plotted in Fig. 5a–f for GE511 and MT908.

For both GE511 and MT908, Orientation 1 has the highest mean stiffness value, followed by Orientations 2 and 3. The difference values between Orientations 1–2 and Orientations 2–3 for GE511 is 1418 and 737 N/m, while for MT908 is 947 and 2091 N/m, respectively. When comparing MT908 Orientations 2–3, a significant difference was observed, whereas the ultimate strength did not differ significantly. The enormous difference may be a result of the MT908 Orientation 3 failure that occurred early on.

The stiffness degradation curves reveal similar responses for the two balls for the three orientations. Both balls show most of the degradation happened during the first four millimeters compression. The degradation then steadily degraded until the fracture except for Orientation 3. In Orientation 2, the stiffness rose after 49 mm of compression and then failed while still rising. In Orientation 1, the stiffness value rose in a cyclic lookalike and then the ball failed. The trendline is also illustrated in Fig. 5a–f with their equation.

4 Conclusion

This paper has presented an overview of the three orientations which can influence the mechanical ultimate force and stiffness of two types of takraw balls which are made from woven synthetic polymeric fibers. The ball stiffness is essential in determining the head injury experienced by a sepak takraw player. This led to the future development of the modern takraw balls as well as the players' head protection.

Orientation 1 shows the highest value of the ultimate force that occurred as well as the stiffness followed by Orientations 2 and 3. The mean value of the ultimate force of GE511 for Orientations 1–3 is 0.66, 0.47, and 0.37 kN respectively, and the mean stiffness is 11606.15 N/m, 10188.14, and 9450.88 N/m respectively, while the mean ultimate force for MT908 for Orientations 1–3 is 0.97, 0.64, and 0.52 kN respectively, and the mean stiffness is 15058.17, 14110.9, and 12,020.09 N/m respectively.

Acknowledgements The authors would like to thank the Universiti Malaysia Pahang for providing financial support under Product Development Grant (PDU203226), Post Graduate Research Grant Scheme (PGRS210344) and to Ministry of Higher Education under Hadiah Latihan Persekutuan (HLP 2021-2023).

References

1. Hanafi H, Brahmana Rangga P (2016) Perbandingan Latihan Bola Digantung Dan Bola Dilambungkan Terhadap Hasil Belajar Sepak Mula Dalam Permainan Sepak Takraw Pada Siswa Putra Kelas X-Is 1 Sma 17 Agustus 1945 Surabaya. Wahana 67(2):71–76. https://doi.org/10.36456/wahana.v67i2.504
2. Hasanuddin I et al (2015) Investigation of the head impact power of a Sepak Takraw ball on Sepak Takraw players. Move Heal Exerc 4(2):45–56. https://doi.org/10.15282/mohe.v4i2.21
3. Lhee SH et al (2021) Sports injury and illness incidence among South Korean elite athletes in the 2018 Asian games: a single-physician prospective study of 782 athletes. BMJ Open Sport Exerc Med 7(1). https://doi.org/10.1136/bmjsem-2019-000689
4. Lin JH, Yu ZC, Zhang JF, Lou CW (2014) Manufacturing techniques and functional properties of the bamboo charcoal/antibacterial/stainless steel metal composite woven fabric. Adv Mater Res 910:238–241. https://doi.org/10.4028/www.scientific.net/AMR.910.238
5. Mubin ASA, Ahmad N, Mamat A, Taha Z, Hasanuddin I (2015) CFD study of drag and lift of Sepak Takraw ball at different face orientations. Adv Mech Eng 7(1). https://doi.org/10.1155/2014/369137
6. Van Den Bosch HLA, Wismans JSHM, Nijmeijer H (2006) Crash helmet testing and design specifications
7. Taha Z, Arif Hassan MH, Hasanuddin I (2015) Analytical modelling of soccer heading. Sadhana—Acad Proc Eng Sci 40(5):1567–1578. https://doi.org/10.1007/s12046-015-0383-5
8. Tierney GJ, Power J, Simms C (2020) Force experienced by the head during heading is influenced more by speed than the mechanical properties of the football. Scand J Med Sci Sport (July):1–8. https://doi.org/10.1111/sms.13816
9. Nevins D, Smith L (2013) Influence of ball properties on simulated ball-to-head impacts. Procedia Eng 60:4–9. https://doi.org/10.1016/j.proeng.2013.07.002
10. Adam NM (2014) Experimental analysis of mechanical properties of selected Takraw balls in Malaysia. Malays J Move Heal Exerc 3(1):57
11. ASTM C496/C496M (1996) Standard test method for splitting tensile strength of cylindrical concrete specimens. ASTM Int (March 1996):1–5. ftp://ftp.astmtmc.cmu.edu/docs/diesel/cummins/procedure_and_ils/ism/Archive/ISMProcedure(Draft10).doc
12. Balakumaran V, Alagirusamy R, Kalyanasundaram D (2021) Epoxy based sandwich composite using three-dimensional integrally woven fabric as core strengthened with additional carbon face-sheets. J Mech Behav Biomed Mater 116(December 2020):104317. https://doi.org/10.1016/j.jmbbm.2021.104317

Development of the Table Tennis Robot Launcher

Irlina Jazlin Jamaludin, Ilmam Mumtaz Islah Munjih, Zulkifli Ahmad, and Mohamad Zairi Baharom

Abstract Table tennis is a dual sport in which two teams compete against an opposing side. In training conditions, it needs consistently receive the ball with varying ball spins and angles to improve the athlete's skills. Therefore, this study aims to develop an automatic table tennis robot that can launch the ball in a different way of rotation. Two 12 V DC motors were placed against each other to produce the opposite direction of ball spin. While four types of rotations which are topspin, backspin, right spin, and left spin, were considered in the evaluation. The microcontroller was used to control the system, including the motor speed and launcher's angle itself. Hence, the combination of variables applied can be customised and increasing the difficulties of training level. In addition, the setting of robot movement can be set up via the control board or wirelessly using Android apps. The reliability study was concerned with the consistency of ball bouncing, ball rotation as well as ball launching. The performance of this robot launcher is satisfactory when the error is less than 5% from the entire repetitive testing. In the experimental session, it is shown that the capability of ball shooting distance, the feed rate of a ball launched, and the ability of the robot launcher to do various ball spins are achieved and suitable to the player. Thus, this table tennis robot launcher benefits the athlete's self-training to improve their skills and technique.

Keywords Table tennis · Robot · Launcher · Android app

1 Introduction

Table tennis is an indoor sport in which the player is to hit the ball over the net and bounces on the opponent's half of the table in such a way that the opponent cannot reach it or return it correctly. Small rackets (bats or paddles) are used by the players to propel the lightweight hollow ball back and forth across the net [1]. This sport is gaining popularity and is often classified as a serious sport around

I. J. Jamaludin (✉) · I. M. Islah Munjih · Z. Ahmad · M. Z. Baharom
Faculty of Mechanical and Automotive Engineering Technology, Universiti Malaysia Pahang, 26600 Pekan, Pahang, Malaysia
e-mail: irlinajazlin@gmail.com

© The Author(s), under exclusive license to Springer Nature Singapore Pte Ltd. 2023

S. F. Syed Omar et al. (eds.), *Innovation and Technology in Sports*, Lecture Notes in Bioengineering, https://doi.org/10.1007/978-981-99-0297-2_17

the world [2]. In 2022, the Malaysian Table Tennis team won two silver medals in Birmingham 2022—Commonwealth Games, which are for women's team events and mixed doubles events, while they won one silver medal and four bronze medals in Philippines 2021—31st SEA Games [3–5]. Participation, especially in grand tournaments, becomes challenging because of the many skilled players in serving and spinning the ball technique. These techniques were already reported as important indicators of performance in high-ranking table tennis players [6–8]. Therefore, the players who want to participate in any table tennis tournaments must fully prepare by enhancing their performance and improving their skills in serving and ball-spinning. This means that our national table tennis team requires assistance with skills improvement. The considered way to improve skills is by doing frequent training. Table tennis training cannot be conducted by a single player because this sport needs at least two players.

The problem was created when the training partner was unavailable for the training session to help with serving and spinning the ball as an opponent of the player. Therefore, the table tennis robot launcher was developed to replace the training partner. Furthermore, in table tennis, a group that practices with the latest training gadgets is likely to perform better than the manual training method group and control group [9].

2 Methodology

The research methodology used to find the solution consists of the design concept, design selection, design development, and testing. The overall research methodology is shown in Fig. 1. The first phase of the research methodology is the design concept which is the process of generating several ideas that may be useful in solving the problems encountered. Next, the design selection is the second phase of the research methodology used to select the suitable solution among the ideas generated in the previous phase. After that, the third phase is design development, in which a suitable solution is developed to realise it in a physically touchable form and able to function as required. Finally, the last phase of this research methodology uses a few testing methods to ensure that the developed design is well functioning, capable of resolving the problems, and achieved all the research objectives.

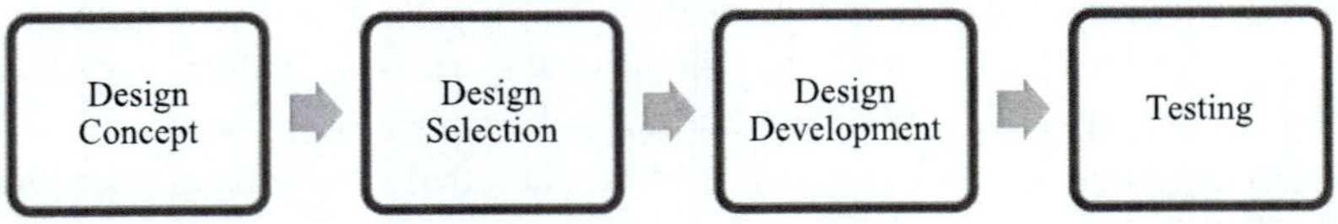

Fig. 1 Methodology flow chart

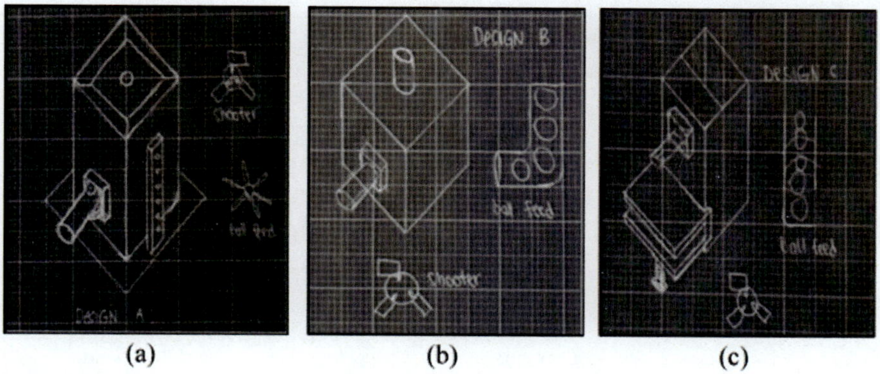

Fig. 2 Design concept that shows **a** Design A, **b** Design B, and **c** Design C

2.1 Design Concept

This research has three unique design concepts, as shown in Fig. 2. Later, the design concepts were discussed.

Design A is designed to be placed on a table. This table tennis robot launcher consists of two detachable parts, which are the body and the base. The body is rectangular to be used as ball storage and features a hole on top to assist the insertion of the balls into it. The body also contains an adjustable shooter that is advantaged to adjust the ball launch projectile by adjusting the height and angle of the shooter. Next, the base is designed to support the body while holding the fan feeder. Design A works when the balls are inserted into the storage (body) and continue to spin by the fan feeder, then pushed to the shooter before being launched.

Design B is invented specifically for the table. This simple table tennis robot launcher contains a rectangular body covering the mechanism, an elbow feeder, and a shooter. Design A functions when the balls are inserted from the top of the elbow feeder and moved down to the shooter before being launched.

Design C is proposed to be clamped on the table. This table tennis robot launcher comprises a rectangular body that helps cover up the mechanism, a clamp in front of the body used to clamp this design on the table, a screw feeder, and a shooter. Design C runs when the balls are inserted from the bottom of the feeder and pushed upward to the shooter by rotating screw before being launched.

2.2 Design Selection

Design selection is performed through screening to manage the complexity of evaluating design concepts. Throughout the screening process, several iterations may be performed. As shown in Table 1, the concepts are first compared with the screening

Table 1 Design selection

Selection criteria	Design A	Design B	Design C
Design uniqueness	+	–	+
Height adjustment	+	0	+
Portability	–	+	–
Ease of manufacture	0	+	–
Stability	0	+	0
Cost of manufacture	0	+	–
Net	1	3	−1
Rank	2	1	3
Continue?	No	Yes	No

criteria. After that, the concept is summed up with the net score and ranking, whether it is continued or combined with the concepts. Finally, Design B is selected as the final design for this research.

2.3 *Design Development*

Design development consists of detailed design, fabrication, app development, and final product. The overall design development phase is discussed as stated.

Detailed Design. The design development starts with a detailed design process which is the technical drawing of the selected design concept produced with CAD software, such as detailed drawings for each part, exploded views, assembly views, and bill of material (BOM). The purpose of the use of detailed design is to ease the fabrication process. Figure 3 is a few detailed designs for this research.

Fabrication. There are a few fabrication methods carried out suited to the material used to achieve the best closest ratio of fabricated product dimension to detailed

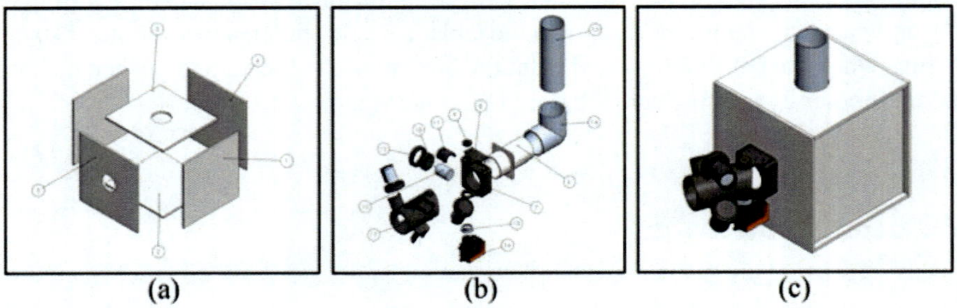

Fig. 3 Detailed design that shows **a** exploded view of the body component, **b** exploded view of shooter component and **c** assembly view of table tennis robot launcher

design dimension. The fabrication methods for this design development are manual cut, gluing, 3D printing, post-processing 3D printing parts, soldering, and fastening. Figure 4 shows several fabrication methods used for this research.

App Development. Arduino Uno is used to control each of the electronic components connected to it. To set Arduino Uno function, lines of code must be embedded into it by connecting it to the computer through Universal Serial Bus (USB). For this research, Arduino ide is the software used to create the coding from scratch. The app was developed by using the MIT App Inventor website, as shown in Fig. 5. Using this website, the customisation of the layout interface, as intended, can be made by adding buttons, sliders, and even pictures. All the components added to it must have their function assigned by arranging their function block.

Then, a simple test was conducted to know whether the app does send the correct data to Arduino and if the code is embedded into Arduino Uno and able to work

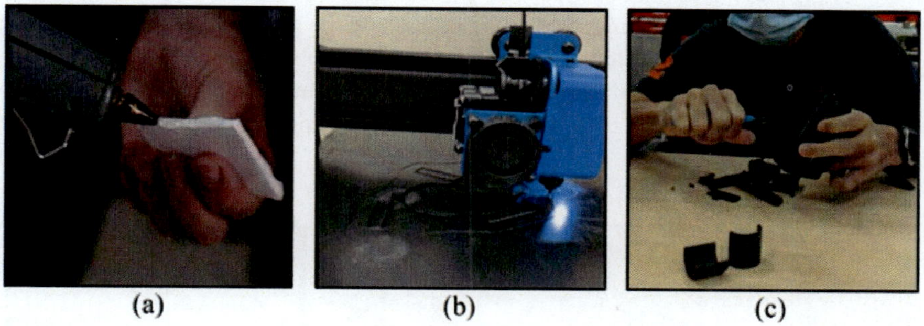

Fig. 4 Fabrication methods that show **a** gluing, **b** 3D printing, and **c** post-processing 3D printing parts

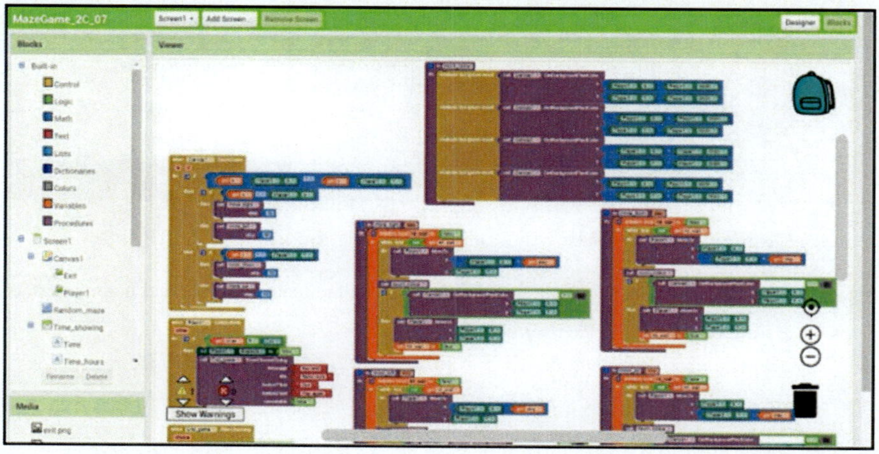

Fig. 5 Coding in MIT app inventor

as intended. There are two ways to verify it, by monitoring the data received by Arduino through the Bluetooth module and observing the behaviour of the electronic components such as the servo motor and DC motor. A feature called "serial monitor" in Arduino ide has been utilised to monitor the data received by Arduino Uno. Both Arduino coding and app are made to complete each other, and they should be able to:

1. Connect the smartphone to the table tennis robot launcher through Bluetooth to communicate; send, and receive data.
2. Control PWM to vary the average power supply to the DC motor to change the rotation speed.
3. Rotate the servo motor that attaches to the shooter to change the direction of the ball launch about 25 degrees to the left or 25 degrees to the right.

Final Product. All 3D printed parts are assembled with other components to produce the final product. The final product has good stability, and the body is able to hold the shooting part without any problem. The smartphone app interface has been built to control the prototype of the developed table tennis robot launcher. Each button is assigned to send different data input to the Bluetooth module attached to Arduino Uno and conduct specific functions. Figure 6 shows a complete product of the table tennis robot launcher.

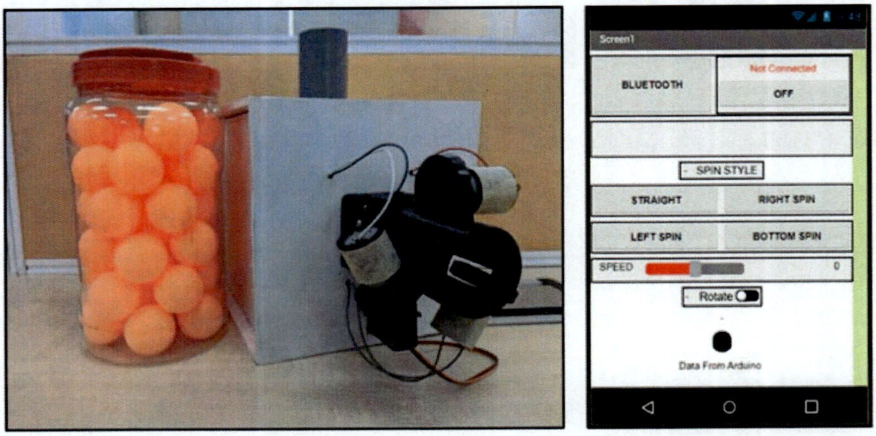

(a) (b)

Fig. 6 The complete product shows **a** full assembly of the fabricated product and **b** app interface

2.4 *Testing*

The result regarding the final product that has been developed is based on the continuous evaluation, observation, and testing made throughout the research, which is done through a series of test procedures, as shown in Fig. 7. Each testing procedure has been made to improve the final product's performance. Several iterations may be performed based on the results of the fitting test, stability test, and shooting capability test to ensure that the final product is well developed.

3 Experimental Results

The experiment was carried out on the final product of the table tennis robot launcher to determine its ball-shooting capabilities. Several parameters are measured by an evaluation which are ball travel distance, ball feed rate, and capability to launch the spin ball. The data recorder is collected by repeatedly experimenting with different PWM values applied as the manipulated variable.

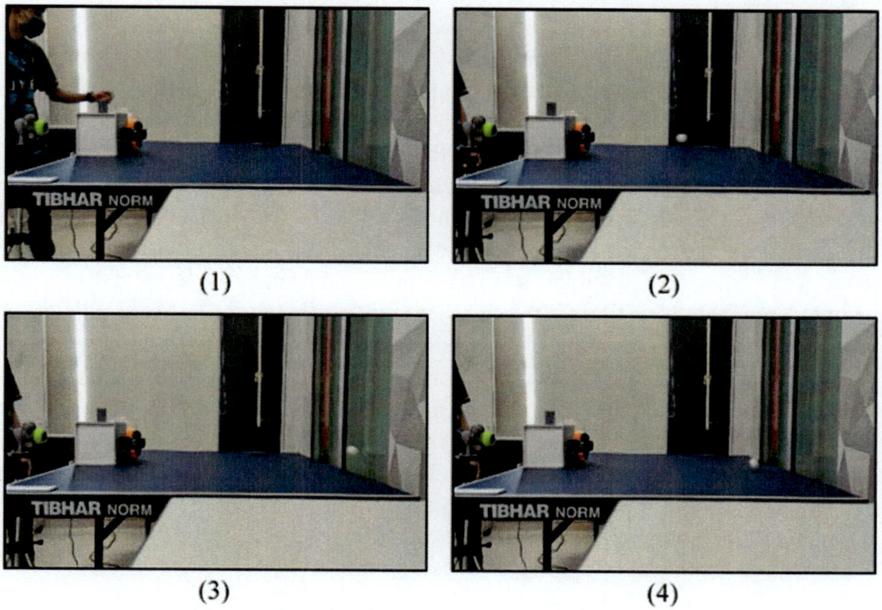

Fig. 7 A series of test procedures for developed Table Tennis Robot Launcher. The (1)–(4) are arranged by the sequences of the test performance

Table 2 Ball travel distance on different PWM values

PWM value	Travel distance (cm)
47–99	20–43 from the end of the table
100–152	43–56 from the end of the table
153–205	57–69 from the end of the table
206–255	69–75 from the end of the table

3.1 Ball Travel Distance

The distance of 40 mm table tennis ball travels on the table tennis table is directly proportional to the value of the PWM set on the smartphone app. The distance is measured from the end of the table tennis table (274 cm × 153 cm) to the first-place ball bounce. The minimum PWM value needed to be applied for the motor to start 47. Table 2 shows the collected data of ball travel distance on different PWM values from the repeatable shooting test.

3.2 Ball Feed Rate

The ball shooting rate ranged from 8 to 15 balls per minute, meaning 2 balls will be launched every 8 s when the rotation mode is not activated. As the ball goes down with the assistance of gravitational force, the feed rate varies with the speed of the motor set. The greater the motor speed, the greater the suction force that pulls the ball towards the wheel at the shooter. Therefore, the disadvantage of this ball feeder mechanism is that it cannot be programmed so that the ball shot in the interval can be minimized through the ball feed to the shooter. It will be difficult for the player to recover from each ball stroke as the ball comes too frequently. Despite that, it can be overcome from time to time through the muscle memory of frequent training. It is suitable to train the awareness of a player to respond quickly to the fast stroke. There is a drastic decrease in the number of balls being launched when rotation mode is activated; only 1–4 balls per minute can be launched. This occurs when a tight angle prevents the ball from moving to the shooter. Furthermore, relying only on gravitational force is not enough to eliminate other force factors that apply to the ball when the shooter rotates to the left and right. The results of the ball feed rate are summarized in Table 3.

Table 3 Ball feed rate

Test	Result
Ball feed rate (shooter static)	8–15 balls per minute
Ball feed rate (shooter rotate)	1–4 balls per minute

3.3 Ball Spin

The table tennis robot launcher did an excellent job of launching table tennis balls with different spin styles, which are topspin and side spin. Sometimes the ball spin is not too significant but does not happen in frequent times. Overall, the table tennis robot launcher can give the sensation and experience for the player to return the spin ball based on their selection spin style.

4 Conclusion

In conclusion, the objective of the research is fulfilled to develop a wireless table tennis robot launcher at an affordable price. Each part of the product is well fabricated through various kinds of fabrication methods. Joined part able to move locally, like the shooter part with the shooter holder. Furthermore, the assembled product is highly stable and resembles the detailed design made in the 3D drawing software. The app development is integrated with the physical part and functions well. According to the experiment result, the capability of ball shooting distance is acceptable for the player to receive the served ball from the table tennis robot launcher. The feed rate of a ball launched is suitable for the player because the ball is shot frequently in an interval of time. The ability of the table tennis robot launcher to apply different spin balls as intended is achieved. For future improvement, the tilt angle adjustment for the shooter can be higher to improve the projection motion of the ball. Next, the speed of the ball coming out of the launcher can be measured in future studies. Other than that, the use of a programmable ball feed mechanism that can adjust the difficulty of the training based on the player's skill can be applied to this product for better functionality. Lastly, the motor with a higher and more stable speed can improve the table tennis robot launcher's performance consistency.

References

1. Table Tennis. https://www.britannica.com/sports/table-tennis. Accessed 22 July 2022
2. Jiangzhou L, Anggina P, Akmal K, Hiroyuki I (2019) Analyzing the improvement process of table tennis using the game refinement theory. In: Proceedings of the Sriwijaya international conference on information technology and its applications (SICONIAN 2019). Atlantis Press, Paris, pp 437–442
3. Birmingham 2022 Commonwealth Games. https://www.ttam.com.my/single-post/birmingham-2022-commonwealth-games. Accessed 28 Aug 2022
4. Sukan Sea Hanoi: Skuad Ping Pong Negara Akur Raih Dua Gangsa. https://berita.rtm.gov.my/index.php/sport/40065-sukan-sea-hanoi-skuad-ping-pong-negara-akur-raih-dua-gangsa. Accessed 28 Aug 2022
5. The Latest SEA Games 2022 Medal Tally for Malaysia. https://www.prestigeonline.com/my/people-events/events/the-latest-sea-games-2022-medal-tally-for-malaysia. Accessed 28 Aug 2022
6. Fuchs M (2018) Table tennis match analysis: a review. J Sports Sci 36(23):2653–2662

7. Zhang H, Zhou Z, Yang Q (2018) Match analyses of table tennis in China: a systematic review. J Sports Sci 36(23):2663–2674
8. Djokic Z, Malagoli I, Katsikadelis M, Straub G (2020) Serve analyses of elite european table tennis matches. Int J Racket Sports Sci 2(1):1–8
9. Purashwani P (2018) A comparative study of different training methods on the performance of table tennis players. Int J Phys Educ Sports Sci 13(5):22–24

Investigation on the Acceleration of Wrist and Waist During a Golf Swing Towards the Ball Trajectory

Kusan Reveendran, Mohd Nadzeri Omar, Nasrul Hadi Johari, Mohd Hasnun Arif Hassan, and Azizul Aziz

Abstract The golf swing involves the movement of numerous body parts and can be viewed in several phases. Understanding how each phase behaves is critical to improving the game. The improvement is commonly quantified in terms of accuracy, which can be determined from the trajectory of the ball. The relationship between golf swings and ball trajectories is normally obtained in a laboratory setting, which is both limited in access and expensive. Golf swings were studied in this study using a wearable device called MetaMotionS (MMS) from MbientLab. The MMS measures the acceleration of a golfer's wrist and waist during a swing. According to the findings, the ball trajectory was influenced by the wrist and waist acceleration. Consequently, the golfer may apply the findings to improve his game.

Keywords Golf swing · Waist motion · Wrist motion · Wrist-waist coordination · Ball trajectory

1 Introduction

The purpose of golf is to get the lowest score possible. Golfers must have a balance of distance, accuracy, and consistency to achieve it. Long-distance hitters have an advantage, especially on a lengthy golf course or hole. Distance alone, however, is insufficient. Golfers must be accurate to bring the golf ball close to the target and must remain consistent throughout the long game.

The golf swing demands complicated multi-body mechanics, especially when distance and accuracy are the goals [1]. A golfer's biomechanics during a golf swing is critical for generating adequate power [2, 3] for distance and controlling the clubface orientation [4, 5] for accuracy. Numerous research has been conducted to investigate

K. Reveendran · M. N. Omar (✉) · N. H. Johari · M. H. A. Hassan
Faculty of Mechanical and Automotive Engineering Technology, Universiti Malaysia Pahang, 26600 Pekan, Pahang, Malaysia
e-mail: nadzeri@ump.edu.my

A. Aziz
Royal Pekan Golf Club, 26600 Pekan,, Pahang, Malaysia

© The Author(s), under exclusive license to Springer Nature Singapore Pte Ltd. 2023

S. F. Syed Omar et al. (eds.), *Innovation and Technology in Sports*, Lecture Notes in Bioengineering, https://doi.org/10.1007/978-981-99-0297-2_18

the biomechanics of body components during a golf swing. The majority of researcher focus on the waist [6], thorax [7], wrist [8], and knee [9]. Coordination between these bodily parts has received little study thus far. Additionally, most of these studies were conducted in a laboratory using an optoelectronic measurement system that included a high-speed camera and sensors attached to the golfer's body. Although it achieves greater accuracy, it has limited accessibility and causes discomfort for the golfer.

One way to determine the accuracy of a golf swing is by looking at the ball's trajectory. Parameters that contribute to the ball trajectory are clubhead velocity, clubface orientation, and ball location on the clubface at contact [10]. Several studies have been conducted to explore the effect of clubface orientation during the impact on the ensuing ball trajectory [10]. The studies demonstrating that the ability to regulate clubface orientation is vital for golfers. However, the studies concentrate on the impact area alone, not on the biomechanics of the golf swing that generates the impact.

In this paper, an investigation was performed to study the coordination between a golfer's wrist and waist during a golf swing. The biomechanics of these body parts were measured using an accelerometer-equipped wearable device. The investigation focuses on the acceleration data, which produces different ball trajectories.

2 Methodology

2.1 Equipment

The acceleration data of two body parts, the wrist, and waist, are the subject of this investigation. The accelerations of these body parts are measured using MetaMotionS (MMS), a wearable Inertial Measurement Units (IMU) device from MbientLab. The MMS includes a BMI270 (6-axis accelerometer and gyroscope), BMM150 (3-axis geomagnetic sensor), BMP280 (temperature and pressure), and ltr-329als-01 (luminosity) integrated sensors. To initiate data collection, store, and export the acquired data, the MetaBase software must be launched on a smartphone that is linked to the MMS via Bluetooth.

2.2 Data Collection

The subject of this research was a mid-handicap armature golfer. The golfer is a male with height and weight measurements of 151 m and 60 kg, respectively. Two MMSs were utilized as shown in Fig. 1, with the MMSs attached to the golfer's left wrist (in the form of an armband) and right waist (in the form of a clip).

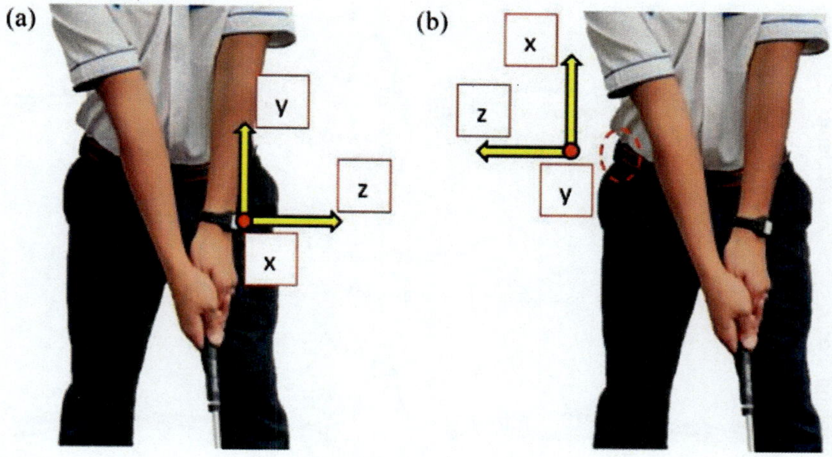

Fig. 1 The location of the IMU and its respected axis. **a** Wrist, and **b** Waist

The data collection was carried out in a driving range. As preparation, the golfer was allowed to do several practices before the data collection. During the data collection, the golfer needs to perform 20 golf swings using a 7-iron and for each swing, he needs to identify the ball trajectory either straight, left, or right. The MMSs are triggered before each swing and deactivated after the swing ends.

3 Results

A data sample that represents the straight, left and right ball trajectories were selected from the collected data. These selected data samples are analyzed according to the wrist-only motion, the waist-only motion, and the wrist-waist motion.

3.1 Wrist Motion

Figure 2 shows the resulting gravitational acceleration (g) of the golfer's wrist that produces a straight, left, and right ball trajectory. Each sample displayed acceleration for the three axes, x, y, and z. In general, the wrist motion of the golfer is consistent which is reflected by the identical shape of the accelerations in all the samples.

When the golfer creates a ball's trajectory to the right, the wrist's acceleration on the x-axis is different. The x-axis peak in Fig. 2c is steep, whereas the peak in Fig. 2a and b is flat. It denotes the amount of time spent transitioning from the backswing to the downswing. In this case (see Fig. 2c), the transition occurred too rapidly which affected the ball's trajectory. Another data sample (that represents the ball's trajectory

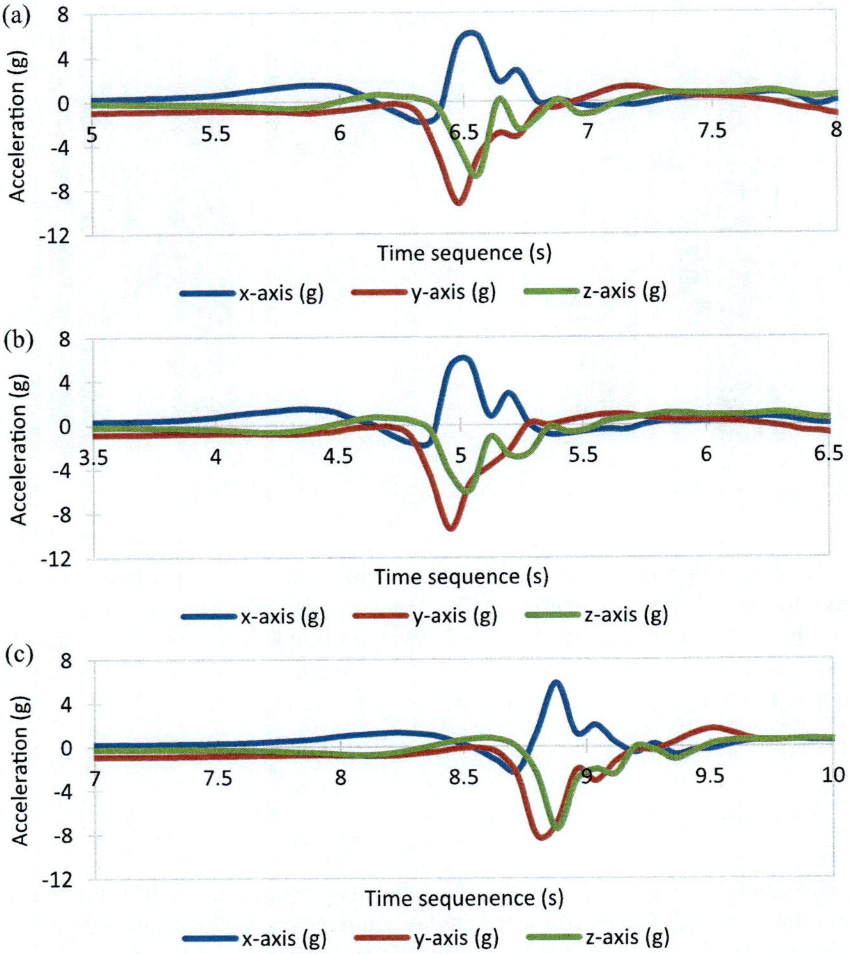

Fig. 2 Wrist motion. **a** Straight trajectory, **b** Left trajectory, and **c** Right trajectory

to the right) was acquired for comparison. As shown in Fig. 3, a similar pattern is exhibited, which verifies the observation.

Another noticeable difference is the acceleration along the z-axis. The lowest acceleration along the z-axis for the straight ball trajectory is close to −6.5 g, as illustrated in Fig. 2a. When compared to the right ball trajectory, Fig. 2c shows that the acceleration along the z-axis is around -1 g higher. It affects the clubface during impact. The golfer is unable to square the clubface during contact due to increased acceleration. As a result, the ball does not launch in a straight trajectory. Meanwhile, when compared to Fig. 3, the same trend is observed, which supports the type of motion that changed the ball's trajectory.

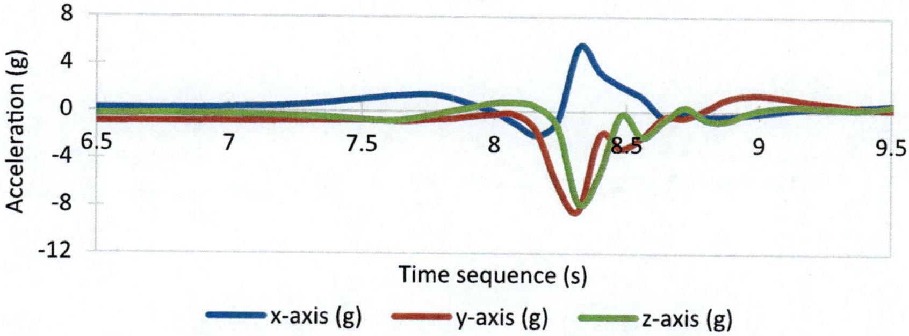

Fig. 3 Another data sample for a right ball trajectory

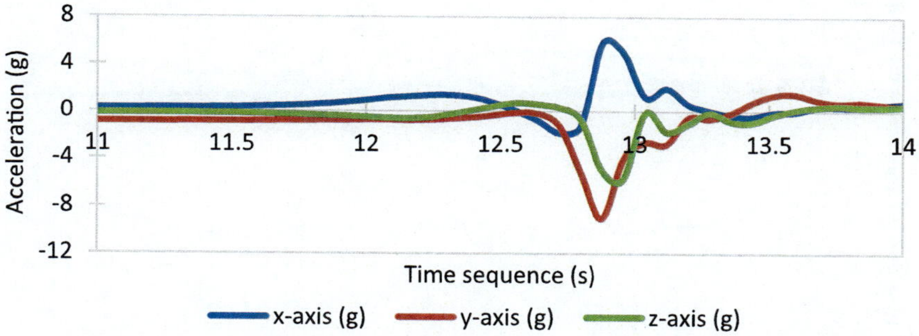

Fig. 4 Another data sample for a straight ball trajectory

Furthermore, in Fig. 2a, the acceleration of the wrist in the z-axis returns to its initial state at some point during the swing execution. This circumstance does not occur in the other samples. To validate, another sample that represents a straight ball trajectory was collected. As shown in Fig. 4, a similar observation is found in which at some point during the execution, the acceleration in the z-axis returns to its initial state. It demonstrates that when the golfer can regulate his wrist motion back to its initial position, indicating that the clubface is square, the possibilities of a straight ball trajectory are increased.

3.2 Waist Motion

Figure 5 depicts the acceleration of the waist during a golf swing, which results in three distinct trajectories. The difference between the samples is seen to be insignificant. It emphasizes the waist motion alone; it does not affect the ball's trajectory.

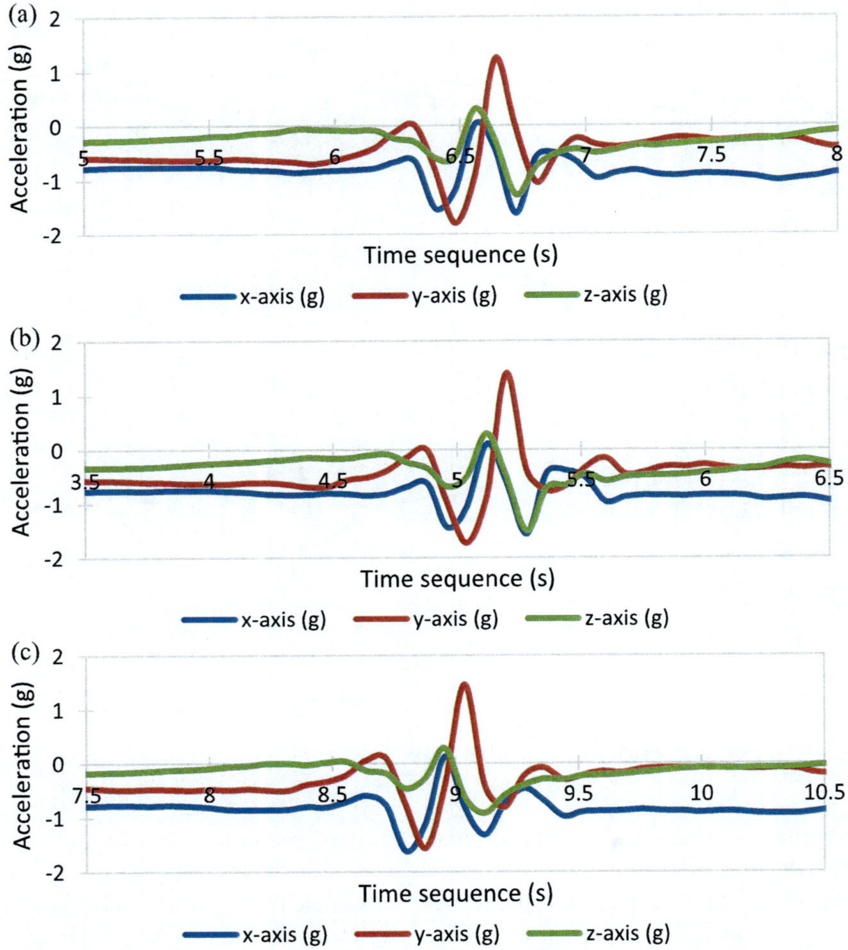

Fig. 5 Waist motion. **a** Straight trajectory, **b** Left trajectory, and **c** Right trajectory

3.3 Coordination of Wrist and Waist

The wrist and waist coordination was examined by observing the behavior of both body parts simultaneously. Figures 6, 7 and 8 depict the coordination of both body parts for a straight, left, and right ball trajectory.

The main distinction is in the waist movement. When examined concurrently with the wrist, the waist motion becomes significant. Using the impact point as a reference, it is shown in Fig. 6 that the initial peak of acceleration of the waist along the x-axis occurred before the impact. Meanwhile, for the left and right trajectories (see Figs. 7 and 8), the first peak appeared just after the impact. The occurrence of the initial peak early in the downswing suggests that the golfer moves his waist during

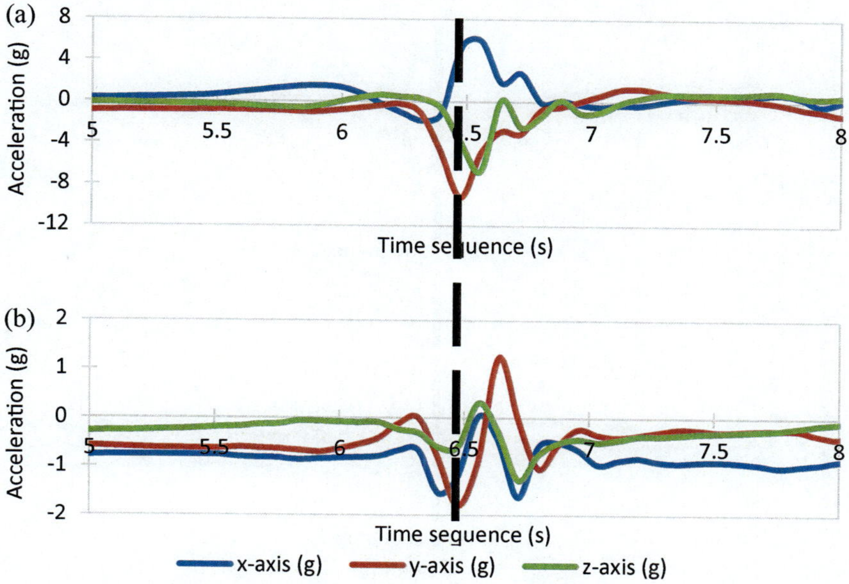

Fig. 6 Coordination of the **a** wrist and **b** waist acceleration for a straight ball trajectory

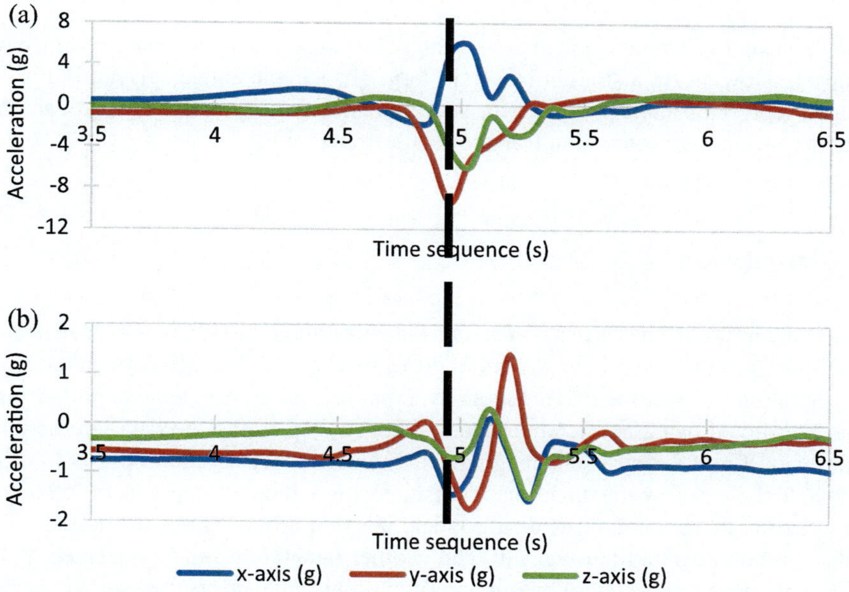

Fig. 7 Coordination of the **a** wrist and **b** waist acceleration for a left ball trajectory

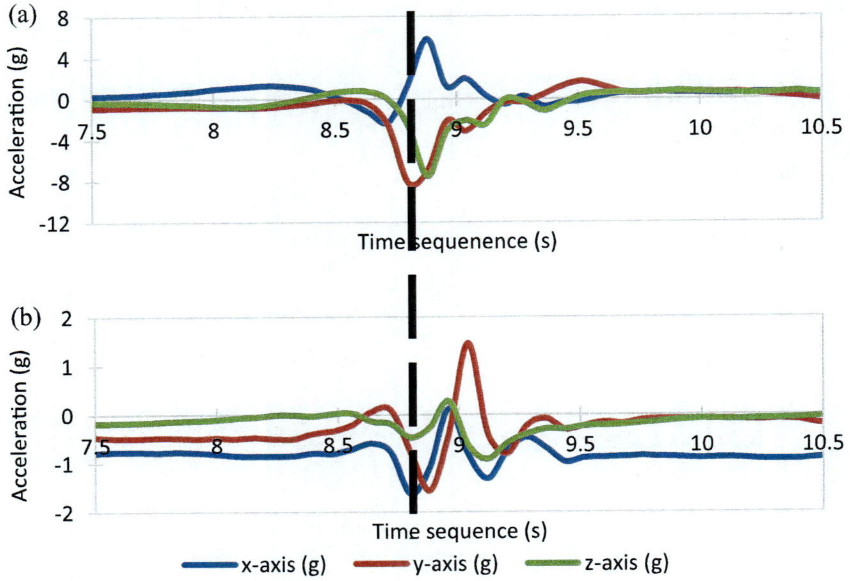

Fig. 8 Coordination of the **a** wrist and **b** waist acceleration for a right ball trajectory

the downswing. The movement assists the golfer in achieving a square clubface on impact, allowing for a straight ball trajectory. The sample data displayed in Fig. 4, which represents another dataset for straight ball trajectory, is used to validate the finding. A similar observation is made, as illustrated in Fig. 9.

4 Discussion

There are multiple phases in a golf swing. Understanding the behavior of each phase, as shown in Fig. 8, can help golfers improve their game. The behavior in terms of acceleration was investigated in this study. It has been demonstrated that golf swings can be studied using acceleration data. Although, separating the data into distinct phases of a golf swing is challenging. The measured acceleration does not correspond to the actual swing direction or sensor position. It only represents the recorded movement. To obtain the true direction and position of the sensor, the acceleration data can be integrated into velocity and another integration for displacement [11]. However, the double integration might result in substantial errors, necessitating error filtration [12]. As a result, other parameters, such as gyroscope properties, maybe a preferable option.

The acceleration data was acquired using the MMS's embedded accelerometer. Although the MMS has been calibrated, the resultant data still require further calibration. For example, some starting values of acceleration are discovered to be different

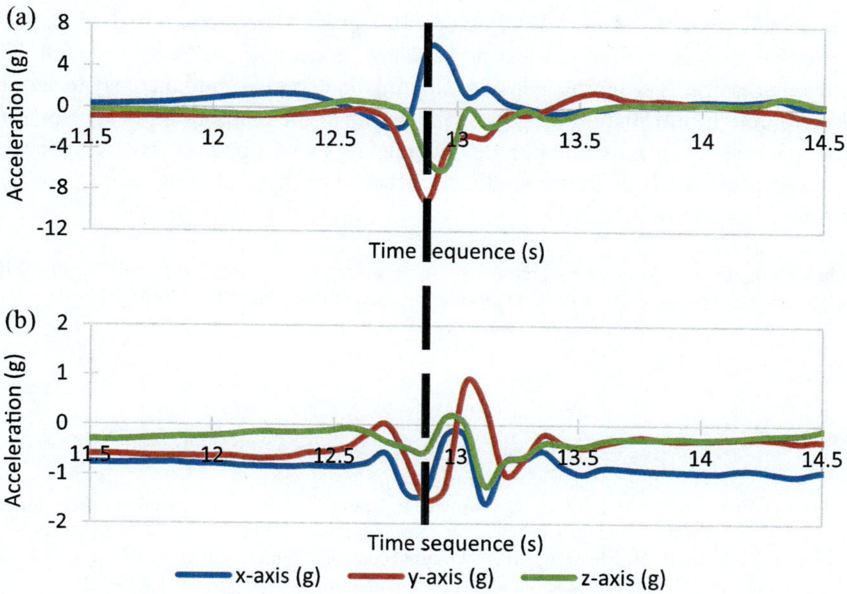

Fig. 9 Another data sample that shows coordination of the **a** wrist and **b** waist acceleration for a straight ball trajectory

even though the data was taken from the same person. Hence, further calibration is essential to determine the true magnitude of the acceleration. The calibration was disregarded in this investigation because it focuses on the acceleration pattern.

The majority of previous studies used optoelectronic technology to analyze golf swings in a laboratory setting. In this study, golf swings were analyzed using a wearable device. The findings indicate the potential of the wearable device. Furthermore, wearable devices have received a lot of interest in a wide range of investigations [13]. In the instance of golf swing analysis, the wearable gadget gives more accessibility at a reduced cost.

It is well-known that each golfer has his type of swing [14]. The presented findings are limited to a single golfer, implying the observations here may not apply to the other golfers. However, the study shows that any type of golf swing for a particular golfer can be identified using the same setup, but it requires individual analysis.

5 Conclusion

The goal of this study is to investigate the acceleration of a golfer's wrist and waist during a golf swing towards the ball trajectory. The wrist and waist accelerations were measured using a wearable device called the MetaMotionS (MMS). According to the findings, when both body parts were examined separately, the wrist acceleration contributed the most to the ball trajectory. However, when the two body parts were investigated simultaneously, waist acceleration is determined to be essential. The study emphasizes the significance of evaluating body parts simultaneously to acquire more thorough information on golf swings. Due to the restrictions of acceleration, other parameters such as Euler angle, velocity, or gyroscope properties can be used as a substitute for it in future studies. Furthermore, by including more golfers, a broader view can be attained.

Acknowledgements Authors are grateful to Universiti Malaysia Pahang for laboratory facilities and the financial support under Product Development Research grant PDU223202.

References

1. Bourgain M, Rouch P, Rouillon O, Thoreux P, Sauret C (2022) Golf swing biomechanics: a systematic review and methodological recommendations for kinematics. Sports 10(6):91
2. Brown C, McNally W, McPhee J (2020) Optimal control of joint torques using direct collocation to maximize ball carry distance in a golf swing. Multibody SysDyn 50(3):323–333
3. Balzerson D, Banerjee J, McPhee J (2016) A three-dimensional forward dynamic model of the golf swing optimized for ball carry distance. Sports Eng 19(4):237–250
4. Langdown BL, Bridge M, Li F-X (2012) Movement variability in the golf swing. Sports Biomechanics 11(2):273–287
5. Corke TW, Betzler NF, Wallace ES, Otto SR (2018) A novel system for tracking Iron Golf Clubheads. Proc Inst Mech Eng Part P: J Sports Eng Technol 233(1):59–66
6. Lamb PF, Pataky TC (2018) The role of pelvis-thorax coupling in controlling within-golf club swing speed. J Sports Sci 36(19):2164–2171
7. Mitchell K, Banks S, Morgan D, Sugaya H (2003) Shoulder motions during the golf swing in male amateur golfers. J Orthop Sports Phys Ther 33(4):196–203
8. Kim YJ, Kim KD, Kim SH, Lee SG, Lee HS (2017) Golf swing analysis system with a dual band and motion analysis algorithm. IEEE Trans Consum Electron 63(3):309–317
9. Frontani F, Prenassi M, Paolini V, Luciant P, Marceglia S, Poliscastro F (2022) Knee kinematic during the golf swing: a cross-sectional analysis between groups of different handicaps. J Phys Educ Sport 22(1):250–255
10. Takagi T, Yokozawa T, Inaba Y, Matsuda Y, Shiraki H (2017) Relationships between clubshaft motions and clubface orientation during the golf swing. Sports Biomech 16(3):387–398
11. Hampson R, Dobie G, West G (2019) Elasticity measurement of soft tissues using hybrid tactile and MARG-based displacement sensor systems. IEEE Sens J 19(22):10262–10270
12. Omar MN, Hassan MHA (2022) Displacement measurement of soft material indentation using light intensity. J Mech Eng 19(1):21–38
13. Seneviratne S, Hu Y, Nguyen T, Lan G, Khalifa S, Thilakarathna K, Hassan M, Seneviratne A (2017) A survey of wearable devices and challenges. IEEE Commun Surv Tutor 19(4):2573–2620
14. Nesbit SM (2005) A three dimensional kinematic and kinetic study of the golf swing. J Sports Sci Med 4:499–519

Badminton Player's Shot Prediction Using Deep Learning

Farzeen Ashfaq, N. Z. Jhanjhi, and Naveed Ali Khan

Abstract The study of object tracking has substantially advanced thanks to the development of deep learning visual recognition and tracking methods. However, because to the additional difficulties they provide, such as the difficulty in tracking small, swiftly moving objects like a ball or shuttlecock due to the fast camera movement and the existence of swings and spins, sports videos are still understudied. To access these massive archives of sports video data and automatically tag and analyse its properties, such as player performance and stroke and shot analysis, an effective end-to-end solution is needed. The aim of this research is to create a complete deep learning based model that can do object detection and tracking in sports movies as well as classify the played stroke. We employed the SF-YOLOv5 model, a lightweight model for the identification of swiftly moving small objects, for this. Then, we utilised the Deep-Sort algorithm and zero shot learning to follow the objects that had been detected. Finally, we classified the played shot using the CNN classifier.

Keywords Visual object tracking · Player stroke prediction · Zero Shot learning · Sports analytics

---

Khan: The author contributed equally to this work.

F. Ashfaq (✉) · N. Z. Jhanjhi (✉) · N. A. Khan
School of Computer Science, SCS, Taylor's University Lakeside Campus, 47500 Subang Jaya, Malaysia
e-mail: farzeen.ashfaq@sd.taylors.edu.my

N. Z. Jhanjhi
e-mail: noorzaman.jhanjhi@taylors.edu.my

N. A. Khan
e-mail: navidali.khan@taylors.edu.my

© The Author(s), under exclusive license to Springer Nature Singapore Pte Ltd. 2023
S. F. Syed Omar et al. (eds.), *Innovation and Technology in Sports*, Lecture Notes in Bioengineering, https://doi.org/10.1007/978-981-99-0297-2_19

1 Introduction

The process of locating a specific object or objects in each frame of a video clip or stream is known as visual object tracking. It is a well-known topic of study in the domain of computer vision. Deep learning-based visual object tracking models and algorithms have found applications in a variety of fields, such as autonomous driving and traffic monitoring [1, 2], security and surveillance systems [3, 4], human-computer interaction [5, 6], games and sports [7, 8] and medicine [9, 10].

There are many benefits to using visual object tracking in sports analytics. In many games, players and sporting goods like balls, bats, shuttlecocks, and hockey sticks are the primary targets for detection. Tracking one or multiple objects in sports recorded videos not only helps umpires and referees make the right decisions during play. But, during practise and training sessions, motion tracking also helps coaches and players. However, despite being captured for practically every significant national or international championship, the sports video data is usually retained and remain inaccessible for large-scale data mining and data analytics. The reason is that it involves manual labelling or inspection, both of which take time and effort. Deep learning algorithms have recently been used in image and video analysis, considerably improving accuracy, automation, and cost. Figure 1 shows some of the most popular deep learning based algorithms used for visual object detection, tracking and classification algorithms.

As a result, in this study, we classify the player's stroke using video analysis of a badminton match while keeping in mind the advantages of deep learning algorithms in the aforementioned domain.

1. Use the compact SF-Yolov5 to detect the player hand and badminton shuttle in the target video frame when they are moving quickly.
2. Track the discovered items in the following several frames using zero shot visual object tracking with DeepSORT.

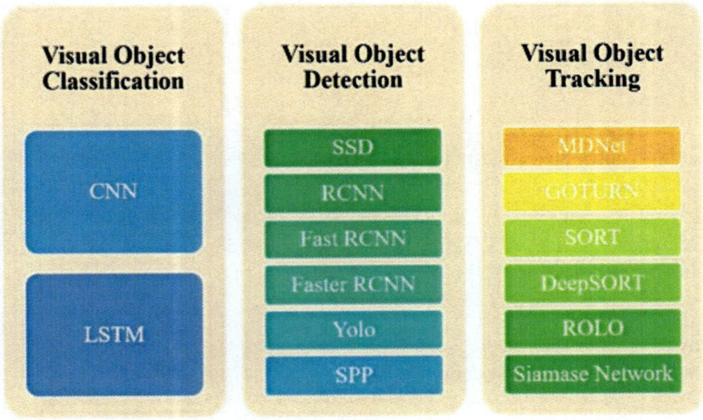

Fig. 1 A list of popular deep learning models

3. Classify the played stroke as a drop, drive, lift, or smash using the trajectory data from numerous frames and the CNN classifier.

The remainder of the paper is structured as follows. The current state of the research on visual object tracking in sports is presented in Sect. 2. Our suggested model shuttle detection and tracking in badminton videos is discussed in Sect. 3. Results are compiled in Sect. 4 along with a brief explanation. Section 5 finishes the paper and offers suggestions for the future.

2 Related Work

In this section, we give a quick overview of current developments in deep learning and vision-based sports analytics models and systems using broadcast video data. A summary view of the existing methods is shown in Table 1.

Zhang et al. [7] used Yolov4 to detect the running athletes and moving football players and then tracked their information in consecutive frames by using DeepSort algorithm. They evaluated their work on NBA and world cup dataset. Buric et al. [8] also used DeepSort algorithm to track the football players in indoor court after detecting them with pre trained Yolov2. They also studied the influence of detection score on tracking. By contrasting various arrangements of detectors and trackers, [42] attempted to solve the Multi Object Tracking (MOT) issue in order to create intelligent basketball courts. According to their findings, FasterRCNN is a more accurate item than others. The JDE approach outperforms the other when Yolov3 detector was used [43]. Some other studies which involved DeepSort algorithm for player or object tracking includes [43–46]. Table 2 shows summary of literature relating to the task of player shot prediction and classification.

Table 1 Summary of existing studies done in sports analytics using broadcast videos

Paper	Performed task
[11–16]	Player detection
[13, 17–21]	Event recognition
[22–26]	Object (ball) detection and tracking
[27–31]	Player movement/pose tracking
[32–36]	Player stroke/shot classification
[37–41]	Sports highlight generation

Table 2 Summary of literature relating with player's shot prediction

Year	Paper	Method	Dataset
2018	[47]	Fusion of Inertial Measurement Unit (IMU) and audio sensor data embedded in a wrist-worn wearable	Video IMU
2019	[48]	AlexNet (CNN)	UCF-101
2020	[49]	LSTM & BiLSTM	Publicly available videos
2020	[50]	Shallow learning and deep learning algorithms to classify the strokes used to analyse the stance	Data collected from IMU sensors
2021	[51]	Stroke detection and classification	TTStroke-21
2021	[52]	3D motion capture to obtain kinemetics of stick data	Cue sports videos
2021	[53]	Multimodal data and neural networks	33 sessions and 510 recorded strokes

3 Proposed Method

For the classification of badminton stroke we define a three stage process as illustrated in Fig. 2. Each of the stage is discussed in detail in the following subsections.

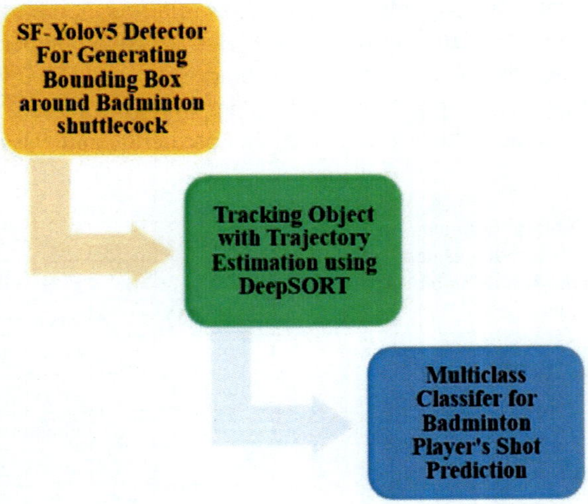

Fig. 2 Components of our proposed model

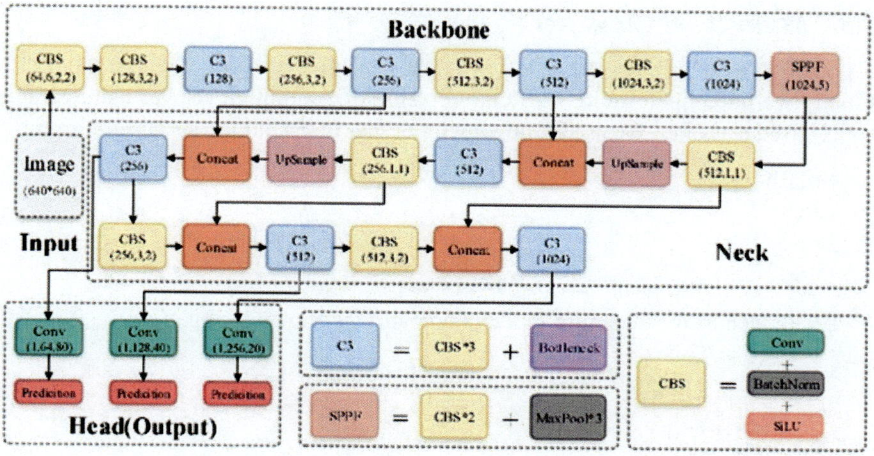

Fig. 3 Yolov5 architecture [54]

3.1 *Object Detector*

For the detection of objects we used [54] as baseline model. The model uses Yolov5 as object detector. Figure 3 shows the overall architecture of Yolov5. As can be seen from the diagram the detector consists of three main parts, backbone, neck and head. in our detector we chose CSPDarknet layer for feature extraction.

3.2 *Object Tracking*

DeepSORT is an extension of SORT (Simple Online Realtime Tracking), which is made up four main components.

1. Detection
2. Estimation
3. Association Identity

In our scenario, the detector has already been covered. Once the detection is complete, we transfer it from the first to the next from using the linear velocity model. The velocity components in SORT and DeepSORT are solved using a Kalman filter. The bounding box shape for each target is then determined by extrapolating its location from the previous frame. The distinction in Deep-SORT is that appearance features from the object bounding box photos are extracted using CNN to create a deep association matrix.

3.3 Multi-label Classifier

Finally we implement multi label classification of badminton strokes using simple CNN based classifier. We used following four types of basminton strokes for training.

1. Smash
2. Lift
3. Drive
4. Drop

The flowchart for the entire model's operation is shown in Fig. 4.

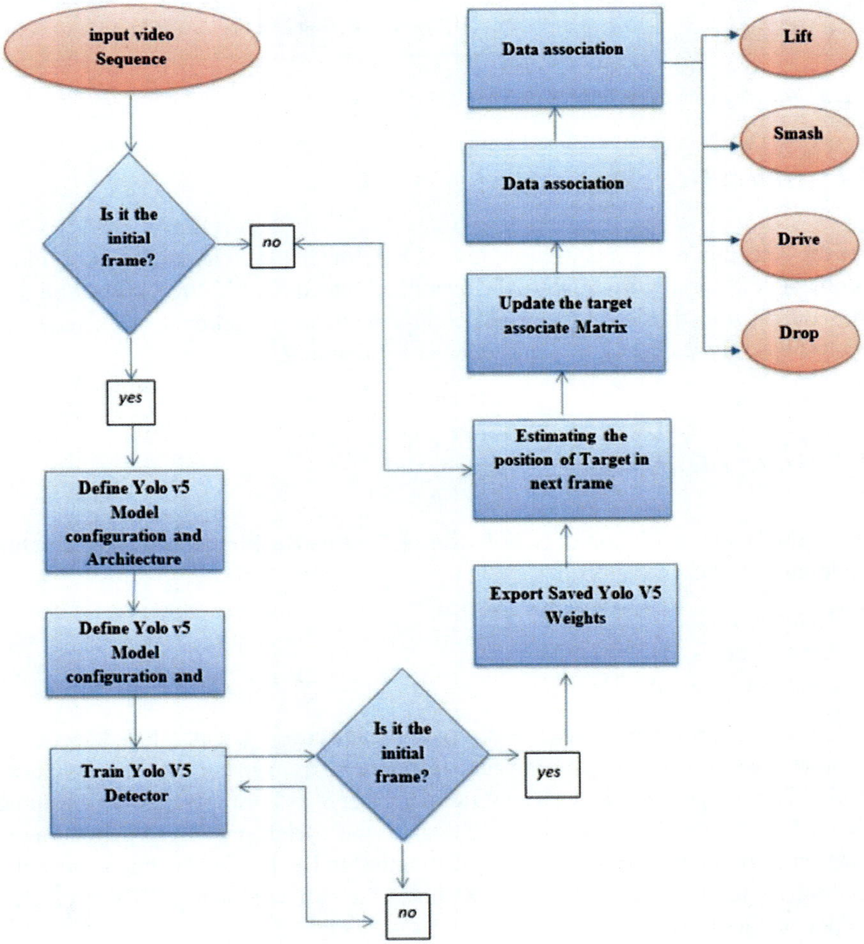

Fig. 4 The flowchart of our proposed model

4 Experiments and Results

The experiment was conducted on Google Colab free NVIDIA Tesla T4 GPU and tensorflow. The custom object detector was trained on Shuttlecock Trajectory Dataset [55]. 26 broadcast videos make up the dataset. Videos have a 1280 × 720 resolution and a 30 frames per second frame rate, respectively.

The match starting and finishing frames were removed during the initial pre-processing because they didn't have any playing moments. Figure 5a–d shows sample bounding box tracking of badminton shuttlecock and player hand movement in four consecutive frames. For the evaluation of our proposed model we selected three multi object trajectory evaluation metrics, including MOTA (Multiple Object Tracking Precision), HOTA (Higher Order Tracking Accuracy) and IDF1 score. MOTA uses two values for calculation the first is the ground truth value and the second is the predicted detection value. The metric calculate the errors in series of frames by using the formula shown in Eq. 1. Also it calculates the accuracy of both the detector and tracer.

$$MOTA = 1 - \frac{\sum_t fN_t + fP_t + IdS_t}{\sum_t G_t} \quad (1)$$

where,

fN_t = False Negative for Predictions,

fP_t = False Positive for Predictions,

G_t = Ground Truth.

The HOTA can be created by combining three IoU scores. It uses an IoU (intersection over union) formulation to calculate scores for each of the three subtasks of tracking evaluation—detection, association, and localization. The IDF1 ratio measures how many accurately identified detections there are compared to the average number of computed and ground-truth detections. Equation 2 calculates identification recall, Eq. 3 calculates identification precision, and Eq. 4 calculates the detection F1 score. The Fig. 6 shows the results of our training in 10000 epochs.

$$detection\_recall = \frac{detection\_TP}{detection\_TP + detection\_FN} \quad (2)$$

$$detection\_precision = \frac{detection\_TP}{detection\_TP + detection\_FN} \quad (3)$$

$$detection\_F1\_score = \frac{detection\_TP}{detection\_TP + 0.5 * detection\_FN + 0.5 * detection\_FP} \quad (4)$$

(a) Frame_01 (b) Frame_02

(c) Frame_03 (d) Frame_04

Fig. 5 An example showing the detection and tracking of shuttlecock and player arm movement in four consecutive frames

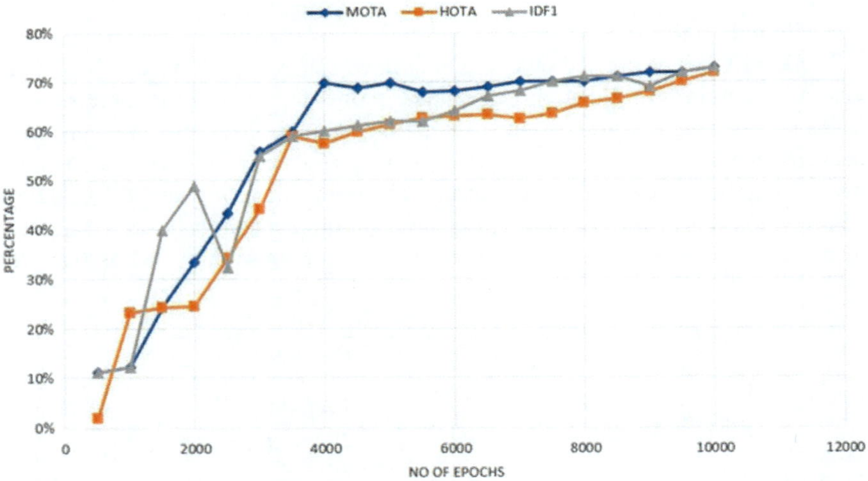

Fig. 6 The performance scores of our proposed model

5 Conclusion

This paper describes a comprehensive technique for classifying player strokes from broadcast video data. Object detection, object tracking over numerous frames, and multiple label categorization are all features of this system. For recognising player hand movement and the position of the badminton shuttle-cock in the video frame, we used YOLOv5 (You Only Look Once). Following that, we tracked both of their positions throughout the course of four successive frames using the trajectory estimate data. Using the information from the frames, we divided the player's shot into four categories: smash, lift, drive, and drop. Over 73% MOTA, 72% HOTA, and 73% IDF1 scores were achieved by our model. Future studies will concentrate on categorising more difficult shots that mostly include occlusions.

References

1. Humayun M, Ashfaq F, Jhanjhi NZ, Alsadun MK (2022) Traffic management: Multi-scale vehicle detection in varying weather conditions using yolov4 and spatial pyramid pooling network. Electronics 11(17):2748
2. Zhang J, Feng W, Yuan T, Wang J, Sangaiah AK (2022) Scstcf: spatial-channel selection and temporal regularized correlation filters for visual tracking. Appl Soft Comput 118:108485
3. Nair SP, Abhinav Reddy K, Alluri PK, Lalitha S (2021) Face recognition and tracking for security surveillance. J Intell Fuzzy Syst 41(5):5337–5345
4. Sarhan A, Elshennawy NM, Diab GM (2021) Human detection/tracking system for video surveillance with noise removal. In: Privacy and security challenges in location aware computing. IGI Global, pp 72–108
5. Zhang Y, Wang C, Wang X, Liu W, Zeng W (2022) Voxeltrack: Multi-person 3d human pose estimation and tracking in the wild. IEEE Trans Pattern Anal Mach Intell
6. Yang Y, Ren Z, Li H, Zhou C, Wang X, Hua G (2021) Learning dynamics via graph neural networks for human pose estimation and tracking. In: Proceedings of the IEEE/CVF conference on computer vision and pattern recognition, pp 8074–8084
7. Zhang Y, Chen Z, Wei B (2020) A sport athlete object tracking based on deep sort and yolo v4 in case of camera movement. In: 2020 IEEE 6th international conference on computer and communications (ICCC), pp 1312–1316. IEEE
8. Buric M, Ivasic-Kos M, Pobar M (2019) Player tracking in sports videos. In: 2019 IEEE international conference on cloud computing technology and science (CloudCom). IEEE, pp 334–340
9. Khairandish M, Sharma M, Jain V, Chatterjee J, Jhanjhi N (2021) A hybrid cnn-svm threshold segmentation approach for tumor detection and classification of mri brain images. IRBM
10. Ghani NBA, Hamid S, Ahmad M, Saadi Y, Jhanjhi N, Alzain MA, Masud M (2022) Tracking dengue on twitter using hybrid filtration-polarity and apache flume. Comput Syst Sci Eng 40(3):913–926
11. Chan A, Levine MD, Javan M (2021) Player identification in hockey broadcast videos. Expert Syst Appl 165:113891
12. Vats K (2022) Player tracking and identification in broadcast ice hockey video
13. Wang T, Li T (2022) Deep learning-based football player detection in videos. Comput Intell Neurosci
14. Naik BT, Hashmi MF (2022) Yolov3-sort: detection and tracking player/ball in soccer sport. J Electron Imaging 32(1):011003

15. Haq MA, Tarashima S, Tagawa N (2022) Heatmap visualization and badminton player detection using convolutional neural network. In: 2022 international electronics symposium (IES). IEEE, pp 627–631
16. Naik BT, Hashmi MF (2021) Ball and player detection & tracking in soccer videos using improved yolov3 model
17. Vats K, Fani M, Clausi DA, Zelek J (2021) Puck localization and multi-task event recognition in broadcast hockey videos. In: Proceedings of the IEEE/CVF conference on computer vision and pattern recognition, pp 4567–4575
18. Vats K, Fani M, Walters P, Clausi DA, Zelek J (2020) Event detection in coarsely annotated sports videos via parallel multi-receptive field 1d convolutions. In: Proceedings of the IEEE/CVF conference on computer vision and pattern recognition workshops, pp 882–883
19. Shingrakhia H, Patel H (2020) Emperor penguin optimized event recognition and summarization for cricket highlight generation. Multimed Syst 26(6):745–759
20. Morra L, Manigrasso F, Lamberti F (2020) Soccer: Computer graphics meets sports analytics for soccer event recognition. SoftwareX 12:100612
21. Khan A, Lazzerini B, Calabrese G, Serafini L (2018) Soccer event detection. In: 4th international conference on image processing and pattern recognition (IPPR 2018). AIRCC Publishing Corporation, pp 119–129
22. Zhang X, Zhang T, Yang Y, Wang Z, Wang G (2020) Real-time golf ball detection and tracking based on convolutional neural networks. In: 2020 IEEE international conference on systems, man, and cybernetics (SMC). IEEE, pp 2808–2813
23. Shah SSA, Khalil MA, Shah SI, Khan US (2018) Ball detection and tracking through image processing using embedded systems. In: 2018 IEEE 21st international multi-topic conference (INMIC). IEEE, pp 1–5
24. Kamble PR, Keskar AG, Bhurchandi KM (2019) A deep learning ball tracking system in soccer videos. Opto-Electron Rev 27(1):58–69
25. Najeeb HD, Ghani RF (2021) A survey on object detection and tracking in soccer videos. MJPS 8(1)
26. Kukleva A, Khan MA, Farazi H, Behnke S (2019) Utilizing temporal information in deep convolutional network for efficient soccer ball detection and tracking. In: Robot world cup. Springer, pp 112–125
27. Ivasic-Kos M, Host K, Pobar M (2021) Application of deep learning methods for detection and tracking of players. In: Deep learning applications. IntechOpen
28. Song C, Zarar S (2019) Sensor fusion for learning-based tracking of controller movement in virtual reality. In: 2019 27th European signal processing conference (EUSIPCO). IEEE, pp 1–5
29. Arbués-Sangüesa A, Ballester C, Haro G (2019) Single-camera basket-ball tracker through pose and semantic feature fusion. arXiv:1906.02042
30. Cuiping, C.: Badminton video analysis based on player tracking and pose trajectory estimation. In: 2021 13th international conference on measuring technology and mechatronics automation (ICMTMA). IEEE, pp 471–474
31. Ren Y, Wang Z, Tan S, Chen Y, Yang J (2021) Winect: 3d human pose tracking for free-form activity using commodity wifi. Proc ACM Interact Mob Wearable Ubiquitous Technol 5(4):1–29
32. Semwal A, Mishra D, Raj V, Sharma J, Mittal A (2018) Cricket shot detection from videos. In: 2018 9th international conference on computing, communication and networking technologies (ICCCNT). IEEE, pp 1–6
33. Ghosh I, Ramamurthy SR, Chakma A, Roy N (2022) Decoach: deep learning-based coaching for badminton player assessment. Pervasive Mob Comput 101608
34. Ebner CJ, Findling RD (2019) Tennis stroke classification: comparing wrist and racket as imu sensor position. In: Proceedings of the 17th international conference on advances in mobile computing & multimedia, pp. 74–83
35. Ghosh I (2020) Deepminton: analyzing stance and stroke to rank badminton players. UMBC Student Collection

36. Kulkarni KM, Shenoy S (2021) Table tennis stroke recognition using two-dimensional human pose estimation. In: Proceedings of the IEEE/CVF conference on computer vision and pattern recognition, pp 4576–4584
37. Khan AA, Rao Y, Shao J (2022) Enet: event based highlight generation network for broadcast sports videos. Multimed Syst 1–12
38. Shukla P, Sadana H, Bansal A, Verma D, Elmadjian C, Raman B, Turk M (2018) Automatic cricket highlight generation using event-driven and excitement-based features. In: Proceedings of the IEEE conference on computer vision and pattern recognition workshops, pp 1800–1808
39. Merler M, Mac K-NC, Joshi D, Nguyen Q-B, Hammer S, Kent J, Xiong J, Do MN, Smith JR, Feris RS (2018) Automatic curation of sports highlights using multimodal excitement features. IEEE Trans Multimed 21(5):1147–1160
40. Gunawardena P, Amila O, Sudarshana H, Nawaratne R, Luhach AK, Alahakoon D, Perera AS, Chitraranjan C, Chilamkurti N, De Silva D (2021) Real-time automated video highlight generation with dual-stream hierarchical growing self-organizing maps. J Real-Time Image Proc 18(5):1457–1475
41. Gupta A, Karel A, Sakthi Balan M (2019) Discovering cricket stroke classes in trimmed telecast videos. In: International conference on computer vision and image processing. Springer, pp 509–520
42. Fu X, Zhang K, Wang C, Fan C (2020) Multiple player tracking in basketball court videos. J Real-Time Image Proc 17(6):1811–1828
43. Host K, Ivasic-Kos M, Pobar M (2020) Tracking handball players with the deepsort algorithm. In: ICPRAM, pp 593–599
44. Sun P, Zhao X, Zhao Y, Jia N, Cao D (2022) Intelligent optimization algorithm of 3d tracking technology in football player moving image analysis. Wirel Commun Mob Comput
45. Pan Z, Zhu J, Bao X, Lin J, Ming J () Research on volleyball players tracking based on improved deepsort. In: 2022 4th international conference on communications, information system and computer engineering (CISCE). IEEE, pp 591–595
46. Ferede S, Xie X, Zhang C, Du J, Shi G (2020) Small ball tracking with trajectory prediction. In: 2020 IEEE 5th international conference on signal and image processing (ICSIP). IEEE, pp. 215–219
47. Sharma M, Anand A, Srivastava R, Kaligounder L (2018) Wearable audio and imu based shot detection in racquet sports. arXiv:1805.05456
48. Minhas RA, Javed A, Irtaza A, Mahmood MT, Joo YB (2019) Shot classification of field sports videos using alexnet convolutional neural network. Appl Sci 9(3):483
49. Bandara I, Bačić B (2020) Strokes classification in cricket batting videos. In: 2020 5th international conference on innovative technologies in intelligent systems and industrial applications (CITISIA). IEEE, pp 1–6
50. Ghosh I, Ramamurthy SR, Roy N (2020) Stancescorer: a data driven approach to score badminton player. In: 2020 IEEE international conference on pervasive computing and communications workshops (PerCom Workshops). IEEE, pp 1–6
51. Martin P-E, Calandre J, Mansencal B, Benois-Pineau J, P´eteri R, Mascarilla L, Morlier J (2021) Sports video: Fine-grained action detection and classification of table tennis strokes from videos for mediaeval 2021. arXiv:2112.11384
52. Kong PW, Pan JW, Komar J, Yang C (2021) Shot performance and cue stick kinematics of top spin shots in cue sports players of various playing levels. ISBS Proc Arch 39(1):49
53. Mat Sanusi KA, Mitri DD, Limbu B, Klemke R (2021) Table tennis tutor: forehand strokes classification based on multimodal data and neural networks. Sensors 21(9):3121
54. Liu H, Sun F, Gu J, Deng L (2022) Sf-yolov5: a lightweight small object detection algorithm based on improved feature fusion mode. Sensors 22(15):5817
55. Sun N-E, Lin Y-C, Chuang S-P, Hsu T-H, Yu D-R, Chung H- Y, Ik T-U (2020) Tracknetv2: efficient shuttlecock tracking network. In: 2020 international conference on pervasive artificial intelligence (ICPAI). IEEE, pp 86–91

Football Analytics for Goal Prediction to Assess Player Performance

Danish Javed, N. Z. Jhanjhi, and Navid Ali Khan

Abstract Machine learning techniques are often used for sports analytics, such as player health prediction and avoidance, appraisal of prospective skill or market worth, and predicting team or player performance. This reshapes the sports performance and helps in coaching the teams and individuals. This research focuses on football analytics, which can help football managers and coaches for reshaping the performance of players to target the goal with higher accuracy and precision. The match results depend on the successful number of goals; any minor mistake may lead to failure. Other statistics, like shots on target and game possessions, have been gaining popularity in recent years. Several attributes are utilized to train an anticipated goal model formed by monitoring football data to evaluate the chance of a shot being a goal. Using historical data and advanced analytics, a credible prediction of a goal, as well as player and team performance, can be deduced. Furthermore, we address the identification and recording of personal talents and statistical categories that distinguish an exceptional goal scorer from the worst goal scorer through football analytics. Feature selection, data size, and parameters used may impact the results of the model. Our research proposes a Goal Prediction Model (GPM) with player analysis trained on data from 9,074 games, including 941,009 events from Europe's top 5 leagues containing the information of five seasons. Our model will explain the observations on expected goals through football analytics and monitor the performance of the players with respect to anticipated goals. This research could benefit football team managers and coaches by reshaping the performance of players.

D. Javed · N. Z. Jhanjhi (✉) · N. A. Khan
School of Computer Science, SCS, Taylor's University Lakeside Campus, 47500 Subang Jaya, Malaysia
e-mail: noorzaman.jhanjhi@taylors.edu.my

D. Javed
e-mail: danish.tulc@gmail.com

N. A. Khan
e-mail: NavidAli.Khan@taylors.edu.my

© The Author(s), under exclusive license to Springer Nature Singapore Pte Ltd. 2023

S. F. Syed Omar et al. (eds.), *Innovation and Technology in Sports*, Lecture Notes in Bioengineering, https://doi.org/10.1007/978-981-99-0297-2_20

Keywords Goal prediction · Football analytics · Machine learning · English Premier League · Spanish La Liga · Italian Serie A · French League 1 · German Bundesliga

1 Introduction

Football analytics is a branch of sports analytics where we utilize past records and sophisticated statistics to assess the performance of a player or a club, make choices, and forecast outcomes and results in order to obtain a competitive edge [1]. The most typical task in sports analytics is performance prediction. Sports analysts use data analysis on players and teams with the purpose of predicting game results, league winners, or club and player efficiency [2]. Predictions might be for short-time or long-time occurrences. As a result, several methodologies and algorithms have been used [3, 4] Clubs collect and analyze data provided by football athletes throughout their training time and games using sophisticated equipment and software. They analyze this information in order to make short-term decisions and long-term organizational growth. Machine learning can be an optimal choice for such applications [5]. Furthermore, substantial data analysis is required for betting firms. Finally, football fans are fascinated with sophisticated statistics and their impact on the game. Due to the aforementioned reasons, the usage of football analytics has grown in recent years [6]. Therefore, football was chosen for our research because of the number of statistical attributes and past data available, its popularity, and the simplicity of its regulations and national championship forms. In contrast, there are unique challenges that make football's long-time projection tough [7].

The proliferation of internet football data is a benefit, but it requires careful preprocessing to predict club and player performance. Regrettably, this is not always straightforward [8]. Furthermore, certain scenarios that are not shown in the data collected might change club and player performance. For example, a club might be rated higher than it should be if its competitors underperform. When a player returns from a significant injury, he or she may have a poor rating performance [9].

Finally, because of the nature of football, statistical collection of match occurrences, as well as player and club ratings, is an unclear procedure. In the past decade, the exponential rate of advancement in technology supporting data gathering, storage, and analysis has had a transformative influence on football analytics and many other industries [10]. Because data is easily accessible, there is a tremendous opportunity to provide multiple critical performance indicators that measure various parts of the game like pass assessment, quantifying controlled areas, assessing hits on goals, and goal-scoring possibilities via possession statistics based on the area of the pitch [11]. Figure 1 presents the general partition of the football pitch to calculate possession statistics.

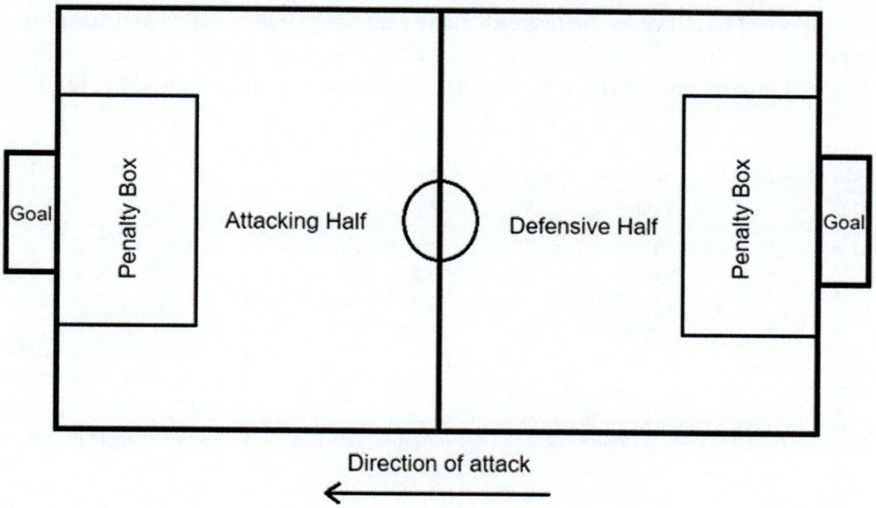

Fig. 1 Areas of the football pitch

1.1 Problem Definition and Contributions

Team and player performance prediction are areas that need to be explored. Sports agents, coaches, and bookies are concerned with the performance of clubs and players from one season to the next. The context of this topic is explained, and the research goals are also established. Goal predictions are difficult due to the unique characteristics of football matches as not many goals are scored in every game [12]. Furthermore, there is no distinct transition between the immediate shift of ball possession and the movement between offensive and defensive moves. Furthermore, player locations and strategies are not set, and the match has a constant flow, which makes capturing match events difficult [13]. It is also worth noting that not only teams are interested in forecasting the result of a match but also the people who like sports betting throughout the world would benefit from such predictions as well [14].

Our research focuses on prior season statistics and historical data and presents the following contributions:

- Proposing a model for football goal analysis and prediction using machine learning.
- Presenting evaluation for player's performance through exploratory analytics.

2 Related Literature

It is widely known that a lot of football clubs are attempting to invest in some type of football analytics to enhance their results by obtaining even a little advantage over their opponents. However, there aren't many teams who will acknowledge using such approaches. Furthermore, it is not easy for a team to set aside the necessary finances to hire sports scientists who will assist them in drawing conclusions [15] (Table 1).

Table 1 Brief literature review of football analytics research

Cite	Purpose	Positives	Findings
Green [16]	Proposes an xG model to calculate the likelihood of a shot being the goal	It is now the most prevalent in football talk shows on TV and end-of-match statistics	Creation of such a statistic is motivated by the need to offer a metric that shows low-scoring character of football rather than other sports
Cardoso et al. [17]	xG model is employed as a meaningful scoring indication about the goals scored in the match	From a statistical stand- point, it may be de- scribed as the average of a significant number of independent samples of a random variable	Aside from being an excellent representation of the score, it is also a solid indication for pre-dicting future team's success
Spearman and Spearman Hudl [18]	Assess the efficiency of off-ball positioning before attempts that may result in goals	It can identify the opportunities during the match and help with opposition analysis. It can automate talent recognition for upcoming players	It could be useful to predict which players, if any, are likely to score at any moment through- out the game, as well as where on the field they are likely to score
Cavus and Biecek [19]	Gives a more representative measure of club and player performance that also suits the low-scoring aspect of game through explainable model	Several attributes are utilized to train an anticipatedgoalmodel by monitoringfootball data to evaluate the chance of a shot being a goal	The choice of these attributes, the amount and date of the data, and the model parameters may impact the model's performance
Patrōn et al. [20]	Trials were conducted onfootball data via appropriate algorithms to forecast a player's position on the pitch as well as goal scoring likelihood	It forecasts a player's goal scoring performance in the next season. It also predicts the number of shots a player takes in each match for goal scoring prediction	Results were impressive, displaying good accuracy, especially because the anticipated number of goals was extremely close to the real number

3 Methodology

In this section, we organize the data such that it may be used as input for our model. We will discuss the dataset used in this research and provide details on our proposed model in Fig. 2.

3.1 Dataset Description

The properties of the data used to train the models must be considered prior to training the models. From the perspective of football fans, it is obvious that the playing style shifts with time and differs throughout leagues. Research [21] provide a response to the question, "How can it be established if this circumstance happened or not?" They performed a comprehensive experimental investigation to evaluate commonly questioned data-related topics such as "How much data is required to train an appropriate model?" "Are these models league-specific?" and "Does data become out of date?" that may impact a model's performance. The study concluded with the answer that data consisting of five seasons would be enough to check the performance of the model. Therefore, we have chosen a dataset that meets these requirements.

The dataset used in this research is "Football events". It consists of 9,074 games, including 941,009 events from the Europe's top 5 leagues containing the information of five seasons. Furthermore, dataset contains 229,135 shots, where 24,441 are goals and others are missed opportunities. We have 28 separate attributes that define each shot. As all 28 qualities are binary, they simply say Yes or No to various aspects of the shot.

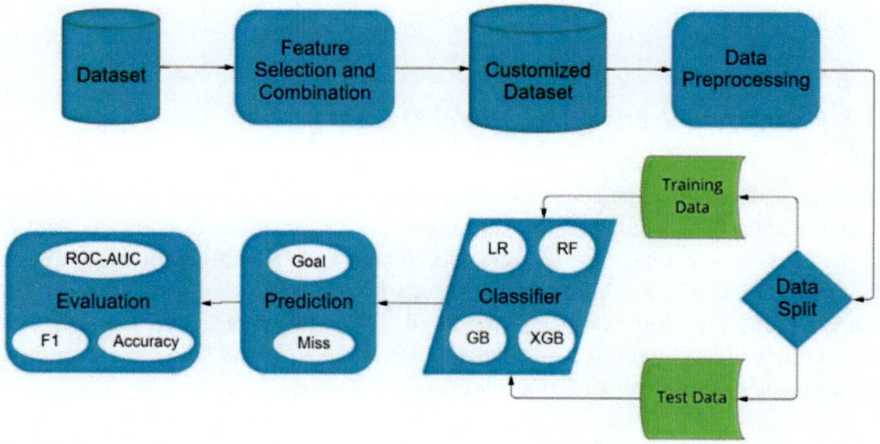

Fig. 2 Anticipated goal model

3.2 Proposed Model

Our model will try to calculate the likelihood that a given shot will result in a goal (see Fig. 2). We may use this measure to examine what transpired in a match besides how many goals each club scored. Because the match is won through goals, and goals are only scored through shots, therefore evaluation measure can only be produced using data from shots. In other words, everything else that may impact how many "anticipated goals" a team scores must occur as a result of having more shots in the first place. For example, if team X suffers three red cards, team Y is likely to score more goals (an increase in goal anticipation). However, having three more players on the field will result in more (and maybe better) shots, and more shots will result in an increase in predicted goals. As a result, the only factor we must consider are the shots and associated information.

For our model, we will not take into consideration individual qualities or talents of the participants in the event. Although it could be argued that if a model should account for this, it might be more accurate. Of course, if an impressive player is one-on-one with the goalkeeper, his odds of scoring are better than those of any other player. Similarly, if the goalkeeper is impressive, the chances of it becoming a goal decrease. However, that is not the purpose of the measure. We're attempting to normalize the likelihood of every player scoring from a certain position in a specific circumstance using thousands of data samples. If the player is an exceptional scorer, he will most likely score more goals than projected, which is fine. If we begin to consider the talents of the people involved, we believe that it will be going too far in the construction of this statistic, and it will become less significant.

For our problem we will set the X and Y sets as follows:

- Target feature 'Y' will just include every shot in the database, regardless of whether it was a goal or a miss (1 or 0). It is our primary variable.
- Feature set 'X' will contain all the pertinent details about the shot that we have in our database.

3.3 Feature Selection and Model Training

Our model does not include pre-processing processes such as missing data imputation and encoding but includes feature selection and combination. Initially the dataset consists of two separate '*csv*' files. League, country and date information is collected from one file, and it is combined with the other to create a new merged dataset that contains all the information regarding '*shots*'. Then different features are selected based on our tasks. We train our model with various machine learning algorithms. To train and verify the models, we adopt an 70-30 train-test split.

4 Results and Experimentation

The performance of our proposed model is studied in terms of numerous metrics under various sampling techniques in this section. The initial stage in model investigation is generally connected to model performance [22]. F1, accuracy, and AUC are some of measurements that can be employed. Unfortunately, just a few articles in the literature connected to the model have applied adequate approaches to address this issue. As mentioned in this study [23], selection of machine learning algorithms can have huge impact in terms of performance of the model, therefore we will utilize various ML algorithms to test the performance of our model. We limited the research in this work to a comparison of model performance between train and test data. Table 2 shows the performance metrics for the various ML algorithms employed. Figure 3(a) and (b) presents confusion matrix for two best performing algorithms.

Table 2 Classification results

Classifier	Accuracy (%)	F1-score	ROC-AUC (%)
Logistic regression (LR)	91.3	0.76	82.1
Gradient boost (GB)	91	0.76	82
SVM	90	0.75	81.5
Random forest	90	0.75	81.5
XGB	89.5	0.74	81
Decision tree	89	0.74	81

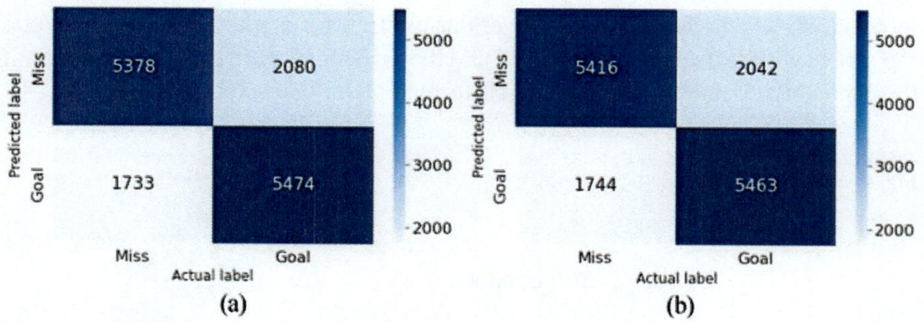

Fig. 3 **a** Confusion matrix for Gradient Boost classifier **b** Confusion matrix for Linear Regression classifier

5 Analytics and Discussion

In this section, we will discuss the various visual analytics related to 'shots' and 'players' and few analytics related to teams and seasons. Figure 4 provides details on short outcomes. When a football player takes a hit towards the net to score a goal, it is considered a shot. That hit can result in various places thus leading to different outcomes as follows:

- If a hit is blocked by the opponent, it is labeled as a 'blocked shot'.
- If a hit strikes the bar of the net, it is labeled as a 'bar strike'.
- If a hit is directed towards the net without being blocked, it is considered a 'target shot', even if it actually does not result in a goal.
- If a hit is directed outside the range of the net, it is considered a 'missed shot'.

We explored the shot placement, we can see in the Fig. 5 that most of the shots were blocked while the shots that are not blocked are either directed at the center of the net or miss to the left or right side but less shots are missed to the bottom side of the net.

We will investigate something vital for our model by checking what percentage of shots result in goals. We'll separate this part into leagues and years to see whether there are any variances in the correlations across various areas of the world or different years. As we can see in the Fig. 6, there is only a small difference between top five leagues in terms of goals and missed opportunities.

We also examined the gap between total goals scored and total predicted goals: This will provide details about the players that scored the most goals in comparison to what was anticipated of them. This analytic can be useful as it provides us the details of whether a player is exceeding their expectations or not. Clubs can use this information when placing bid on player during 'transfer market'. Figure 7 provides top scorer with respect to goals scored and the goals they were expected to score. If the gap is bigger that means a player performed better than expected.

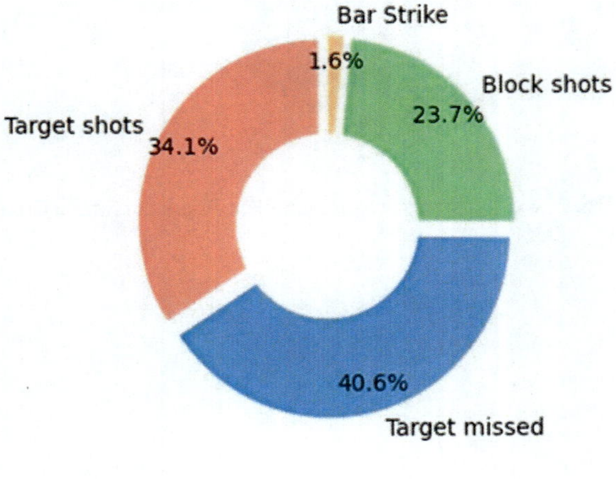

Fig. 4 Shot outcome

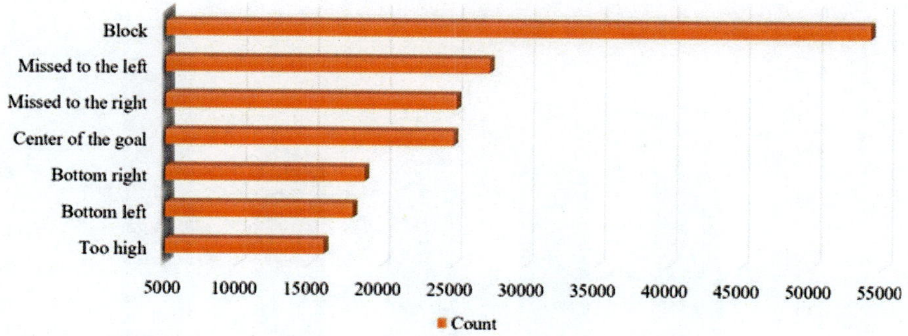

Fig. 5 Top 7 shot placement

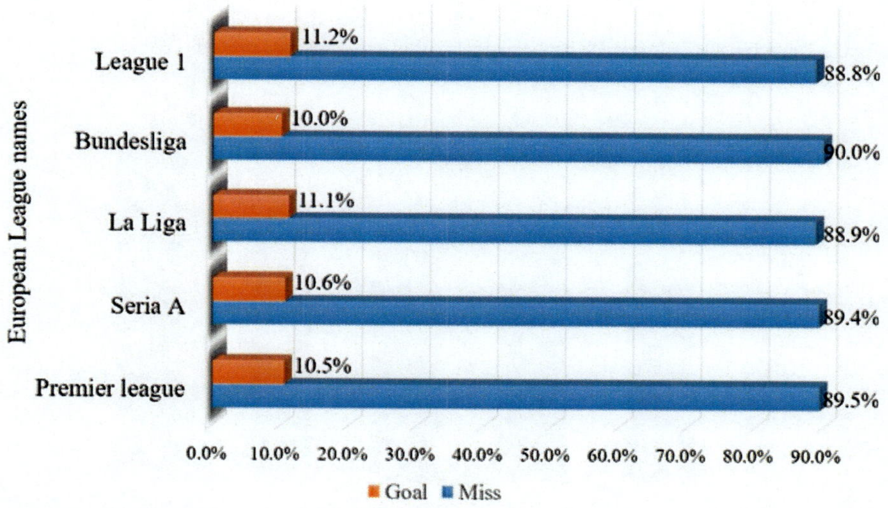

Fig. 6 Goal versus Miss comparison of top 5 leagues

We can examine the ratio of total goals accomplished to total projected goals. This will give us the exact relationship between how many goals the players really scored and how many goals they were projected to score. Figure 8 lists the top five players with respect to a goal scored per one anticipates goal. If the ratio is higher than 1, it implies that the player performed better than anticipated. In this metric the list of the players changed as we analyzed the actual vs anticipated goals based on ratio of expectations per one goal.

Similarly, we can also analyze the worst finishers in the league by finding the ratio between goal scored and the number of goals the player was anticipated to score. If the ratio is less than 1, it implies that the player performed worse than anticipated. Figure 9 shows the worst five performing players.

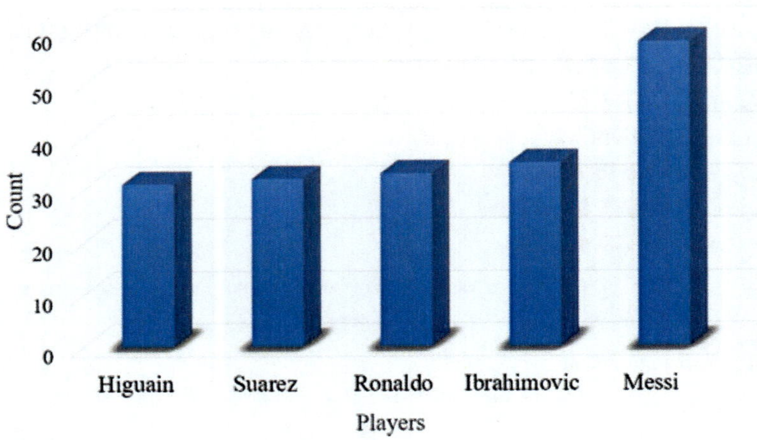

Fig. 7 Top 5 finishers–difference between goal scored and goals anticipated

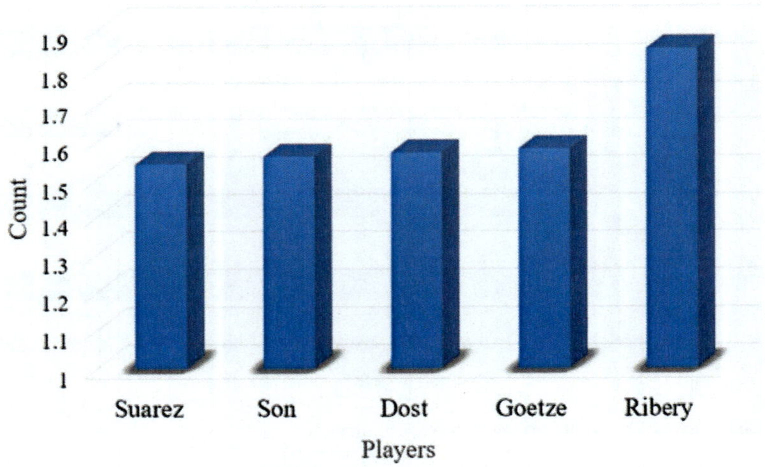

Fig. 8 Top 5 finishers-goals scored per one anticipated goal

By analyzing the stats of the player who made the pass that came before the shot, we can assess how much anticipated goals a player made from their passing. We analyzed the mean anticipated goals of these passes to check the top five passers. Figure 10 lists the top players by passing criteria.

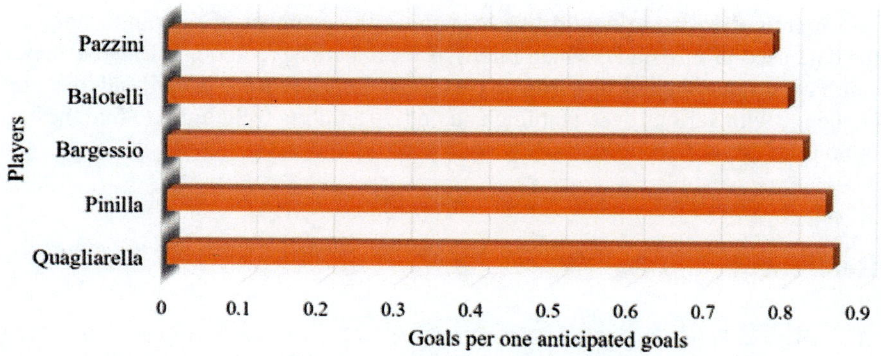

Fig. 9 Worst 5 finishers-goals scored per one anticipates goals

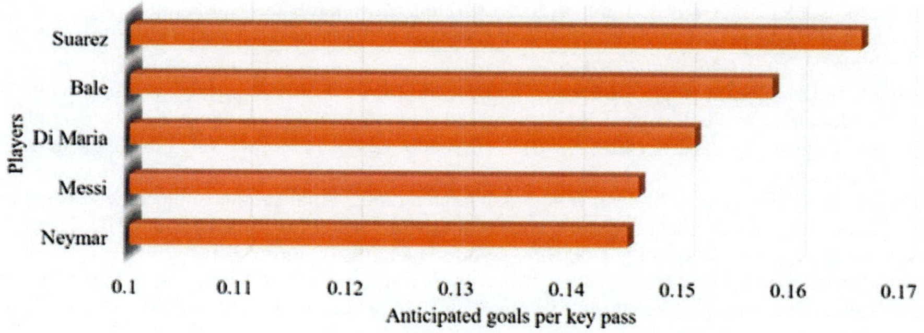

Fig. 10 Top 5 passers-anticipated goals per key pass

6 Conclusion

Our research looked at two core examples of football analytics: goal prediction and player performance with respect to goals. The purpose of the first experiment was to forecast whether a shot will end up in goal or not based on various features. The output of our model has been used to evaluate the performance of a team or player. They give information based on the disparity between actual and predicted goals for assessing a team's or a player's defensive and offensive efficiency. However, we concentrated on using the behavior of our model to discern the link between characteristics and reaction, that is the anticipated goal score. Performance of a player or a club can be enhanced by adjusting strategy depending on factors that affect the anticipated goal value, such as distance to goal, angle to goal, and others. In order to achieve this, we developed a model trained on data from five seasons of top-five European leagues using various ML algorithms. We also provided detailed player analytics to find the best and worst performing players with respect to anticipated goal value. Our model performs with great accuracy to predict the outcomes as a goal or miss with an accuracy of 91.3%.

For future work, we would like to address the problem of data imbalance for the data used in this study, which might result in better predictions. Data balancing can have a huge impact in terms of performance. Therefore, we will address this challenge with various oversampling and under-sampling techniques to find the best option.

References

1. Mendes L, Fontanive S (2021) Football analytics: a literature analysis from 2010 to 2020. Instituto Superior de Estatística e Gestão de Informação, Universidade Nova de Lisboa
2. Rastegari H, Haghighat M, Nourafza N (2013) A review of data mining techniques for result prediction in sports. In: Advances in computer science. https://www.researchgate.net/publication/262560138
3. Stübinger J, Mangold B, Knoll J (2020) Machine learning in football betting: prediction of match results based on player characteristics. Appl Sci (Switzerland) 10(1). https://doi.org/10.3390/app10010046. (Jan 2020)
4. Herold M, Goes F, Nopp S, Bauer P, Thompson C, Meyer T (2019) Machine learning in men's professional football: current applications and future directions for improving attacking play. Int J Sports Sci Coach 14(6):798–817. https://doi.org/10.1177/1747954119879350. (SAGE Publications Inc., 01 Dec 2019)
5. Simon CGK, Jhanjhi NZ, Goh WW, Sukumaran S (2022) Applications of machine learning in knowledge management system: a comprehensive review. J Inf Knowl Manag 21(2). https://doi.org/10.1142/S0219649222500174. (World Scientific, 01 Jun 2022)
6. Behravan I, Razavi SM (2021) A novel machine learning method for estimating football players' value in the transfer market. Soft Comput 25(3):2499–2511. https://doi.org/10.1007/s00500-020-05319-3. (Feb 2021)
7. Cefis M (2022) Football analytics: a bibliometric study about the last decade contributions. Electron J Appl Stat Anal 15(1):232–248. https://doi.org/10.1285/i20705948v15n1p232
8. Baboota R, Kaur H (2019) Predictive analysis and modelling football results using machine learning approach for English Premier League. Int J Forecast 35(2):741–755. https://doi.org/10.1016/j.ijforecast.2018.01.003
9. Claudino JG, de Capanema DO, de Souza TV, Serrão JC, Machado Pereira AC, Nassis GP (2019) Current approaches to the use of artificial intelligence for injury risk assessment and performance prediction in team sports: a systematic review. Sports Med Open 5(1). https://doi.org/10.1186/s40798-019-0202-3. (Springer, 01 Dec 2019)
10. Pariath R, Shah S (2018) Player performance prediction in football game
11. Eldridge D, Pulling C, Robins M (2013) Visual exploratory activity and resultant behavioural analysis of youth midfield soccer players. J Hum Sport Exerc 8(3). https://doi.org/10.4100/jhse.2013.8.Proc3.02
12. Bunker RP, Thabtah F (2019) A machine learning framework for sport result prediction. Appl Comput Inf 15(1):27–33. https://doi.org/10.1016/j.aci.2017.09.005
13. Pantzalis VC, Tjortjis C (2020) Sports analytics for football league table and player performance prediction. https://doi.org/10.1109/IISA50023.2020.9284352. (Jul 2020)
14. Miljkovic D, Gajic L (2010) The use of data mining for basketball matches outcomes prediction. In: IEEE 8th international symposium on intelligent systems and informatics, 10–11 Sept 2010, Subotica, Serbia
15. Corscadden J, Eastman R, University of Virginia, Institute of Electrical and Electronics Engineers (2018) Developing analytical tools to impact U.Va. Football performance
16. Green S (2012) Assessing the performance of premier league goalscorers. In: OptaPro Blog. https://www.statsperform.com/resource/assessing-the-performance-of-premier-league-goalscorers/. Accessed 24 Sept 2022

17. da Cardoso FSL, González-Víllora S, Guilherme J, Teoldo I (2019) Young soccer players with higher tactical knowledge display lower cognitive effort. Percept Mot Skills 126(3):499–514. https://doi.org/10.1177/0031512519826437. (Jun 2019)
18. Spearman W, Spearman Hudl W (2018) Beyond expected goals. https://www.researchgate.net/publication/327139841
19. Cavus M, Biecek P (2022) Explainable expected goal models for performance analysis in football analytics. http://arxiv.org/abs/2206.07212. (Jun 2022)
20. Patrōn P, Peloponnēsou P, Peiraiōs P, Institute of Electrical and Electronics Engineers (2019) Sports analytics algorithms for performance prediction
21. Brefeld U, Davis J, van Haaren J, Zimmermann A (eds) Machine learning and data mining for sports analytics, vol 1324. Springer International Publishing, Cham. https://doi.org/10.1007/978-3-030-64912-8
22. Naif Almuayqil S, Humayun M, Zaman Jhanjhi N, Fahaad Almufareh M, Javed D (2022) Framework for improved sentiment analysis via random minority oversampling for user tweet review classification, vol 11, pp 3058. https://doi.org/10.3390/electronics11193058
23. Dogra V, Verma S, Jhanjhi N, Ghosh U, Le D-N (2022) A comparative analysis of machine learning models for banking news extraction by multiclass classification with imbalanced datasets of financial news: challenges and solutions

Detection of Localized Muscle Fatigue by Using Wireless EMG Among Track and Field Athletes

Muhammad Zulhilmi Chek Azman, Mohammad Azzeim Mat Jusoh, and Nurul Syuhada Khusaini

Abstract Muscle exhaustion is one of the most common injury types affecting most athletes and even ordinary people. This study aims to understand more about athletes' muscular fatigue, especially track and field athletes such as sprinters. The electromyography (EMG) shall capture the muscle activation signals, specifically on the leg muscles, as both gastrocnemius muscles quickly became fatigued during the calf raise exercise. The electricity flow through the calf muscle can be measured via EMG, reflecting the amount of fatigue experienced. In this research, three selected athletes were recruited as the test subject. The EMG sensor was attached to the calf muscle of the athlete during the calf raise exercise and the subject was instructed to maintain composure until fatigue set in. Using the ProEMG software, the data was transferred to the computer for further analysis. The preliminary result found that the average median frequency dropped to 6.13 $\pm$ 0.03%, and the root mean square amplitude increased by 11.98 $\pm$ 0.02% during fatigue conditions. The endurance time for each subject varies, depending on the physical properties of the body and performance experience. The understanding from the findings is pivotal for achieving optimal muscular strength and endurance development and reducing the prospect of training-related injuries to sport-person.

Keywords EMG · Gastrocnemius muscle · Calf raise exercise

1 Introduction

1.1 Research Background

Muscle fatigue is a decrease in maximal force during contraction [1]. The capacity to lift or move is limited by extreme weariness. Various research was conducted to study muscular exhaustion [2]. Many detection methods are already used on muscle

M. Z. C. Azman · M. A. Mat Jusoh (✉) · N. S. Khusaini
School of Mechanical Engineering, College of Engineering, Universiti Teknologi MARA (UiTM), 40450 Shah Alam, Selangor, Malaysia
e-mail: m_azzeim@uitm.edu.my

© The Author(s), under exclusive license to Springer Nature Singapore Pte Ltd. 2023

S. F. Syed Omar et al. (eds.), *Innovation and Technology in Sports*, Lecture Notes in Bioengineering, https://doi.org/10.1007/978-981-99-0297-2_21

signals to identify fatigue [3]. Surface electromyography (sEMG) is the primary approach for recording and studying muscle activities, as it records the electrical signal of the muscles. Other elements can contribute to fatigue, such as muscle fibre composition, ionic control in the plasma, energy supply, neurological processes, and many others. According to research, muscular tiredness is linked to musculoskeletal injuries during sports training and competition [2].

The gastrocnemius muscle (located above the soleus muscle) is a larger calf muscle that accounts for 50% of the calf muscle [4]. It extends from behind the knee to the Achilles tendon at the heel. The gastrocnemius muscle has two heads: the lateral and the medial, which attach to the Achilles tendon and the calcaneus. Injuries to the gastrocnemius often occur at the distal muscle–tendon junction in the medial head of the muscle. In high-performance athletes, calf muscle strain can negatively impact their running and is also a significant reason for their absence from competition [5]. The frequent strain on the medial head of the gastrocnemius muscle during sprinting can lead to an injury. Athletes who suffer from this condition reported severe pain and soreness at the back of their calf, comparable to a ball impact directly in that spot. Without consistent training, calf muscles cannot maintain the necessary strength and fitness level to support athletic activity, which increases the risk of injury. Stronger calf muscles may benefit some sports, like volleyball and basketball.

The goal of this research is to monitor the different persistence of the calf raise exercise, which affects the athletes' performance level. This study provides an insight into the EMG activity of the most dominant calf muscles; the gastrocnemius lateralis (GSL) and gastrocnemius medialis (GSM) muscles, of both legs by using the sEMG. The increment of EMG signal amplitude could suggest an increase in muscle activation [6]. The features from the frequency and time domains were selected to extract relevant information from the EMG signal and thus analyze muscle fatigue. The current study is limited in sample size, as only three test subjects enrolled for the qualitative research.

2 Methodology

The experiment was performed to achieve the objectives. It was designed to ensure that no obstacles affect the experiment's progression. During the experiment, the subject's gastrocnemius muscle was tested by performing the calf raise exercise.

The investigation focused primarily on the two head positions of the gastrocnemius muscle: the gastrocnemius medialis and gastrocnemius lateralis. These specific muscles were chosen since they control the running's stride and pattern. To accelerate their pace, sprinters required extensive calf activity during running.

Table 1 General information of test subject

	Gender	Age	Height (cm)	Weight (kg)	BMI (Kg/m^2)
Volunteer 1	Male	24	173	69	23.1
Volunteer 2	Male	25	160	58	22.7
Volunteer 3	Male	23	173	60	20

2.1 Participant

In this project, three selected athletes actively engaged in athletic sports were recruited as subjects (age = 24 ± 1, weight = 62.33 ± 5.86 kg, height = 168.67 ± 7.51 cm). Before agreeing to participate, these subjects were briefed about the motive and the study's procedure. For findings to be approved, subjects must be free from muscle injuries and soreness. Detailed information regarding the participants are summarized in Table 1.

2.2 Experimental Setup

The surface EMG data were recorded using a 4-channel wireless Myon 320 sEMG system at FSR Gymnasium, UiTM Shah Alam, for all the volunteers. Before data collection, volunteers' skin was shaved to remove hairs and then cleansed with an alcohol swab to reduce skin impedance, thus optimizing signal acquisition. The sEMG sensor was designed to be placed on the gastrocnemius muscle, following guidelines established by the European project organization SENIAM (Surface Electromyography for the Non-Invasive Assessment of Muscles) [7]. According to a rehabilitation physician's recommendation, conductive adhesive hydrogel disposable foam electrodes (Kendall Product) with one and 12-inch diameters and a 2-cm inter-electrode spacing were applied in parallel to the muscle fibres of the gastrocnemius of each volunteer [8] (Fig. 1).

The test subjects performed five minutes of static stretching before the actual examination. After the preparation, subjects were given the task to complete the calf raise exercise for 2 min, according to a metronome beat of 60 beats per minute. The challenge for the subjects was raising their heels as high as possible while keeping their knees straight. After completing the first repetition, the subjects took a complete recovery break as much as they wanted to perform the second repetition without pausing to take a break in between. The procedure was continued for a total of three repetitions. To avoid crosstalk effects, the sampling frequency of the data processing software (ProEMG) was set to 1000 Hz. The surface EMG signal was recorded every 40 s segment.

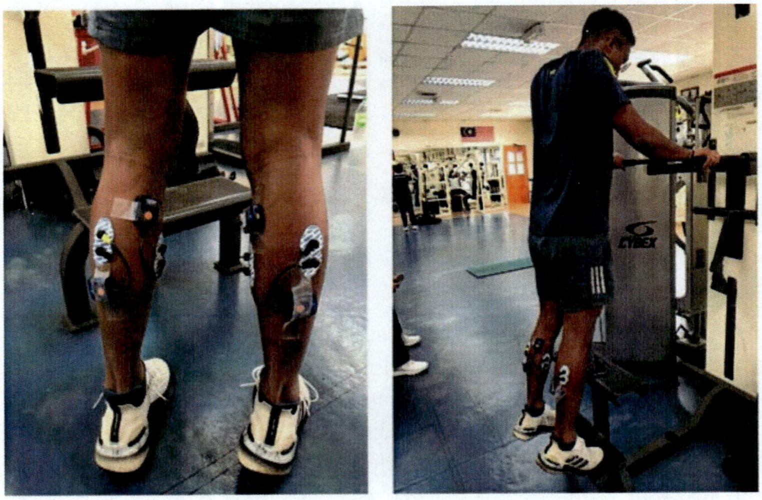

(a) (b)

Fig. 1 Visuals of **(a)** the electrode and transmitter placement on gastrocnemius muscle and **(b)** the actual condition of the participant performing the test

2.3 *Data Acquisition*

The raw EMG signals from the different subjects were quantified and processed into the ProEMG 2.0 software. As shown in Fig. 4, the signal processing approach was used on raw EMG data and divided into two parts which the preprocessing and feature extraction. The EMG signal was recorded at a sampling rate of 1000 Hz with a band-pass filter ranging from 10–400 Hz. Using the Fast Fourier Transform (FTT) analysis pipeline included in ProEMG 2.0, the power frequency spectrums of the raw data were initially discovered. It was done to examine the artefact noises' characteristics [9]. After the notch filter was set at 50 Hz, a Butterworth high-pass filter at 400 Hz and a Butterworth low-pass filter at 10 Hz were applied. From the filtered EMG signal, time-domain and frequency-domain feature was extracted to root mean square (RMS), median frequency (MDF) and mean frequency (MNF) for each, respectively. The selection of these parameters was made because previous research has shown that it is most commonly used to identify shifts in muscle performance associated with fatigue [10] (Fig. 2).

The processing approach that was frequently used in conjunction with frequency analysis is amplitude analysis. This is commonly performed by integrating the signal to calculate the area beneath the signal's curve or by utilizing this equation to get the root mean square. It is an advanced function used to evaluate EMG signals. RMS is an amplitude Gaussian random process correlating to non-fatiguing contraction with

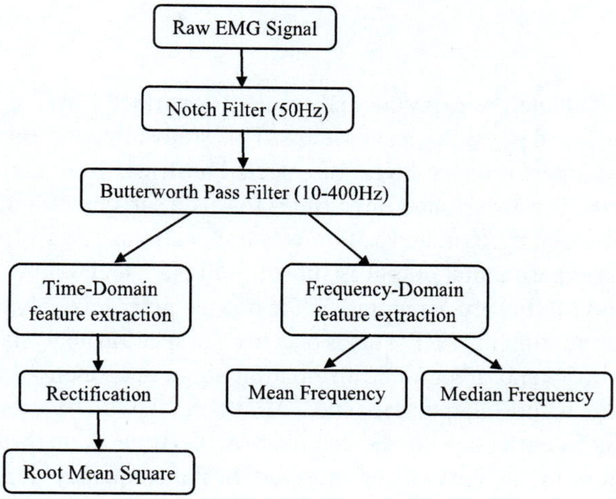

Fig. 2 The EMG signal processing steps

constant force [11]. The level of muscle exhaustion can be assessed by monitoring the rate of RMS's change of the signal's amplitude [12], where T refers to time and f2(t) refers to the rectified signal as shown below:

$$Root\ Mean\ Square\ (RMS) = \sqrt{\frac{1}{T}\int_{0}^{T} f^2(t)dt} \tag{1}$$

Finding the mean and median frequency is one of EMG's most popular processing methods. The mean and median frequencies both have similar tendencies toward the same direction. In the frequency spectrum of a signal, fast-twitch fibres represent high frequencies, and slow-twitch fibres represent low frequencies [13]. Despite the high levels of exhaustion experienced by fast twitch fibres, a decrease in the mean and median frequency typically indicates a shift toward the recruitment of slow twitch fibres [14]. A change in emphasis toward fast-twitch recruitment is characterized by an increase in the frequency with which it occurs. The MNF and MDF are shown as follows:

$$\sum_{m=1}^{M}(f_m P_m)/\sum_{m=1}^{M}(P_m) \tag{2}$$

$$\sum_{m=1}^{MDF} P_m = \sum_{m=MDF}^{M} P_m = \frac{1}{2}\sum_{m=1}^{M} P_m \tag{3}$$

3 Result

It is crucial to evaluate the experimental results for surface EMG signals to understand the physical and physiological systems. This study aimed to examine muscular fatigue at varying performance levels of selected athletes. To ensure the reliability of the sEMG data, it was repeated three times for each subject with the same experimental setup. It took approximately 30 min to finish all three trials for each subject. A sample of the experimental output is shown in Figs. 3, 4 and 5.

The MDF and MNF were generated on the power spectrum analysis of the sEMG signals that emerge from the FFT. This is because the spectral analysis of the data can be more trustworthy and give more information about the functions of the muscles when compared to the other approaches. However, it will be more reliable and apparent if it is integrated with the features of the time domain to obtain more information about the activity of the muscles. In the frequency domain, fatigue is related to a shift toward lower frequencies, as shown in Fig. 5.

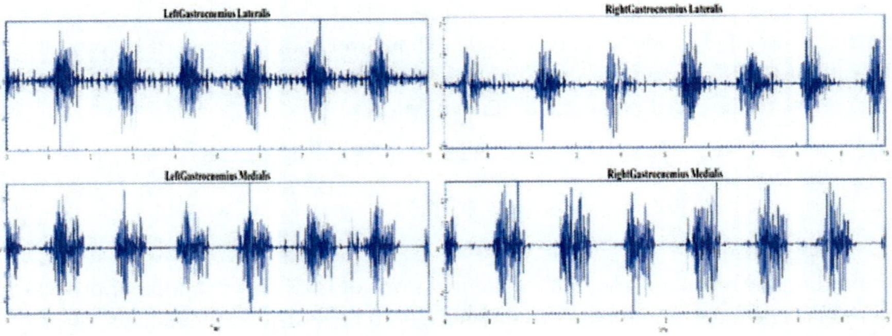

Fig. 3 The raw sEMG signal cropped in 10 s from the experiment for both legs

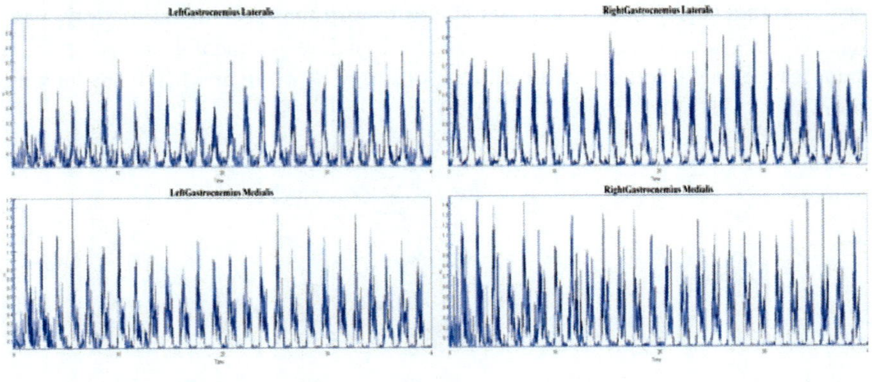

Fig. 4 The result of the RMS signal in 40 s from the experiment for both legs

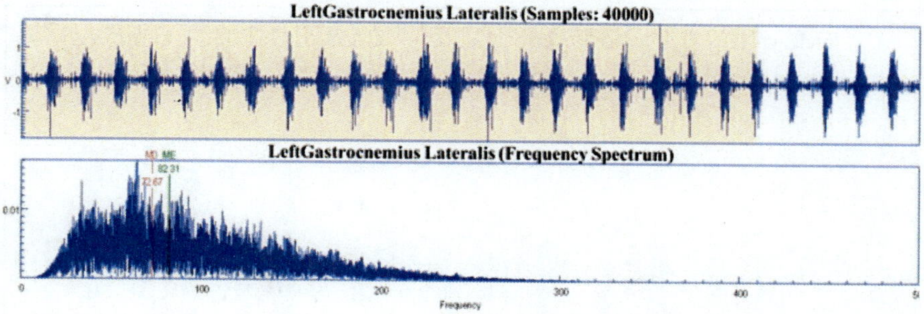

Fig. 5 Sample of FFT of sEMG signal GSL(LL) on 4000 samples

The RMS values of the GSL and GSM muscles of the left leg (LL) and right leg (RL) were summarized in Fig. 6. The figure illustrates the difference in performance between the subjects regarding time-domain properties for the muscles in both legs. According to the previous research, during the calf raise exercise, the higher performance test subject had a more significant mean value for IEMG and RMS in terms of performance difference along with time progression in both muscles of the individual during calf raise exercise [4]. In this experiment, a signal was possibly corrupted for electrode monitoring GSL muscles of the right leg of subject one because the electrode slightly detached from the subject's skin after the trial. This was either caused by excessive sweating or the electrode was not adequately attached from the start of the experiment. For this study, RMS amplitude for every subject and muscle increased, except for subject 1's GSL of the right leg and the inconsistent trend for GSM of the left leg's subject two and subject three.

The mean frequency value is displayed in Fig. 7 for the GSL and GSM muscles on both legs of each subject. For this study, the mean frequency for all muscles in both legs of each subject was decreased except for the GSL muscle of the right leg's subject 1. Experiment results show that their ability to activate their GSL and GSM muscles throughout the test phase differs according to their current performance level of endurance.

4 Discussion and Conclusion

This study showed a difference in the EMG activity of the left and right leg, from muscle to muscle, and in the time-domain and frequency-domain parameters retrieved. During the whole experiment, there was no significant difference in the EMG activity level of some muscles between the subjects. It was found that the left leg was used more for the calf raise trial. However, it was seen that the surface EMG activity of the right and left leg and the activity of each muscle varied between the test subjects.

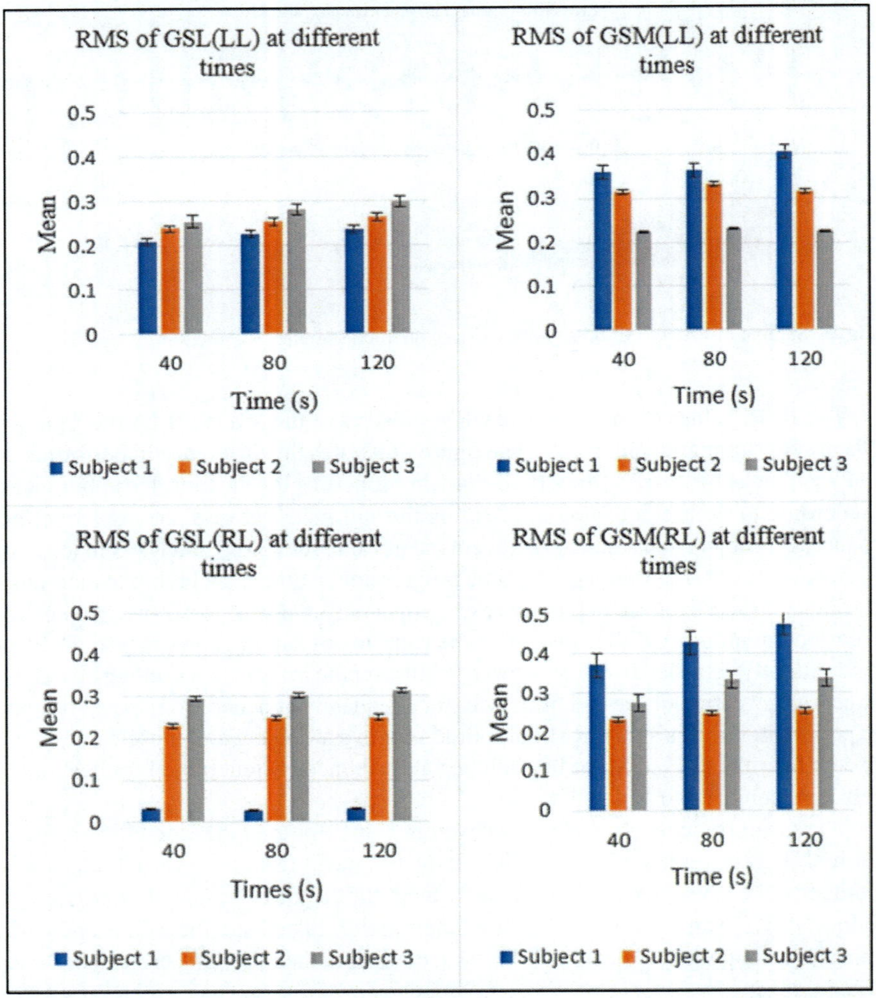

Fig. 6 Comparison of Time-domain features (RMS) in terms of performance level between subjects with standard error

To conclude, quantitative findings for surface EMG signal were extracted while monitoring the gastrocnemius muscle for muscle fatigue. The current limitation of the study was due to the limited sample size, as only three people enrolled in this experiment as qualitative research. In future, a higher test subject enrolment could ensure reliable results. Application-wise, athletes need to be aware of the findings to attain optimal muscle strength and strength endurance development while lowering the chance of training-related injuries.

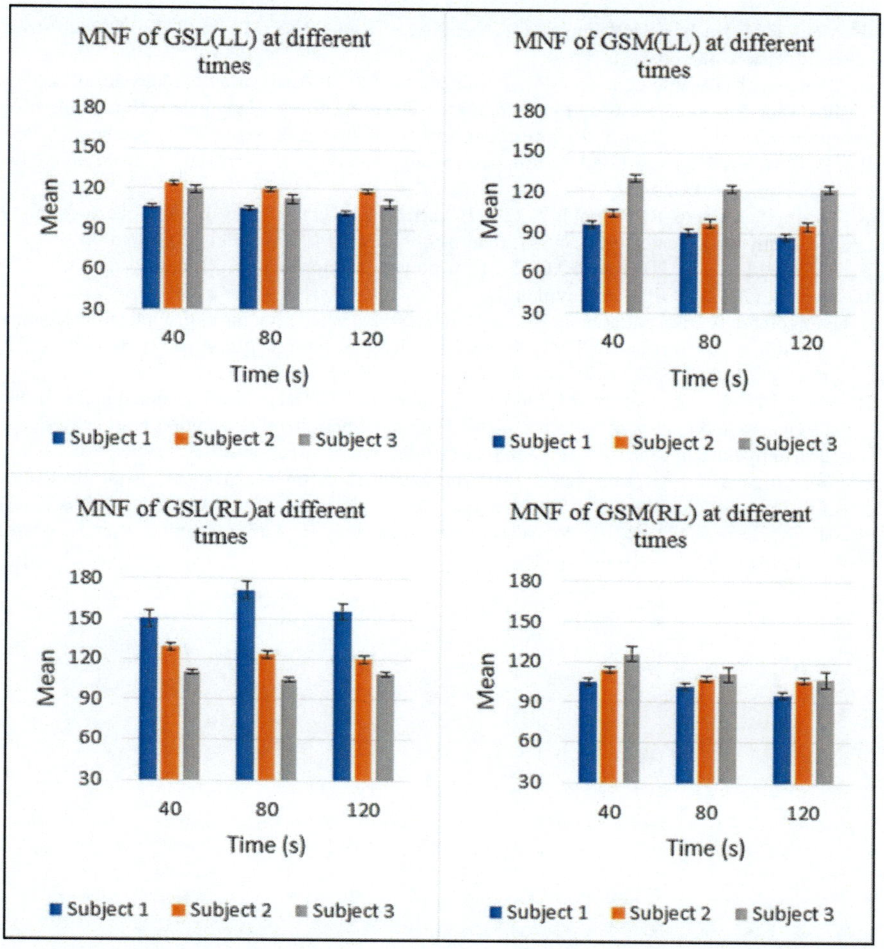

Fig. 7 Comparison of Frequency-domain features (MNF) in terms of performance level between subjects with standard error

References

1. Wallmann HW (2007) Muscle fatigue. In: Sports-specific rehabilitation. Churchill Livingstone, pp 87–95
2. Yousif HA et al (2019) Assessment of muscles fatigue based on surface EMG signals using machine learning and statistical approaches: a review. In: IOP conference series: materials science and engineering, vol 705, no 1. https://doi.org/10.1088/1757-899X/705/1/012010
3. Al-Mulla MR, Sepulveda F, Colley M (2011) A review of non-invasive techniques to detect and predict localized muscle fatigue. Sensors 11(4):3545–3594. https://doi.org/10.3390/s110403545
4. Mandal S, Thakur M, Thakur K, Singh BK (2021) Comparative investigation of different classification techniques for epilepsy detection using EEG signals

5. Meek WM, Kucharik MP, Eberlin CT, Naessig SA, Rudisill SS, Martin SD (2022) Calf strain in athletes, vol 0, no 3, pp 1–9
6. Mcmanus L, De Vito G, Lowery MM, Mcmanus L (2020) Analysis and biophysics of surface EMG for physiotherapists and kinesiologists: toward a common language with rehabilitation engineers, vol 11, no, pp 1–25. https://doi.org/10.3389/fneur.2020.576729. (October)
7. HJ Hermens HJ et al (1999) European recommendations for surface ElectroMyoGraphy. Roessingh Res Dev 8–11
8. Mesin L, Merletti R, Rainoldi A (2009) Surface EMG: the issue of electrode location. J Electromyogr Kinesiol 19(5):719–726. https://doi.org/10.1016/j.jelekin.2008.07.006
9. Released LR et al (2013) proEMG 2.0-How do you do? no, pp 1–51. (March)
10. Freed A (2012) Design, prototyping
11. Phinyomark A, Phukpattaranont P, Limsakul C (2012) Expert systems with applications feature reduction and selection for EMG signal classification. Expert Syst Appl 39(8):7420–7431. https://doi.org/10.1016/j.eswa.2012.01.102
12. Mesin L, Cescon C, Gazzoni M, Merletti R, Rainoldi A (2009) A bi-dimensional index for the selective assessment of myoelectric manifestations of peripheral and central muscle fatigue. J Electromyogr Kinesiol 19(5):851–863. https://doi.org/10.1016/j.jelekin.2008.08.003
13. Wakeling JM, Uehli K, Rozitis AI (2006) Muscle fibre recruitment can respond to the mechanics of the muscle contraction. J R Soc Interface 3(9):533. https://doi.org/10.1098/RSIF.2006.0113
14. MSEC, Lewis AD (2020) Visualizing muscular fatigue during running exercises with electromyography-derived variables

An Evaluation of Different Input Transformation for the Classification of Skateboarding Tricks by Means of Transfer Learning

Muhamad Amirul Abdullah, Muhammad Ar Rahim Ibrahim, Muhammad Nur Aiman Shapiee, Mohd Azraai Mohd Razman, Rabiu Muazu Musa, Noor Azuan Abu Osman, Muhammad Aizzat Zakaria, and Anwar P. P. Abdul Majeed

Abstract This study aims to investigate the effect of different input images, namely raw data (RAW) and Continuous Wavelet Transform (CWT) towards the discriminating of street skateboarding tricks, i.e., Ollie, Kickflip, Shove-it, Nollie and Frontside 180 through a variety of transfer learning with optimised k-Nearest Neighbors (kNN) pipelines. Six amateur skateboarders participated in the study, executed the aforesaid tricks five times per trick on an instrumented skateboard where six time-domain signals were extracted prior it was transformed to RAW and CWT. It was shown from the study that the CWT-InceptionV3-optimised kNN pipeline could attain an average test and validation accuracy of 90%.

Keywords Classification · k-Nearest Neighbor · Machine learning · Skateboarding · Transfer learning

1 Introduction

Skateboarding is a type of extreme sport that has now been acknowledged through its Olympic debut at the 2020 Summer Olympics [1]. One of the famous skateboarding

M. A. Abdullah · M. A. R. Ibrahim · M. N. A. Shapiee · M. A. Mohd Razman · M. A. Zakaria · A. P. P. Abdul Majeed
Innovative Manufacturing, Mechatronics and Sports Laboratory, Faculty of Manufacturing and Mechatronic Engineering Technology, Universiti Malaysia Pahang, Pahang, Malaysia

R. M. Musa
Centre for Fundamental and Continuing Education, Department of Credited Co-Curriculum, Universiti Malaysia Terengganu, Terengganu, Malaysia

N. A. Abu Osman
Faculty of Engineering, Universiti Malaya, Kuala Lumpur, Malaysia

A. P. P. Abdul Majeed (✉)
School of Robotics, XJTLU Entrepreneur College, Xi'an Jiaotong Liverpool University, Taicang, China
e-mail: Anwar.Majeed@xjtlu.edu.cn

© The Author(s), under exclusive license to Springer Nature Singapore Pte Ltd. 2023

S. F. Syed Omar et al. (eds.), *Innovation and Technology in Sports*, Lecture Notes in Bioengineering, https://doi.org/10.1007/978-981-99-0297-2_22

styles that have gained traction over the years is street skateboarding that requires the skateboarders to execute different tricks such as jumps (Ollies), flips and mid-air spins amongst others.

It is important to mention at this juncture that at present, the evaluation of the execution of the tricks in a sporting event is carried out by a few selected professional judges manually. However, it is been reported that judging is indeed a rather difficult task in a skateboarding event. This is primarily due to the involvement of many factors, for instance, style, speed, difficulty, consistency, trick selection and originality, amongst others [2].

In order to mitigate the aforesaid issues, different methods have been proposed by researchers with regards to activity recognition for different types of sporting events [3–12]. Nevertheless, limited studies have been investigated on the recognition of skateboarding tricks [13–17]. Amongst the pioneers of investigating the classification of skateboarding (along with similar nature sports, viz. snowboarding) are Groh et al. [8, 14, 15, 18, 19]. Utilising four different machine learning (ML) classifiers, Groh et al. [14] investigated the classification of six different skateboarding tricks, and it was demonstrated that the Naïve Bayes (NB) model based on selected features could yield appreciable classification accuracy (CA) rate.

In an extended investigation, Groh et al. [15] attempted at classifying thirteen skateboarding trick classes and it was shown that the Radial Basis (RB)-Support Vector Machine (SVM) model with its associated features could gain a CA of up to 89.1%. Other researchers have also attempted to classify skateboarding tricks, for example, Correa et al. [16] constructed an Artificial Neural Network (ANN) architecture to classify five tricks. In contrast, Anlauff et al. [13] examined the effectiveness of Linear Discriminant Analysis (LDA) in discriminating three classes, and in a more recent study, Abdullah et al. [17] examined the ability of Logistic Regression (LR), SVM, kNN, ANN, NB, and RF, by considering features extracted statistically from the time-domain signals acquired.

Therefore, it is obvious from the scant literature that was published, the use of machine learning to classify skateboarding tricks showed appreciable classification accuracy [13–17]. Nevertheless, it is worth noting that statistically derived time-domain signals may eliminate important information from the signal. Conversely, the use of transfer learning, features are automatically derived from the altered signals, eliminating the requirement for handcrafted feature extraction from the signals. In order to classify skateboarding tricks, this paper intends to evaluate the impact of input transformation through the Continous Wavelet Transformation (CWT), against taking the data as is (RAW) and on the effectiveness of different transfer learning models coupled with optimised kNN architecture.

2 Methods

The instrumented inertial measurement unit (IMU) device was used to acquire the time-domain signals from the skateboarding tricks investigated, i.e., Ollie (O), Nollie

Frontside Shove-It (NFS), Frontside 180 (FS180), Pop Shove-It (PS), and Kickflip (KF), respectively. Six amateur skateboarders aged 20 ± 7 who had at least five years of expertise executed the tricks. Every trick that was carried out successfully was collected five times. For detailed accounts of the ruses and the construction of the instrumented IMU device, the readers are referred to Abdullah et al. [17].

In transforming the input image, as per the successful trick from the IMU, an average of six time-domain-based raw signals were acquired from the instrumented IMU device. The signals are the angular velocities (°/s) along the x-axis (ΩX), y-axis (ΩX), and z-axis (ΩZ), respectively, as well as the linear accelerations (m/s2) along the x-axis (aX), y-axis (aY), and z-axis (aZ). Based on the default image size from Table 1, all six signals were combined to create a single image that represented a single skateboarding trick. Thus, 150 synthesised pictures in total were used in this study. The base of the input image is the raw transformation (RAW), in which the individual signals are stacked in a single image. Whilst the other form of input image is the transformation is converting the RAW image to a scalogram-based image via the Continuous Wavelet Transform (CWT).

In the current study, a total of five pre-trained Convolutional Neural Network (CNN) models for transfer learning were used: VGG16, VGG19, Inception V3, DenseNet201, and MobileNetV2. The Keras library's arguments were used to set the transfer learning (TL) models [20]. It is important to note that the fully connected layers of the TL models is replaced with the kNN model in this study, while the convolutional weights of the models remained frozen. Through the aforementioned convolutional layers [21], the features are automatically extracted. Table 1 lists the TL models along with the settings that were used for each.

k-Nearest Neighbors (kNN) is an instance-based learning model and one of the simplest machine learning models that have been developed. Owing to its simplicity, it is also often dubbed as lazy learning [22]. In the present study, the effect of different selection of kNN hyperparameters were evaluated based on its ability to classify the skateboarding tricks from the extracted features attained via the TL models. The hyperparameters tuned for kNN are the number of neighbors (varied from 1 to 25), the weight function (between uniform and distance), as well as the distance metric (i.e., Minkowski, Euclidean, Cosine and Manhattan).

The exhaustive grid search optimisation approach, along with the three-fold cross-validation technique, was employed to obtain the best hyperparameters amongst the

Table 1 TL models specifications

No	Model	Flatten reshape	Input image size	
			Height	Width
1	Densenet201	7*7*1920	224	224
2	InceptionV3	8*8*2048	299	299
3	MobilenetV2	7*7*1280	224	224
4	VGG16	7*7*512	224	224
5	VGG19	7*7*512	224	224

200 developed kNN models for the extracted features per transfer learning model and per image input. Therefore, 4200 pipelines were evaluated in the present investigation. Out of the 150 synthesised images, 100 was used for training, whilst the remaining 50 were evenly split for validation and testing purposes. The accuracy score is used to appraise the different pipelines developed. The pipelines were developed via a Python 3.7-based IDE, Spyder 3.3.6, along with its associated libraries, i.e., Keras 2.3.1: Tensorflow 1.14.0 and scikit-learn 0.22.121.

3 Results and Discussion

The transformed signals produced from the FS180 trick are shown in Fig. 1. The different accuracies evaluated for the various pipelines formulated are tabulated in Table 2. It is evident from Fig. 2 that the CWT-based input image, in general generates improved CA on the test and validation dataset as compared to the RAW-based input image, across the various TL. It was demonstrated that neither the transfer learning model nor any of the input images produce 100% train accuracies. However, it should be noted that nearly half of the TL models achieved a train accuracy score of more than 90%.

It could be seen that for the CWT-based input images, the InceptionV3-optimised kNN pipeline could yield an average test and validation CA of 90%. Although the DenseNet 201, MobileNetV2 as well as VGG19 provided higher training CA, nonetheless, a reduced classification performance was detected on its ability to classify on the test and validation dataset, demonstrating a slight overfitting behavior. The best or optimised hyperparameter combination of the kNN for the CWT-InceptionV3 pipeline is 1, uniform and cosine for the number of neighbors, weight function and

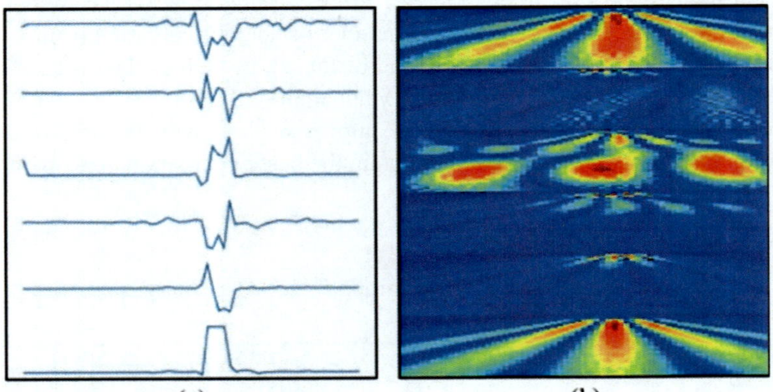

Fig. 1 Synthesised input image transformation. **a** RAW. **b** CWT

Table 2 The classification accuracy of transfer learning model with the optimised kNN for different input image

No	Input image	Transfer learning model	Accuracy		
			Train	Validate	Test
1	RAW	DenseNet201	0.92	0.88	0.76
2	RAW	InceptionV3	0.89	0.72	0.76
3	RAW	MobileNetV2	0.95	0.88	0.84
4	RAW	VGG16	0.92	0.80	0.80
5	RAW	VGG19	0.96	0.88	0.84
6	CWT	DenseNet201	0.95	0.92	0.80
7	CWT	InceptionV3	0.92	0.92	0.88
8	CWT	MobileNetV2	0.94	0.88	0.88
9	CWT	VGG16	0.92	0.88	0.80
10	CWT	VGG19	0.93	0.88	0.84

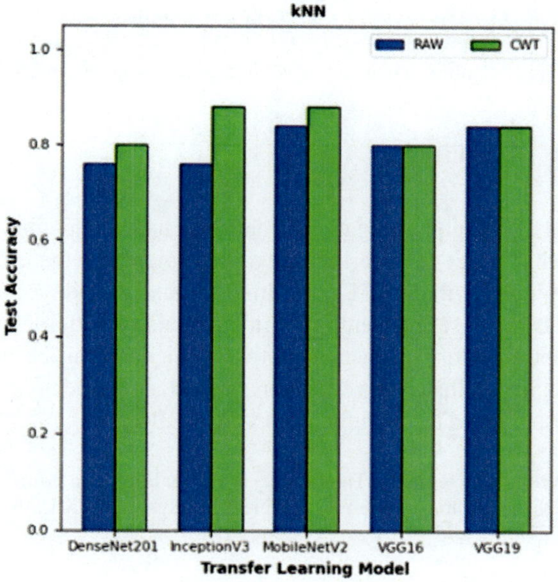

Fig. 2 Test CA of five different pipelines developed for RAW and CWT input images

distance metric, respectively. Figure 3 shows the confusion matrix of the CWT-InceptionV3-optimised kNN pipeline on the test dataset. It could be that the misclassification arises from NFS as well as PS. In which two of the NFS trick were misclassified as PS180, whilst one of the PS tricks was misclassified as NFS.

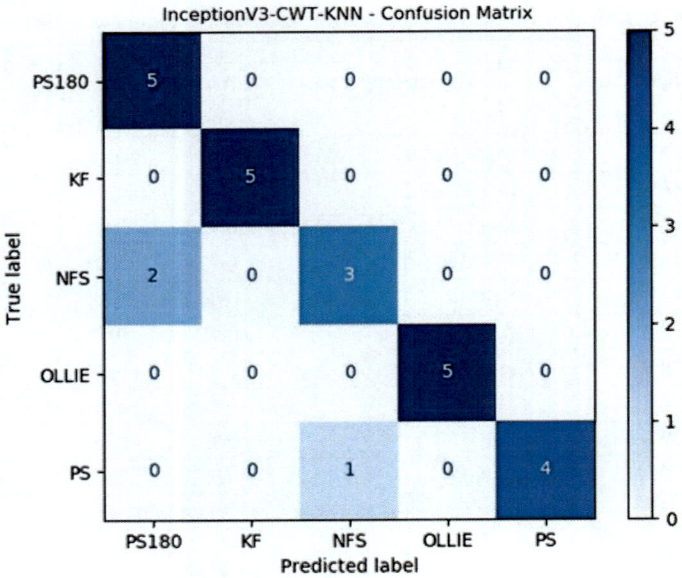

Fig. 3 Confusion matrix of CWT-InceptionV3-optimised kNN pipeline on the test dataset

4 Conclusion

The present investigation establishes that the input image transformation does somewhat affect the discrimination performance of the tricks evaluated. From the study, it appears that CWT input image could provide a better inference to the different TL models. Different input image transformation methods and the hyperparameter optimisation via metaheuristic techniques shall be investigated in the future. It is expected that a more objective-based assessment on skateboarding tricks could be established through the proposed architecture.

Acknowledgements The Ministry of Higher Education (MoHE), Malaysia and Universiti Malaysia Pahang is acknowledged for funding the study via FRGS/1/2019/TK03/UMP/02/6 (RDU1901115).

References

1. TOKYO2020/About this sport/Skateboarding (n.d.)
2. Stein M, Janetzko H, Lamprecht A, Breitkreutz T, Zimmermann P, Goldlücke B, Schreck T, Andrienko G, Grossniklaus M, Keim DA (2018) Bring It to the Pitch: Combining Video and Movement Data to Enhance Team Sport Analysis. IEEE Trans Vis Comput Graph 24:13–22
3. K. Rangasamy, M.A. As'ari, N.A. Rahmad, N.F. Ghazali, Hockey activity recognition using pre-trained deep learning model, ICT Express. (2020).

4. Hong Y, Ling C, Ye Z (2018) End-to-end soccer video scene and event classification with deep transfer learning.In: 2018 International conference on intelligent systems and computer vision ISCV. pp 1–4
5. Brock H, Ohgi Y, Lee J (2017) Learning to judge like a human: Convolutional networks for classification of ski jumping errors. In: Proceedings of 2017 ACM international symposium on wearable computers ISWC. Part F1305, pp 106–113
6. Anand A, Sharma M, Srivastava R, Kaligounder L, Prakash D (2017) Wearable motion sensor based analysis of swing sports. In: 2017 16th IEEE International conference on machine learning and applications. IEEE, pp 261–267
7. Büthe L, Blanke U, Capkevics H, Tröster G (2016) A wearable sensing system for timing analysis in tennis. In: 2016 IEEE 13th International Conference wearable implant body sense Networks. IEEE, pp 43–48
8. Groh BH, Fleckenstein M, Eskofier BM, Wearable trick classification in freestyle snowboarding, in, (2016) IEEE 13th Int. Conf. Wearable Implant. Body Sens. Networks, IEEE 2016:89–93
9. Adelsberger R, Tröster G (2013) Experts lift differently: Classification of weight-lifting athletes. In: 2013 IEEE International Conference Body Sensing Networks. IEEE, pp 1–6
10. Connaghan D, Kelly P, O'Connor NE, Gaffney M, Walsh M, O'Mathuna C (2011) Multi-sensor classification of tennis strokes. In: SENSORS, 2011 IEEE. IEEE, pp 1437–1440
11. Parkkari J, Kujala UM, Kannus P (2001) Is it possible to prevent sports injuries? Review of controlled clinical trials and recommendations for future work, Sports Med. 31:985–995
12. Gordon D, Hanne J.-H, Berchtold M, Miyaki T, Beigl M (2012) Group activity recognition using wearable sensing devices, Mob Ubiqui Syst Comp Network Serv 104:350–361
13. Anlauff J, Weitnauer E, Lehnhardt A, Schirmer S, Zehe S, Tonekaboni K (2010) A method for outdoor skateboarding video games. In: ACM International conference proceeding series, pp 40–43
14. Groh BH, Kautz T, Schuldhaus D (2015) IMU-based trick classification in skateboarding, KDD work. Large-Scale Sport Anal
15. Groh BH, Fleckenstein M, Kautz T, Eskofier BM (2017) Classification and visualisation of skateboard tricks using wearable sensors. Pervasive Mob Comput 40:42–55
16. Corrêa NK, de Lima JCM, Russomano T, dos Santos MA (2017) Development of a skateboarding trick classifier using accelerometry and machine learning. Res Biomed Eng. 33:362–369
17. Abdullah MA, Ibrahim MAR, Bin Shapiee MNA, Mohd Razman MA, Musa RM, Abdul Majeed APP (2020) The classification of skateboarding trick manoeuvres through the integration of IMU and machine learning. In: Lect Notes Mechanical Engineering. Springer, pp 67–74
18. Groh BH, Fleckenstein M, Eskofier BM (2016) Wearable trick classification in freestyle snowboarding, In: 2016 IEEE 13th International conference on wearable and implantable body sensor networks (BSN),pp 89–93
19. Groh BH, Flaschka J, Wirth M, Kautz T, Fleckenstein M, Eskofier BM (2016) Wearable real-time skateboard trick visualization and its community perception. IEEE Comput Graph Appl 36:12–18
20. Keras API reference/Keras applications (n.d.)
21. KumarJLM, Rashid M, Musa RM, Razman MAM, Sulaiman N, Jailani R, Abdul Majeed APP (2021) The classification of EEG-based wink signals: A CWT-Transfer Learning pipeline. ICT Express
22. Muazu Musa R, Abdul Majeed APP, Taha Z, Abdullah MR, Husin Musawi Maliki AB, Azura Kosni N (2019) The application of artificial neural network and k-nearest neighbour classification models in the scouting of high-performance archers from a selected fitness and motor skill performance parameters. Sci Sport

Ball-Oriented Soccer Simulation (BOSS)

Muhammad Hamdan, Raihana Sharir, Wee Kian Yeo, Zulkifli Mohamed, Sapto Adi, and Raja Mohammed Firhad Raja Azidin

Abstract The Ball-Oriented Soccer Simulation (BOSS) is an innovation from soccer match-play simulations previously used in scientific studies that aim to replicate soccer match-play demands. Several shortcomings from the previous simulations may include the utility of unidirectional treadmill running, large spatial requirement, or lack of individual ball-handling actions. With the utility of the multidirectional overground running, compact, BOSS in soccer studies, much of these shortcomings may be easily overcome in a safe, controlled, laboratory setting. The ball-handling activities in the BOSS were designed to closely follow match-play frequencies and distances to create an ecologically valid soccer match-play simulation. Mean heart rate and rating of perceived exertion during the BOSS was found to be similar to previous overground soccer match simulations (HR: 155±6 bpm; RPE: 14±1). Practical applications of the BOSS include its utility in multiple research disciplines such as exercise physiology, nutrition, sports psychology, sports biomechanics as well as skills and motor control. The BOSS may also be used in real-world settings such as using the protocol on injury risk screening during pre-season or as a return-to-play biomechanical assessment protocol for players following post-injury or post-surgical rehabilitation. In the future, perhaps the BOSS may be used as a blueprint for the development of different sport-specific simulation protocol for the betterment of sports research as well as sports equipment production.

M. Hamdan · R. Sharir · R. M. F. Raja Azidin (✉)
Faculty of Sports Science and Recreation, UiTM, Shah Alam, Malaysia
e-mail: firhad@uitm.edu.my

W. K. Yeo
National Sports Institute of Malaysia, Kuala Lumpur, Malaysia

Z. Mohamed
College of Engineering, UiTM, Shah Alam, Malaysia

Z. Mohamed · R. M. F. Raja Azidin
Sports Engineering and Artificial Intelligence Center, UiTM, Shah Alam, Malaysia

S. Adi · R. M. F. Raja Azidin
Department of Sports Science, Universitas Negeri Malang, Jawa Timur, Indonesia

R. M. F. Raja Azidin
Selangor Football Club, Shah Alam, Malaysia

© The Author(s), under exclusive license to Springer Nature Singapore Pte Ltd. 2023
S. F. Syed Omar et al. (eds.), *Innovation and Technology in Sports*, Lecture Notes in Bioengineering, https://doi.org/10.1007/978-981-99-0297-2_23

Keywords BOSS · Soccer · Simulation · Assessment

1 Introduction

Previous epidemiological research has shown that soccer matches can have a higher injury rate than training sessions [1, 2]. More importantly, in a study [2], it was made evident that the injury incidences tend to happen more during the later stages of a match (i.e. the last 15 min of both halves). Numerous researchers saw this as a side effect of the repetitive physical strain of soccer games, which diminished athletic performance. Apart from increased injury rate over time, studies have noted reductions in work rate such as distance covered and high-intensity running [3, 4], as well as physiological (i.e. muscle strength; [5–7] and biomechanical [8–11] parameters.

Studies have since incorporated physical activity protocols into testing sessions to replicate the exertions observed in soccer matches. Several studies have come close to replicating the demands of soccer match-play. Studies by Small, McNaughton [7], Williams, Abt [12], Bendiksen, Bischoff [13], and Raja Azidin, Sankey [9] were particularly of interest with regards to this current study for their own reasons with regard to soccer match-play demands. Small, McNaughton [7] utilized a 20-m, multidirectional soccer simulation to replicate soccer specific running profiles such as walking, jogging, striding and sprinting; whereas Raja Azidin, Sankey [9] used a 15-m variation to the simulation to replicate both physiological and physical loading [14] of actual soccer match-play. The soccer match-play simulation by Williams, Abt [12] incorporated ball shooting actions and Bendiksen, Bischoff [13] included dribbling and passing actions in their simulation. It is rather unfortunate, however, that the frequencies and distances covered during ball handling actions were questionable as they were not reported in the two studies. The replication of ball handling action is crucial in replicating soccer match-play, especially since research has depicted that a majority of injuries in soccer occurred during an offensive game-play (during ball possession) [2].

In actual soccer matches, a single player may dribble the ball an accumulative range of from 120 to 280 m [15] interceded by a series of roughly 120 ball handling actions consisting of shooting, passing, and heading [16] while trying to regain or maintain possession, or trying to score a goal. In order to ecologically replicate actual soccer demands in match-play, there is a need to incorporate the frequencies and distances covered in possession of the ball with the current existing soccer match simulation. Recently, a simulation that fulfills the demand have been introduced [17, 18], however, the report included a small number samples (7 participants) in their study which only suited the analysis of physical parameters in the study, and not the physiological parameters due to the design and aim of the study. Thus, the physiological responses of the simulation remained debatable. The aim of this study was to investigate the physiological responses induced by soccer match simulations which incorporates ball handling actions.

2 Methods

2.1 Study Design

Ethical approval was granted for the human trials in this study. Participants were required to attend two (one familiarization and one testing) sessions. Each participant was required to complete a 90 min a match-play simulation with a ball. Throughout the simulation, the participants' heart rate and rate of perceived exertion were monitored. All participants completed the simulation at the same time of day to account for circadian variation [19]. The participants were tested in a 3 h' post absorptive state with no vigorous physical exertions or any alcohol and caffeine consumption 24 h prior to testing.

2.2 Participants

Eighteen (n = 18) male participants volunteered for this study. Their mean age, height, and body mass were 25.3 ± 4 y, 1.67 ± 0.05 m, and 65.0 ± 9.23 kg respectively. Sample size calculation was conducted a-priori using G*Power (version 3.1.9.2, Universität Kiel, Germany) and it was determined that to achieve 80% statistical power and an estimated moderate effect size *f* of 0.25, a minimum number of 10 participants is required for this study. A written informed consent form was be provided, and the study was performed in accordance with the university ethics committee guidelines.

2.3 Soccer-Match-Play Simulations

The Ball-Oriented Soccer Simulation (BOSS) was based on the overground soccer match simulation (OSMS) [20] which has been found to elicit similar physiological responses as during actual match-play [4, 9] and imply similar physical loading of the player during an actual soccer match [14]. Ball handling skills such as ball passing, running with a ball, ball shooting and ball heading was instructed at selected positions throughout the course as illustrated in Fig. 1. The movement intensity and activities performed by the participants whilst completing the BOSS was maintained via verbal cues on a recorded audio file. The audio cues consisted of a 15 min activity profile (Table 1) repeated three times to mimic one half of a soccer match.

The distribution of the ball handling skills was done in accordance with the individual ball action frequency provided by Link and Hoernig [15] and was based on another research by Ruiz, Power [21]. While the BOSS includes a total of 120 ball handling abilities over a period of 90 min, previous study reported a range of 117–126 ball actions per player each match [15]. The simulation's overall distance

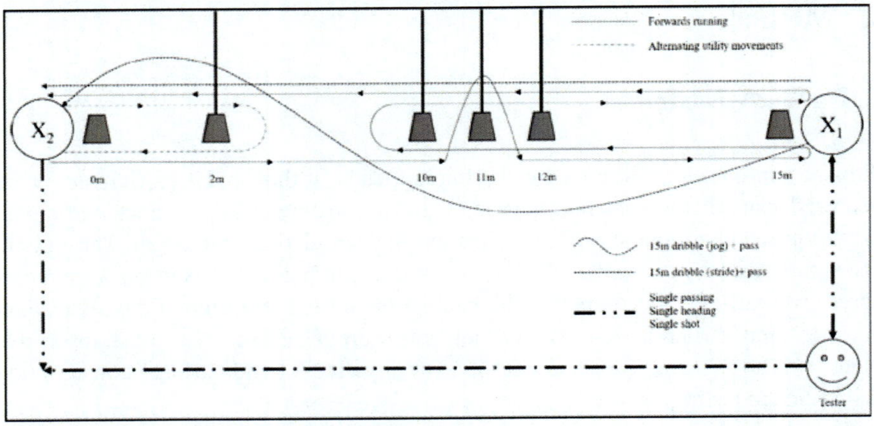

Fig. 1 A schematic diagram of the layout of the BOSS

travelled when handling a ball was around 270 m, which is in line with the ranges that players often reach while in possession of the ball [16].

2.4 Heart Rate and Rate of Perceived Exertion

Throughout the BOSS, heart rate (Polar heart rate system, Electro, Finland) and rate of perceived exertion (RPE, 20-point Borg scale) were recorded every five minutes.

2.5 Statistical Analyses

Statistical analyses were conducted using statistical software package SPSS (Version 25; SPSS Inc., USA). Mean and standard deviation were calculated for each variable. A one-way repeated measures analysis of variance was performed to compare differences in physiological changes (heart rate and RPE) over time (every 5-min interval). Mauchly's test of sphericity was used to check for equality of variance between simulations and different times. Post-hoc analysis was conducted using Bonferroni procedures. Time was treated as the independent variable, whereas heart rate and RPE were treated as dependent variables. Alpha was set at 0.05.

Table 1 Activity profiles during the BOSS

0 – 5 min		5 – 10 min		10 – 15 min	
Activity	Time (s)	Activity	Time (s)	Activity	Time (s)
Stand	4	Upjog	10	Jog	10
Upjog single pass	10	Walk 10	17	Sidestride single shot	7
Walk	17	Sidejog	10	Stride	7
Sidestride	7	Sprint single shot	6	Upjog single pass	10
Walk	17	Stand	4	Walk	17
Upjog	10	Upjog	10	Sidejog	10
Sprint single shot	6	Walk	17	Stand	4
Stand	4	Sidejog	10	Walk	17
Sidejog	10	Walk	17	Upjog single pass	10
Walk	17	Upstride single shot	7	Walk	17
Upjog single pass	10	Walk	17	Sidestride	7
Walk	17	Sidejog	10	Walk	17
Sidestride	7	Walk	17	Upjog single header	10
Walk	17	Upjog	10	Stand	4
Stand	4	Jog and 15 m Dribble	10	Walk	17
Upjog single header	10	Sidejog	10	Sidejog	10
Walk	17	Stand	4	Walk	17
Sidejog	10	Walk	17	Upjog	10
Jog and 15 m dribble	10	Upjog single pass	10	Sprint single shot	6
Upjog	10	Walk	17	Stand	4
Walk	17	Sidestride	7	Sidejog	10
Sidejog	10	Walk	17	Walk	17
Walk	17	Upjog	10	Upjog single pass	10
Upstride single shot	7	Walk	17	Walk	17
Stand	4	Sidejog	10	Sidestride	7
Walk	17	Walk	17	Walk	17
Sidejog	10	Stand	4	Upjog	10
Walk	17	Upjog	10	Stride and 15 m Dribble	7

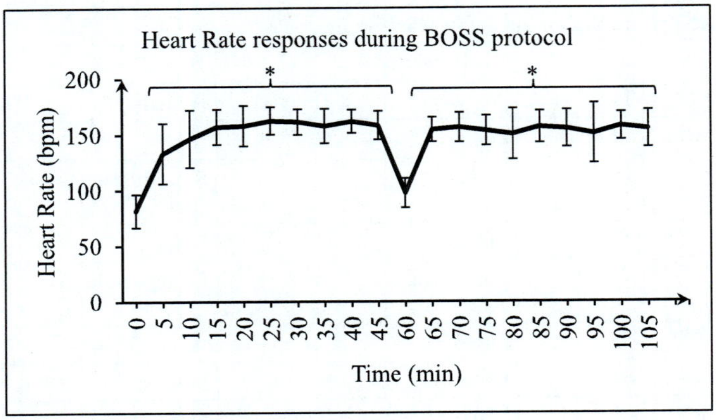

Fig. 2 Heart Rate responses during BOSS protocol. * Indicates significant differences compared to pre-exertion levels

3 Results

3.1 Heart Rate

Throughout the BOSS, the mean heart rates maintained between 155 ± 6 bpm. There was a significant main effect of time on heart rate ($F_{19,323} = 37.835$, $p < 0.001$, partial $\eta^2 = 0.690$). Post-hoc analysis revealed that the BOSS induced significantly elevated heart rates compared to pre-test conditions ($p < 0.001$; Fig. 2).

3.2 Rate of Perceived Exertion

The mean rating of perceived exertions maintained between 14 ± 1. There was a significant main effect of time on RPE across the two simulations ($F_{19,323} = 36.476$, $p < 0.001$, partial $\eta^2 = 0.682$). Post-hoc analysis revealed that the BOSS induced significantly elevated heart rates compared to pre-test conditions ($p < 0.05$; Fig. 3).

4 Discussion

This study was conducted to investigate the physiological responses induced by soccer match simulations that incorporates ball handling actions. The key findings in this study suggest that match-play simulations with ball-handling activities elicited increased heart rates and ratings of perceived exertions.

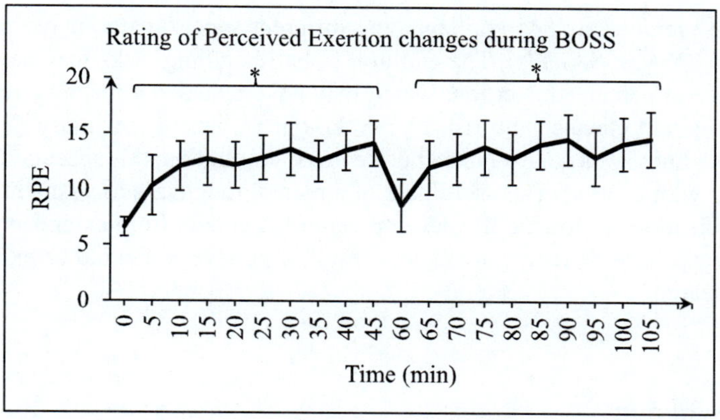

Fig. 3 Rating of Perceived Exertion changes during BOSS. * Indicates significant differences compared to pre-exertion levels

Our results show that the BOSS induced similar physiological response to the OSMS protocol from previous studies [9]. It has been demonstrated that the heart rate responses during the OSMS protocol were consistent with heart rates during actual match-play [9]. Several differences should still be addressed between the findings in this study and that found from a previous investigation [9]. Mean heart rate and RPE recorded in the previous study [9] were slightly higher in comparison to this study (HR: 160±7 bpm versus. 155±6 bpm; RPE: 15±2 versus 14±1). These variations could be explained by anthropometric variations, such as average height, average body mass, and perhaps even ethnicity. In certain investigations, participants' heart rate responses were found to be quite comparable to those in this study, where participants' features like height (1.6–1.72 m) and body mass (49–80 kg) could be seen [22, 23], whereas the participants in another study [9] were slightly taller (1.74±0.07 m), with more consistent body mass (73±8 kg). These anthropometric variations may point to a range of work rates among the participants, which would explain the variations in heart rates observed in these investigations. However, due to the complexity of the association between anthropometric profiles and work rates as a result of multiple work-rate drivers, this finding should be made with caution [24].

Another finding in this study is that the BOSS protocol induced elevated perception of exertions, that were notably higher than in previous studies using overground soccer match-play simulation without ball-handling activities [9]. However, these differences were only observed during the first 15 min of match-play simulation. For example, after 5 min of BOSS exertion, mean RPE was already above 10, whereas previous studies reported a mean RPE of 8 [9]. These scenarios might be brought on by the extra ball-handling tasks built into the BOSS protocol. Ratings of perceived exertion remained high after further effort, but they were consistent with other investigations. This finding may reveal new follow-up insight on previous findings [7–9] who observed increased biomechanical risk of injury following a passive half-time rest. With previous studies implementing different soccer match-play simulations (treadmill, OSMS, SAFT^{90}), The addition of ball-handling tasks may have caused an increase in perceived exertion, which may have revealed a mental strain on the participants and elevated the risk of psychological hazard for injury [25]. Since mental toughness was not evaluated before this investigation, this speculation should be treated with caution. The usefulness of a mental toughness assessment in injury risk stratifications following the development of tiredness from extended exertions and decreased muscle temperatures as a result of passive rest could be examined in future research.

5 Conclusion

The physiological alterations that occur during match-play simulations with integrated ball-handling activities may be first documented in this work. With higher perceived exertion ratings early in the protocol, the BOSS protocol was able to recreate similar reactions to earlier overground match-play simulations and actual match-play. According to these findings, including ball-handling movements could provide further insight into soccer-specific multifactorial injury risk factors from the biomechanical, physiological, and perhaps even psychological perspectives. For this reason, compared to earlier simulations, the BOSS protocol is thought to depict soccer-specific match-play exertion demands more accurately. Practical applications of the BOSS include its utility in multiple research disciplines such as exercise physiology, nutrition, sports psychology, sports biomechanics as well as skills and motor control. The BOSS may also be used in real-world settings such as using the protocol on injury risk screening during pre-season or as a return-to-play biomechanical assessment protocol for players following post-injury or post-surgical rehabilitation. In the future, perhaps the BOSS may be used as a blueprint for the development of different sport-specific simulation protocol for the betterment of sports research as well as sports equipment production.

References

1. Herrero H, Salinero JJ, Del Coso J (2014) Injuries among Spanish male amateur soccer players: a retrospective population study. Am J Sports Med 42(1):78–85
2. Hawkins RD et al (2001) The associated football medical research programme: an audit of injuries in professional football. Br J Sport Med 35:43–47
3. Mohr M, Krustrup P, Bangsbo J (2003) Match performance of high-standard soccer players with special reference to development of fatigue. J Sport Sci 21(7):519–528
4. Bangsbo J, Nørregaard L, Thorsoe F (1991) Activity profile of competition soccer. Can J Sport Sci 16(2):110–116
5. Delextrat A, Gregory J, Cohen D (2010) The Use of the Functional H: Q Ratio to Assess Fatigue in Soccer. Int J Sports Med 31(3):192

6. Greig M (2008) The influence of soccer-specific fatigue on peak isokinetic torque production of the knee flexors and extensors. Am J Sports Med 36(7):1403–1409
7. Small K et al (2010) The effects of multidirectional soccer-specific fatigue on markers of hamstring injury risk. J Sci Med Sport 13(1):120–125
8. Greig M (2009) The influence of soccer-specific activity on the kinematics of an agility sprint. Eur J Sport Sci 9(1):23–33
9. Raja Azidin RMF et al. (2015) Effects of treadmill versus overground soccer match simulations on biomechanical markers of anterior cruciate ligament injury risk in side cutting. J Sport Sci 33(13):1332–1341
10. Sanna G, O'Connor KM (2008) Fatigue-related changes in stance leg mechanics during sidestep cutting maneuvers. J Clin Biomech 23(7):946–954
11. Small K et al (2009) Int J Sports Med 30(8):573
12. Williams JD, Abt G, Kilding AE (2010) Ball-sport endurance and sprint test (BEAST90): Validity and reliability of a 90-minute soccer performance test. J Strength & Cond Res 24(12):3209–3218
13. Bendiksen M et al (2012) The Copenhagen Soccer Test: physiological response and fatigue development. J Med Sci Sport Exerc 44(8):1595–1603
14. Barreira P et al (2016) Mechanical player load™ using trunk-mounted accelerometry in football: Is it a reliable, task-and player-specific observation? J Sport Sci: 1–8
15. Link D, Hoernig M (2017) Individual ball possession in soccer. PLoS ONE 12(7):e0179953
16. Di Salvo V et al (2007) Performance characteristics according to playing position in elite soccer. Int J Sports Med 28(03):222–227
17. Hamdan M et al (2020) Changes in hamstring eccentric peak torques and angles of peak torque following 90 Minutes of soccer specific exertions. Malays J Mov, Health & Exerc 9(2)
18. Hamdan M, Raja Azidin RMF (2022) Lab-controlled soccer: A review of soccer match-play simulations. Malays J Sport Sci Recreat (MJSSR) 18(2):255–273
19. Reilly T, Brooks G (1986) Exercise and the circadian variation in body temperature measures. Int J Sports Med 7(06):358–362
20. Raja Azidin RMF (2015) Effect of soccer match-play on markers of anterior cruciate ligament injury risk. Liverpool John Moores University
21. Ruiz H et al. (2017) The Leicester City fairytale?: Utilizing new soccer analytics tools to compare performance in the 15/16 & 16/17 EPL seasons. In: Proceedings of the 23rd ACM SIGKDD International conference on knowledge discovery and data mining. https://doi.org/10.1145/3097983.3098121
22. Hamdan M et al. (2018) The Effects of A 90-minute simulated soccer match-play on Knee and hip kinematics. Mov, Health & Exerc (MoHE) 7(1)
23. Hamdan M et al (2018) Knee and hip extension responses to prolonged simulated soccer match-play: A 2D study. Int J Eng Technol (UAE) 7(4):6–14
24. Rienzi E et al (2000) Investigation of anthropometric and work-rate profiles of elite South American international soccer players. J Sports Med Phys Fitness 40(2):162–169
25. Mohd Nor MA et al. (2003) Mental toughness profile as one of psychological predictors of injuries among Malaysian professional football players. In: Seminar, Pan Pacific, Johar Bahru

Prototype of IoT-Based Timing Gate System for Sports Application

Muhammad Farhan Mohd Jamil, Zulkifli Mohamed, Mohd Fauzi Ibrahim, Mohd Hanif Mohd Ramli, and Nurul Syuhadah Khusaini

Abstract The Internet of Things (IoT) now plays an important role in sports especially in tracking athlete's development. IoT offers limitless opportunities to monitor athlete's growth by creating personalized measuring instrument for athletes. In sports, track and field routine, protocol as well as teaching events is one of the areas that could benefit from IoT. Speed and agility training for sport requires a variety of training approaches, much as strength and power training does. A few approaches can be used to measure speed for athletes, including timing gate. However, the cost of developing a timing gate system is on the high side, hence there is a need to develop a cost-effective timing gate that is reliable and serve the same purpose. The development of timing gate device is used in this study to assess and enhance all aspects of athletic training, including speed, acceleration, responsiveness, power, and elevation. The prototype of the timing gate system is successfully developed by combining hardware and software. This project seeks to improve the architecture of a smart timing gate system for sports applications by utilizing ESP32 Wi-Fi module as the microcontroller and infrared sensors as input to achieve high accuracy. All data are transferred to Blynk apps and later displayed on LCD. This paper outlines the design and execution of the timing gate system application in sports.

Keywords Internet of things · Sports application · Time gate system · ESP32 Wi-Fi module

---

M. F. Mohd Jamil · Z. Mohamed · M. H. Mohd Ramli · N. S. Khusaini (✉)
School of Mechanical Engineering, College of Engineering, Universiti Teknologi MARA Shah Alam, Shah Alam, Malaysia
e-mail: nurulsyuhadah2089@uitm.edu.my

Sports Engineering and Artificial Intelligence Centre, Universiti Teknologi MARA Shah Alam, Selangor, Malaysia

M. F. Ibrahim
Malaysia National Sports Institute, Bukit Jalil, Kuala Lumpur, Malaysia

© The Author(s), under exclusive license to Springer Nature Singapore Pte Ltd. 2023

S. F. Syed Omar et al. (eds.), *Innovation and Technology in Sports*, Lecture Notes in Bioengineering, https://doi.org/10.1007/978-981-99-0297-2_24

1 Introduction

In order to improve athlete's performance, modern sports training must undergo the process of data acquisition and data analysis of athlete's performance data. Coaches must design training and routines for each athlete based on his or her specific needs [1]. Collecting and analysing athlete data during practises is critical in modern sports training for improving athlete performance. Also in sports, speed will always be a major factor in an athlete's success rate.

It is vital to test athletes on a continuous basis and acquire empirical data on their performance to ensure that their speed and agility are improving. This, however, is impossible without a precise and reliable timing system. Speed and agility testing accuracy and reliability are critical for providing valid feedbacks to athletes. Failure to meet high standards of reliability by using stopwatches or non-calibrated timing systems jeopardizes athlete performance assessment and may provide inaccurate feedback that is detrimental to the training program. To improve themselves, professional athletes are incorporating IoT devices into their training routines. These devices could be smart wearable devices or any other cutting-edge technology. IoT timing gates have also gained widespread attention due to their ability to communicate real-time data and ensure athletes' proper timing as they pass through the gates. These will help athletes improve their skills and reaction time by integrating training technologies that can exchange information and coordinate decisions in real-time to improve training session efficiency [2]. The Internet of Things (IoT) is a network of physical objects that are embedded with sensors, software, and other technologies that enable them to communicate and exchange data with other devices and systems over the internet. IoT devices ranging from simple household items to complex industrial machinery are being used.

Today, there is more than 7 billion Internet of Things (IoT) connected devices and it has been predicted this figure will expand to 10 billion by 2020 and 22 billion by 2025 [3]. Components and systems in information technology are becoming smaller, more affordable, more impactful, and more widely used. People are now exposed to information technology through widespread information and communication infrastructures that encompass every aspect of our everyday lives [4]. As the concept of the Internet of Things (IoT) has gained traction in the technology world, it has enabled the incorporation of information technologies into a wide range of corporate areas and processes, as well as the growth of numerous innovative technological solutions [5]. As a result, IoT technology offers numerous opportunities for reducing unnecessary labour costs and procedures in order to improve our lives and the environment. The Internet of Things enables a more informed approach to daily occurrences such as finding parking spaces, exercising, monitoring weather conditions, breathing cleaner air, and preparing better meals [7].

Relating IoT application with timing-gate, it is indeed possible to develop a cost-effective timing gate system. For instance, ESP32 microcontroller is suitable for wide deployment capabilities, support for Wi-Fi standards and protocols as well as low-cost solution [8]. As a result, it is suitable for both personal and professional use, and the solution is easily upgradeable and scalable [9]. Furthermore, the ESP32 microcontroller is an appropriate communication platform that can be integrated with third-party applications for real-time data display and monitoring such as Blynk.

The Blynk platform collects and analyses data from IoT sensor nodes. It includes methods for integrating the system into mobile apps, as well as the creation of data and knowledge dashboards for display, making the system easy to use and measure [4]. Sensors can be extremely simple or complicated.

Sensors are classified based on their requirements, conversion method, material used, and the properties observed [10, 11]. Infrared (IR) sensors are frequently used to measure mid-range distances. The infrared (IR) sensor has a higher resolution than the ultrasonic (US) sensor, making it ideal for measuring relatively short distances within their corresponding useful ranges [12]. Sensor selection is an essential aspect of any system design because it has a notable impact on system performance over its lifetime and may even affect prototype quality [13].

There is a wealth of technological equipment for sports monitoring, but little research has been conducted on the time gate system for athlete usage. The current product is so expensive that it cannot be used for sports research and development in our country.

As a result, this paper focuses on creating a timing gate system using the IoT Blynk platform. This research will focus on lowering the cost of the system while maintaining its accuracy and durability for long-term use. This development both supports and validates the use of IoT for sports applications. Using ESP32 as the microcontroller, this project aims to create a smart time gate system that notifies users via Blynk Platform in their smartphones. During the development of this system, the responsiveness and reliability of the used sensors and modules will be evaluated.

2 Methodology

2.1 Software Development

The first stage is to collect all essential information on the selected topic. Analyzing current research papers, articles, journals, and books on the Internet of Things (IoT), sports technology, and the current time gate system is part of the procedure [15, 16]. Before working on the programming code in the Arduino IDE, the circuit's designation and configuration are initiated by creating a functional system flowchart, as shown in Fig. 1. Figure 1 depicts the development prototype's flowchart.

In terms of software development, the Arduino IDE acts as a compiler, compiling the written source code [17]. The prototype requires the creation of code using the Arduino IDE platform, which allows the ESP32 to be programmed, as well as a link to the Blynk app. This code must include the connection requirements between the ESP32 and the application in order for the app to notify the user if the sensor detects

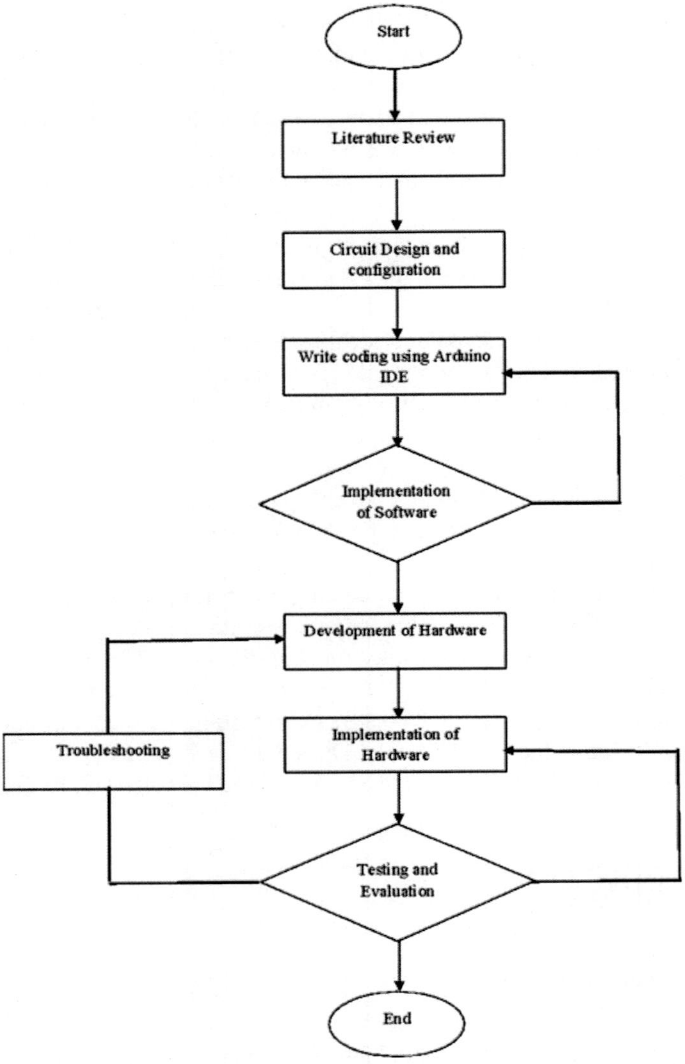

Fig. 1 Study flowchart

motion. If a bug is discovered during compilation, the code must be revised before hardware development can begin.

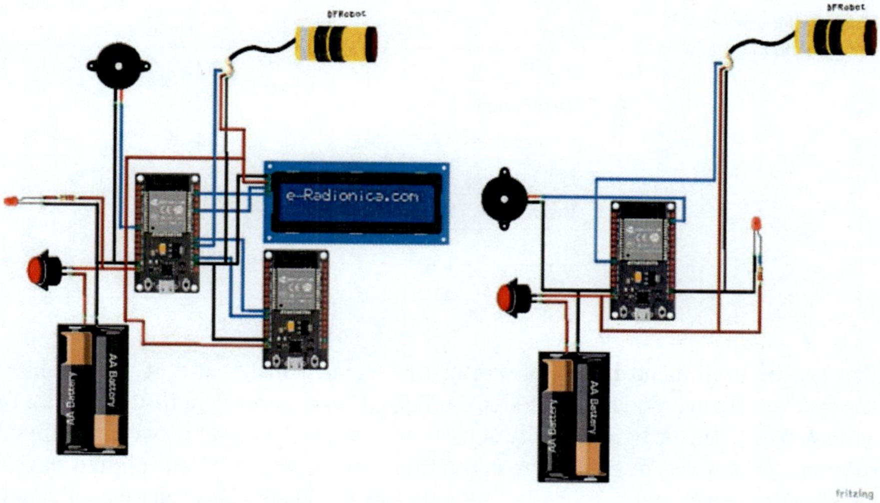

Fig. 2 Wiring circuit diagram

2.2 *Hardware Development*

According to the wiring diagram shown in Fig. 2, all physical components were connected using solder and wire connectors. Figure 2 also shows the prototype's pin out. The Blynk project allows the user to select the hardware model with which to operate. The ESP32 was chosen in this case, and it was connected via Wi-Fi. The ESP32 is one of Espressif's most powerful chips on the market right now.

In many ways, the ESP32 is an improvement over previous chip. In addition to Wi-Fi, the ESP32 has a Bluetooth connection module that allows it to communicate with Bluetooth-enabled devices. It uses the same CPU chip as the ESP8266, but with two more cores and a faster clock speed. This model has more memory and GPIO (general-purpose input/output) pins. The Controller Area Network (CAN) bus connection on the ESP32 is a significant improvement over the ESP8266's lack of this feature [16, 17]. Table 1 depicts the major components used in this project. Knowing that PVC is one of the most used plastics in electronic equipment casing, a PVC box was chosen as the casing for this prototype. These products are frequently lighter, less expensive, and provide numerous performance benefits [18].

2.3 *Calibration*

Calibration of the sensors is required to ensure that their output falls within an acceptable, effective range. The calibration method follows the instructions in the user manual provided by the manufacturer. The SN-E18 includes an infrared sensor

Table 1 Hardware component

Item	Quantity
ESP32	3
Switch button	2
Li-on battery 4800 m-AH	4
LCD display module	1
Buzzer	2
PVC box	2
Stand	2

that can be used as an IR signal barrier reflection. It makes use of a specialized sensor to detect the modulated IR signal reflected from a nearby or distant object. To achieve detection, the light intensity of the transmitter and receiver between the digital infrared sensors can be altered by current. As shown in Fig. 3, an infrared device for distance detection has an adjustable range of 6–80 cm. This sensor application employs a DC 5 V energy source and a current of 100 mA. Because infrared reflection is critical to the device's functionality, the best material for the sensing device is black or any opaque substance that allows less light to pass through. Output Digital TTL Signal; 0 = GND and 1 = 5 V.

As shown in Fig. 3, the range is adjusted using a multi-turn screw on the unit's back. The detection range can be improved by turning the screw counterclockwise CCW). The module can detect the presence of an object within the specified range but cannot estimate its distance. The calibration procedure is depicted in Fig. 4. The development of the prototype needs the writing of instructional code using the Arduino IDE platform, which enables the ESP32 to be programed and linked to the Blynk app. This code must include the specific connection between the ESP32 and Blynk in order to alert the user via the app if the sensor detects motion. If a problem appears during compilation, the code must be revised before hardware development.

As shown in Fig. 4, adjust VR at the end of the Sensor. Examine the described LED change at the sensor's end. If the LED is turned off (OUTPUT = 1), turn the VR

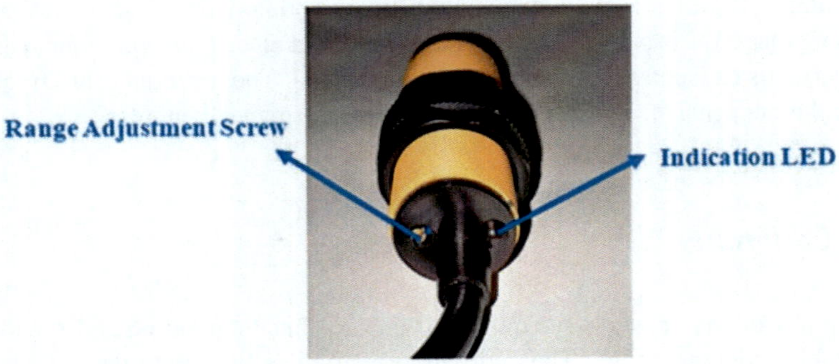

Fig. 3 Sensor E18 range adjustment screw and indication LED

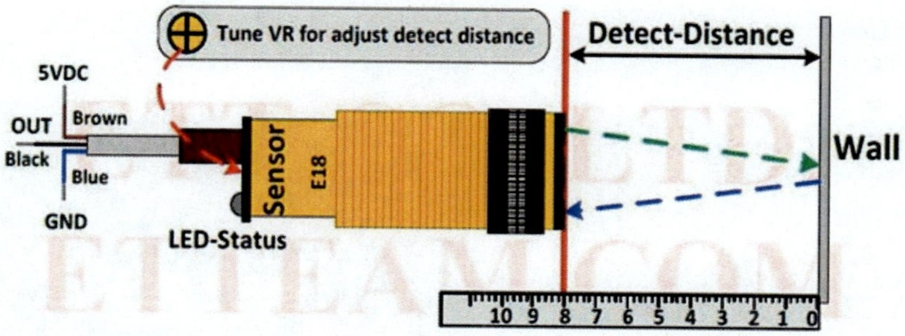

Fig. 4 Calibration detection range for sensor E18

CCW until the LED is turned on (OUTPUT = 0), then stop turning the VR. Distance detection occurs at the point where the LED changes state. This is a conditional statement. If the sensor's distance is less than or equal to the detected distance, the LED Status is ON and the OUTPUT becomes Logic 0. Otherwise, LED Status is set to OFF, and OUTPUT is set to Logic 1. If the LED is turned on (OUTPUT = 0), turn the VR CCW until the LED is turned off (OUTPUT = 1), and then stop turning the VR. Distance detection is performed at the place where the LED changes state. This is a conditional action. If the distance of the sensor is higher than or equal to the distance detected, LED Status is OFF, and OUTPUT becomes Logic 1; otherwise, LED status is ON, and OUTPUT becomes Logic 0 instead.

Moving the sensor allows you to test its functionality. When the sensor's head moves and passes the specified distance detection, the LED illuminates if the sensor's distance is less than or equal to the specified distance detection, but the LED is deactivated if the sensor's distance is greater than or equal to the specified distance detection. It fails to configure distance detection for the sensor if any of the conditional operations are not met. After testing and calibrating the sensor, we discovered that its effective range is 10–70 cm.

2.4 System Testing

Figure 5 shows the prototype being tested on a tarmac road. The surrounding of 0 m elevation has a straight road, indicating that the area is suitable for testing the prototype. The test was conducted in the evening at 1800 h on a sunny day with a temperature of 33 °C.

The prototype's height was set at 110 cm, as shown in Fig. 6, and the distance between the tester and sensor was set at 40 cm, as shown in Fig. 7. For human time reading, a measuring tape and a stopwatch were used. As shown in Fig. 8, this prototype was tested with a different distance between sensors of 2, 4, 6, 8, 10 m. The time recorded for both time-gate prototype and stopwatch is taken three times and yielded the average reading time for both readings. To ensure accurate results, the test subject walks at the same pace as the sensors. Also, the delayed time reading

Fig. 5 The surrounding area where the prototype was tested

displays of the LCD and Blynk apps were also recorded. The smart time gate system was fully operational, and the prototype was successfully tested.

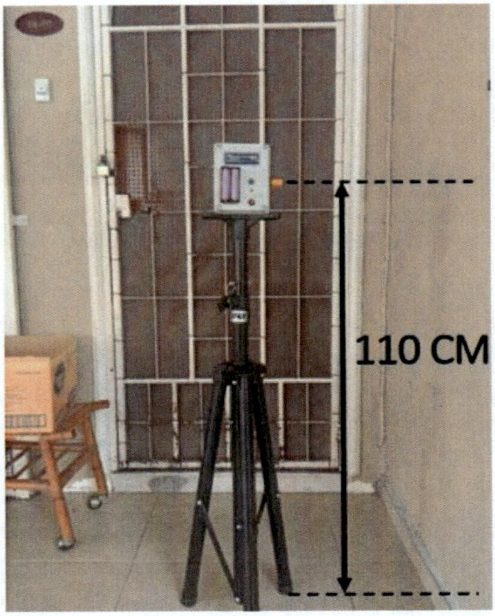

Fig. 6 The height of the prototype was fixed at 110 cm

Fig. 7 The distance between the tester and sensor is 40 cm

Fig. 8 Measurement of different distances between sensors

3 Result and Discussion

Software and hardware have been integrated and are working together. Blynk's interaction with ESP32 was aided by readily available Blynk libraries. They help with the entire process of sending and receiving data inputs. All that was required was the activation of authentication tokens, which recognize the setup device via the Blynk platform, allowing interaction and sensor data transmission from the EPS32 MCU to the cloud. Figure 9 shows how the data is visualized by Blynk's mobile dashboard. Sensor readings can be viewed by the user. Table 2 describes the results of the four variables involved in this study. The first variable is the difference in sensor distances. Followed by the delay time from prototype to Blynk apps. Next is the average reading time on stopwatch and prototype (all time measurements are in seconds) (s). The formula in Equation can be used to calculate the average time (1). However, the delay time from prototype measurement reading to Blynk apps is measured in milliseconds due to small differences in value.

$$\frac{Total\,time\,calculate}{Total\,time\,in\,the\,data} = Average\,time\,measure$$

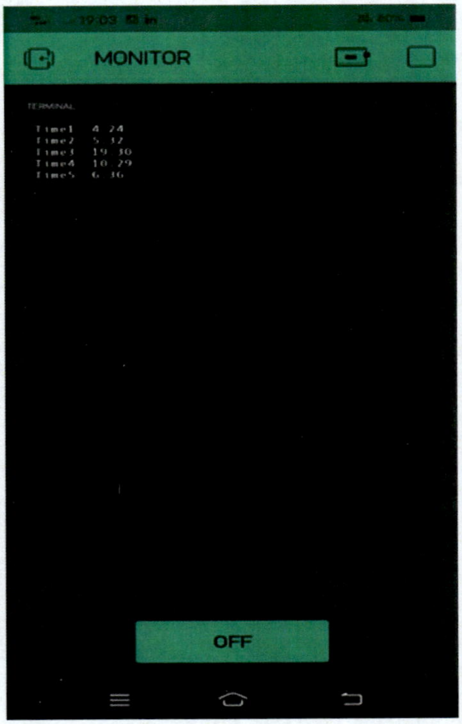

Fig. 9 Blynk display monitor

Table 2 Timing data for stopwatch and prototype

Distance between sensors (m)	Delay time to Blynk apps (milliseconds)	Average stopwatch reading (s)	Average prototype reading (s)
2	112	1.75	1.62
4	127	3.00	2.83
6	141	4.44	4.24
8	168	5.45	5.32
10	173	6.55	6.36

Figure 10 depicts the delay time from the prototype to the Blynk monitor as a bar chart and numerical value. Raw or processed data can be transferred from any MCU-board-connected sensor or actuator using Blynk. The data sent to Blynk went through a Datastream that is compliant with Blynk. Each value is then timestamped and then stored in the Blynk's cloud. The x-axis represents the distance between the sensor or prototype, and the y-axis represents the delay in milliseconds. The delay time recorded was insufficient because the data was transmitted smoothly and efficiently. From observation during experiment, when the distance increase, the delay will increase too because of the electromagnetic wave sent by the connection loses energy as it goes further. This is due to the speed of bits decreases with increasing distance. As the signal loses energy, it becomes more difficult to send data.

Figure 11 depicts the average time measurement reading of the stopwatch and prototype at various distances as a bar chart and numerical value. The test was repeated three times for each distance, and the average time was calculated using Eq. (1). The prototype and the stopwatch used by the test subject have slightly different values. The x-axis represented the distance between the sensor or prototype, while the y-axis represented the average time measurement in seconds for both data sets. The disparity in time measurement between the prototype and the stopwatch (controlled by test subject) is due to human error. Common errors such as failing to respond to prototype's warning buzzer as well as taking too long to stop the stopwatch is somewhat unexpected. However due to minor differences, the reading

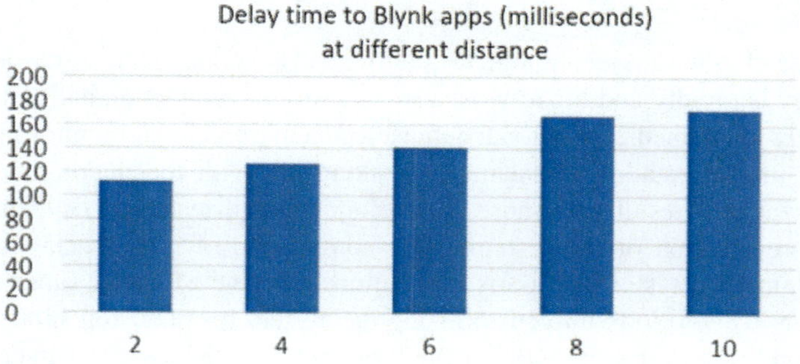

Fig. 10 Delay time data

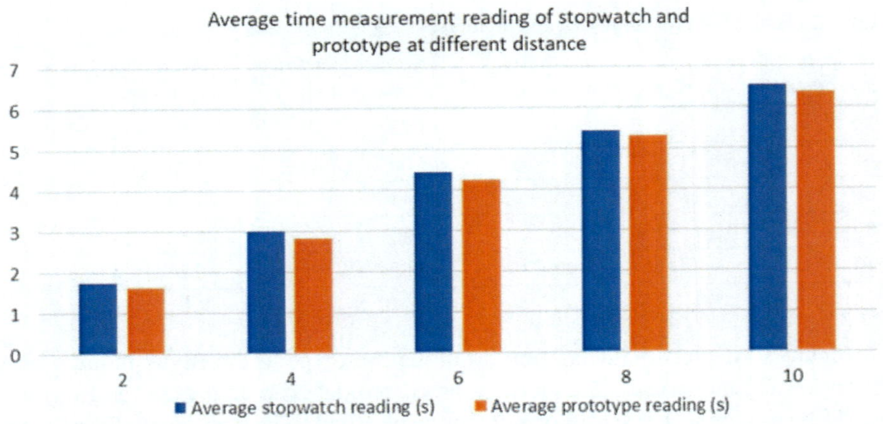

Fig. 11 Average time measurement data

is acceptable and it did demonstrate that the sensor E18 has high accuracy and detects parallel lines.

4 Conclusion and Recommendation

The paper discusses the component selection, implementation strategy, and initial system experiences to develop a prototype of an affordable IoT timing gate system. This smart timing gate system development was successful and fully functional, as evidenced by the data display during the testing phase.

The prototype successfully combined software and hardware, and the system's performance was assessed. Despite a minor delay in data transmission between the prototype and the Blynk app, this system was seamlessly integrated between hardware and software to meet the objectives. Furthermore, the sensor was suitable for this system, and the low-power IoT device for monitoring sports applications based on the ESP32 microcontroller and the Blynk cloud platform performed admirably and met the goal.

Because of its unique technological qualities and benefits, wireless sensor network technology is rapidly and broadly used in a variety of production and living areas, particularly in the medical, national defense, and environmental monitoring sectors.

Direct implementation of technology, alongside with the nation's continuing development will result in a more mature and refined technical development. A long-term evaluation of the system and a comparison of sensor data with established commercial system for sports applications will be the focus of future research. In addition, we will look into customizing the system for precision sports applications. In addition to testing with a variety of users or athletes, the personalization will consider a power supply based on a solar panel and battery that can display battery health in percent (%) on an LCD screen. Hence, it is possible to replicate the prototype in the coming years to create various system applications.

References

1. Ishida K (2019) IoT application in sports to support skill acquisition and improvement. In: Proceedings—2019 IEEE 12th Conference on service-oriented computing and applications, SOCA 2019. pp 184–189. https://doi.org/10.1109/SOCA.2019.00034
2. Wei et al. (2021) Dataflow management in the internet of things: Sensing, control, and security. [Online]. Available https://www.sciopen.com/article_pdf/1496413795440541698.pdf
3. What is IoT. https://www.oracle.com/my/internet-of-things/what-is-iot/
4. Babic D, Jovovic I, Popovic T, Kovac N, Cakic S (2022) An internet of things system for environmental monitoring based on ESP32 and Blynk. https://doi.org/10.1109/IT54280.2022.9743538
5. Zhang C (2021) Intelligent internet of things service based on artificial intelligence technology. In: 2021 IEEE 2nd International conference on big data, artificial intelligence and internet of things engineering, ICBAIE 2021. pp 731–734. https://doi.org/10.1109/ICBAIE52039.2021.9390061
6. SCAD College of Engineering and Technology and Institute of Electrical and Electronics Engineers (2019).In: Proceedings of the international conference on trends in electronics and informatics (ICOEI 2019), pp 23–25
7. Kumar MS, Janaki G, Hariharan S, Priya PS (2021) IoT based warning system from the threat of wild animals. In: Proceedings of the 6th International conference on communication and electronics systems, ICCES 2021. pp 550–553. https://doi.org/10.1109/ICCES51350.2021.9489181
8. Babiuch M, Foltynek P, Smutny P (2019) Using the ESP32 microcontroller for data processing. https://doi.org/10.1109/CarpathianCC.2019.8765944
9. Maier A, Sharp A, Vagapov Y (2017) Comparative analysis and practical implementation of the ESP32 microcontroller module for the internet of things. In: 2017 Internet technologies and applications, ITA 2017–Proceedings of the 7th international conference. pp 143–148. https://doi.org/10.1109/ITECHA.2017.8101926
10. Karnik A, Adke D, Sathe P (2020) Low-cost compact theft-detection system using MPU-6050 and Blynk IoT Platform. In: 2020 IEEE Bombay section signature conference, IBSSC 2020. pp 113–118. https://doi.org/10.1109/IBSSC51096.2020.9332214
11. Kos A, Umek A, Tomazic S, Zhang Y (2018) Identification and selection of sensors suitable for integration into sport equipment: Smart golf club. In: Proceedings - 2016 International conference on identification, information, and knowledge in the internet of Things, IIKI 2016, vol 2018. pp 128–133. https://doi.org/10.1109/IIKI.2016.71
12. Mustapha B, Zayegh A, Begg RK (2013) Ultrasonic and infrared sensors performance in a wireless obstacle detection system. In: Proceedings - 1st International conference on artificial intelligence, modelling and simulation, AIMS 2013. pp 487–492. https://doi.org/10.1109/AIMS.2013.89
13. Yang J, Lv W (2021) Optimization of sports training systems based on wireless sensor networks algorithms. IEEE Sens J 21(22):25075–25082. https://doi.org/10.1109/JSEN.2020.3046290
14. Suresh RP, Shajahan Kavalakkal S, Shereef S, SAS, Sebastian J (2019) IoT based toll gate system using RFID; IoT based toll gate system using RFID
15. Picking R (2017) Internet technologies and applications (ITA) : Proceedings of the seventh international conference : Tuesday 12th–Friday 15th September 2017. Wrexham Glyndŵr University, Wales, UK
16. Sun CI, Huang JT, Weng SC, Chien MF (2019) City marathon active timing system using Bluetooth low energy technology. Electronics (Switzerland) 8(2). https://doi.org/10.3390/electronics8020252

17. Le MT, Tran Ta Thi N, Vo MT (2019) Design and implementation of real-time robot controlling system using upper human body motion detection on FPGA. In: Proceedings–2019 19th International symposium on communications and information technologies, ISCIT. pp 215–220. https://doi.org/10.1109/ISCIT.2019.8905212
18. Torres-Montoya S, Montoya-Goez Y, Tapias L (2014) Comparing different mechanical models for the evaluation of mechanical behavior of reuse in PVC electrosurgical pencils through their hardness and Young's Modulus relation. https://doi.org/10.1109/PAHCE.2014.6849634